T0211492

Advances in Intelligent Systems and Computing

Volume 1123

The series "Advances in Intelligent Systems and Computing" contains publications on theory, applications, and design methods of Intelligent Systems and Intelligent Computing. Virtually all disciplines such as engineering, natural sciences, computer and information science, ICT, economics, business, e-commerce, environment, healthcare, life science are covered. The list of topics spans all the areas of modern intelligent systems and computing such as: computational intelligence, soft computing including neural networks, fuzzy systems, evolutionary computing and the fusion of these paradigms, social intelligence, ambient intelligence, computational neuroscience, artificial life, virtual worlds and society, cognitive science and systems, Perception and Vision, DNA and immune based systems, self-organizing and adaptive systems, e-Learning and teaching, human-centered and human-centric computing, recommender systems, intelligent control, robotics and mechatronics including human-machine teaming, knowledge-based paradigms, learning paradigms, machine ethics, intelligent data analysis, knowledge management, intelligent agents, intelligent decision making and support, intelligent network security, trust management, interactive entertainment, Web intelligence and multimedia.

The publications within "Advances in Intelligent Systems and Computing" are primarily proceedings of important conferences, symposia and congresses. They cover significant recent developments in the field, both of a foundational and applicable character. An important characteristic feature of the series is the short publication time and world-wide distribution. This permits a rapid and broad dissemination of research results.

** Indexing: The books of this series are submitted to ISI Proceedings, EI-Compendex, DBLP, SCOPUS, Google Scholar and Springerlink **

More information about this series at http://www.springer.com/series/11156

M. Hadi Amini
Editor

Optimization, Learning, and Control for Interdependent Complex Networks

Springer

Editor
M. Hadi Amini
School of Computing and Information
Sciences
Florida International University
Miami, FL, USA

Sustainability, Optimization, and Learning
for InterDependent Networks Laboratory
(solid lab)
Florida International University
Miami, FL, USA

ISSN 2194-5357 ISSN 2194-5365 (electronic)
Advances in Intelligent Systems and Computing
ISBN 978-3-030-34096-4 ISBN 978-3-030-34094-0 (eBook)
https://doi.org/10.1007/978-3-030-34094-0

This Springer imprint is published by the registered company Springer Nature Switzerland AG.
The registered company address is: Gewerbestrasse 11, 6330 Cham, Switzerland

Preface

This book focuses on a wide range of optimization, learning, and control algorithms for interdependent complex networks and their application in smart cities infrastructures, intelligent transportation networks, and smart energy systems. This book paves the way for researchers working on optimization, learning, and control spread over the fields of computer science, operations research, electrical engineering, civil engineering, and system engineering. It covers optimization algorithms for large-scale problems from theoretical foundations to real-world applications; learning-based methods to enable intelligence in future smart cities, and control techniques to deal with the optimal and robust operation of complex system. It further introduces novel algorithms for data analytics in large-scale interdependent networks.

Miami, FL, USA M. Hadi Amini, Ph.D., D.Eng.

Contents

About the Editor

M. Hadi Amini is an Assistant Professor at the School of Computing and Information Sciences at Florida International University (FIU). He is also the founding director of Sustainability, Optimization, and Learning for InterDependent networks laboratory (solid lab). He received his Ph.D. and M.Sc. from Carnegie Mellon University in 2019 and 2015 respectively. He also holds a doctoral degree in Computer Science and Technology. Prior to that, he received M.Sc. from Tarbiat Modares University in 2013, and B.Sc. from Sharif University of Technology in 2011. His research interests include distributed machine learning and optimization algorithms, distributed intelligence, sensor networks, interdependent networks, and cyberphysical resilience. Application domains include energy systems, healthcare, device-free human sensing, and transportation networks.

Prof. Amini is a life member of IEEE-Eta Kappa Nu (IEEE-HKN), the honor society of IEEE. He organized a panel on distributed learning and novel artificial intelligence algorithms, and their application to healthcare, robotics, energy cybersecurity, distributed sensing, and policy issues in 2019 workshop on artificial intelligence at FIU. He also served as President of Carnegie Mellon University Energy Science and Innovation Club; as technical program committee of several IEEE and ACM conferences; and as the lead editor for a book series on "Sustainable Interdependent Networks" since 2017. He has published more than 80 refereed journal and conference papers, and book chapters. He has coauthored two books and edited three books on various aspects of optimization and machine learning for interdependent networks. He is the recipient of the best paper award of "IEEE Conference on Computational Science & Computational Intelligence" in 2019, best reviewer award from four IEEE Transactions, the best journal paper award in "Journal of Modern Power Systems and Clean Energy," and the dean's honorary award from the President of Sharif University of Technology. He is dedicated to educating next generation of computer scientists and engineers by developing and teaching a new interdisciplinary course entitled "Optimization Methods for Computing: Theory and Applications" that engages students from Computer Science, Electrical and Computer Engineering, Civil Engineering, Transportation Engineering, as well as Construction, Infrastructure & Sustainability Engineering (homepage: www.hadiamini.com; lab website: www.solidlab.network).

Chapter 1
Panorama of Optimization, Control, and Learning Algorithms for Interdependent SWEET (Societal, Water, Energy, Economic, and Transportation) Networks

M. Hadi Amini

Abstract In this chapter, I first introduce a high level overview of Interdependent SWEET Networks, including societal networks as a pivot in smart cities, water network, energy networks (e.g., power systems and gas network), economic networks that facilitate financial transactions among entities in other networks, and transportation networks. I then explain how optimization, learning, and data analytic can improve the interdependent operation of these networks. This chapter also provides an overview of this book and its two main parts: theoretical algorithms and real-world applications.

Keywords Interdependent networks · Smart city · Complex networks · Optimization · Machine learning · Control · Data analytic

1.1 Introduction

Integration of novel technologies enables the transition from current urban environments towards smart cities [1–3]. In order to deal with the increasing complexity of the underlying infrastructures in future smart cities, there is a need to develop efficient optimization, learning, and control algorithms. This chapter first provides a brief introduction of Interdependent SWEET Networks as subsets of the networks

M. H. Amini (✉)
School of Computing and Information Sciences, Florida International University, Miami, FL, USA

Sustainability, Optimization, and Learning for InterDependent Networks Laboratory (solid lab), Florida International University, Miami, FL, USA
e-mail: amini@cs.fiu.edu; hadi.amini@ieee.org; www.solidlab.network

© Springer Nature Switzerland AG 2020
M. H. Amini (ed.), *Optimization, Learning, and Control for Interdependent Complex Networks*, Advances in Intelligent Systems and Computing 1123,
https://doi.org/10.1007/978-3-030-34094-0_1

in smart cities. It then provides the abstracts of all chapters in the first volume of this book. The key idea is studying smart cities as human-centered network of networks, i.e., societal networks are the main coupling point among all infrastructures that are designed to serve society, e.g., water or energy networks. Figure 1.1 represents a general overview of interdependent SWEET networks, their mutual couplings, and their interaction with societal networks.

One example of the interdependence among smart city infrastructures is interdependent power system and electrified transportation networks [4–10]. I have explored different aspects of the interdependence among power and transportation networks, including physics-based models for electric vehicle energy demand [5], a visionary perspective and algorithms for optimal operation of these two networks [6, 9], and simultaneous integration of electric vehicle parking lots and renewable energy resources [7, 8]. While Fig. 1.1 represents the interdependence among these networks, there is another emerging issue in terms of decision making paradigm change. As each of these networks has a specific objective, there might be conflict

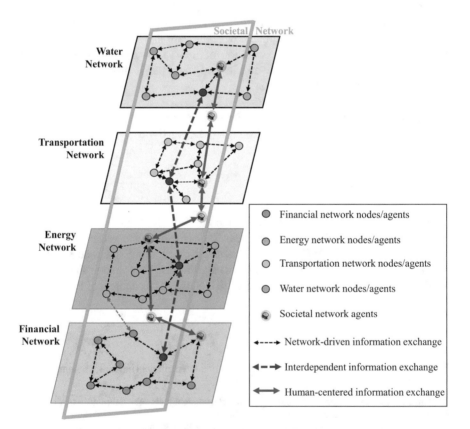

Fig. 1.1 Societal networks as the main coupling network in interdependent SWEET networks

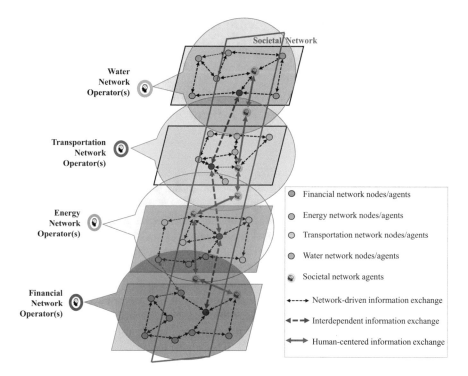

Fig. 1.2 Interdependent decision making in interdependent SWEET networks

while they are formulating their decision making in terms of optimization problems. Figure 1.2 represents an overview of interdependent decision making, where two or more networks need to reach a consensus while finding their locally optimum operating point. This is another area that required attention of researchers from network science, data science, operation research, engineering, and computer science to develop efficient algorithms that are capable of integrating the interdependence among multiple networks.

This book includes two main parts:

Part I: Theoretical Algorithms for Optimization, Learning, and Data Analytics in Interdependent Complex Networks This part covers Chaps. 2–6 of the book.

Part II: Application of Optimization, Learning, and Control in Interdependent Complex Networks This part includes Chaps. 7–12 of the book.

1.2 Part I: Theoretical Algorithms for Optimization, Learning, and Data Analytics in Interdependent Complex Networks

1.2.1 Chapter 2: Promises of Fully Distributed Optimization for IoT-Based Smart City Infrastructures: Theory and Applications

Modern wireless communication and sensor technologies enable ubiquitous sensing through distributed agents in various networks. In the context, collaborative interaction of Internet of Things (IoT) connected end-user devices enables achieving certain goals. In this chapter, authors aim at introducing a holistic distributed framework that enhances operational performance of smart cities infrastructures. To this end, they first introduce a holistic framework that enables distributed coordination of heterogeneous agents such as PEVs. Then, they explain a fully distributed *consensus + innovations* approach for coordinating among agents. Our proposed *consensus + innovations* sits at the core of the proposed framework. In a nutshell, the proposed distributed algorithm achieves a distributed solution for a collaborative decision making problem through iterative agent-based computations and limited inter-agent communications. This algorithm enables fully distributed coordination of agents, plug-and-play capability, and scalability of the solution algorithm for future network expansion.

1.2.2 Chapter 3: Evolutionary Computation, Optimization, and Learning Algorithms for Data Science

A large number of engineering, science, and computational problems have yet to be solved in a computationally efficient way. One of the emerging challenges is how evolving technologies grow towards autonomy and intelligent decision making. This leads to collection of large amounts of data from various sensing and measurement technologies, e.g., cameras, smart phones, health sensors, smart electricity meters, and environment sensors. Hence, it is imperative to develop efficient algorithms for generation, analysis, classification, and illustration of data. Meanwhile, data is structured purposefully through different representations, such as large-scale networks and graphs. Therefore, data plays a pivotal role in technologies by introducing several challenges: *how to present, what to present, why to present*. Researchers explored various approaches to implement a comprehensive solution to express their results in every particular domain, such that the solution enhances the performance and minimizes cost, especially time complexity. In this chapter, authors focus on data science as a crucial area, specifically focusing on a curse of dimensionality (CoD) which is due to the large amount of generated/sensed/collected data, especially large sets of extracted features for a particular purpose. This motivates researchers to think about optimization and apply nature-inspired algorithms, such as meta-heuristic and evolutionary algorithms (EAs) to solve large-scale optimization problems. Building on the strategies of these algorithms, researchers solve large-scale engineering and computational problems with innovative solutions. Although these algorithms look un-deterministic, they are robust enough to reach an optimal solution. To that end, researchers try to run their algorithms more than usually suggested, around 20 or 30 times, then they compute the mean of result and report only the average of 20/30 runs' result. This high number of runs becomes necessary because EAs, based on their randomness initialization, converge the best result, which would not be correct if only relying on one specific run. Certainly, researchers do not adopt evolutionary algorithms unless they face a problem which is suffering from placement in local optimal solution, rather than global optimal solution. In this chapter, authors first develop a clear and formal definition of the CoD problem, next they focus on feature extraction techniques and categories, then they provide a general overview of meta-heuristic algorithms, its terminology, and desirable properties of evolutionary algorithms.

1.2.3 Chapter 4: Applications of Nature-Inspired Algorithms for Dimension Reduction: Enabling Efficient Data Analytics

In Chap. 3, authors have explored the theoretical aspects of feature selection and evolutionary algorithms. In this chapter, they focus on an optimization approach for enhancing data analytic process, i.e., they propose to introduce the applications of the nature-inspired algorithms in data science. Feature selection optimization is hybrid approach leveraging pure feature selection techniques and evolutionary algorithms process to optimize the selected features. Researchers try to iterate this process until to converge to optimal feature subsets. Feature selection optimization is non-specific domain approach which enable scientists to apply this to their data technically.

Data scientists always attempt to find an advanced way to work with data that are successfully conducted in a short time with high computational efficiency and low time complexity, leading to efficient data analytics. Thus, by increasing generated/measured/sensed data from various sources, analysis, manipulation, and illustration of data grow exponentially. Due to the large-scale datasets, curse of dimensionality (CoD) is one of the NP-hard problems in data science. Hence, several efforts have been focused on leveraging evolutionary algorithms (EAs) to address the complexity issues in large-scale data analytics problems. Dimension reduction, together with EAs, lends itself to solve CoD and solve complex problems, in terms of time complexity, efficiently. In this chapter, authors first provide a brief overview of previous studies that focused on solving CoD using feature extraction optimization process. They also discuss practical examples of research studies are successfully tackled some application domains, such as image processing, sentiment analysis, network traffics/anomalies analysis, credit score analysis, and other benchmark functions/datasets analysis.

1.2.4 Chapter 5: Feature Selection in High-Dimensional Data

Today, with the increase of data dimensions, many challenges are faced in many contexts including machine learning, informatics, and medicine. However, reducing data dimension can be considered as a basic method in handling high-dimensional data, because by reducing dimensions, applying many of the existing operations on data is facilitated. Microarray data are derived from tissues and cells considering differences in the gene, which can be useful for diagnosing disease and tumors. Due to the large number of features (genes) and small number of samples in microarray datasets, selecting the most salient genes is a difficult task. Among the many techniques of machine learning, feature selection and data classification play a very important and widespread role in enhancing human life, from detecting voice emotion to detecting illness in the body. In medicine, an effective gene selection can greatly enhance the process of prediction and diagnosis of cancer. After selecting effective genes, the duty of a specific classifier is usually to discriminate healthy people form patients that are suffering from cancer based on their expression of the selected genes. A vast body of feature selection methods has been proposed for high-dimensional microarray data. Traditionally, these methods fall into three categories including filter, wrapper, and hybrid approaches. Furthermore, new techniques such as ensemble methods have recently been developed to improve the process of feature selection and classification. This chapter presents an overview of the most popular feature selection methods to deal with high-dimensional data and analyze their performance under different conditions. The chapter starts with a global overview of the high-dimensional data and feature selection. It then reviews the state-of-the-art methods on filter algorithms. In the next three it further describes the wrapper, hybrid, and embedded methods and the ensemble techniques recently considered by the researchers. It finally presents experimental results of the most significant methods on high-dimensional data.

1.2.5   Chapter 6: An Introduction to Advanced Machine Learning: Meta-Learning Algorithms, Applications, and Promises

In Chaps. 3 and 4, authors have explored the theoretical aspects of feature extraction optimization processes for solving large-scale problems and overcoming machine learning limitations. Majority of optimization algorithms that have been introduced in these chapters guarantee the optimal performance of supervised learning, given offline and discrete data, to deal with curse of dimensionality (CoD) problem. These algorithms, however, are not tailored for solving emerging learning problems. One of the important issues caused by online data is lack of sufficient samples per class. Further, traditional machine learning algorithms cannot achieve accurate training based on limited distributed data, as data has proliferated and dispersed significantly. Machine learning employs a strict model or embedded engine to train and predict which still fails to learn unseen classes and sufficiently use online data. In this chapter, authors introduce these challenges elaborately. They further investigate meta-learning (MTL) algorithm, and their application and promises to solve the emerging problems by answering *how autonomous agents can learn to learn?*.

1.3   Part II: Application of Optimization, Learning, and Control in Interdependent Complex Networks

1.3.1   Chapter 7: Predictive Analytics in Future Power Systems: A Panorama and State-of-the-Art of Deep Learning Applications

The challenges surrounding the optimal operation of power systems are growing in various dimensions, due in part to increasingly distributed energy resources and a progression towards large-scale transportation electrification. Currently, the increasing uncertainties associated with both renewable energy generation and demand are largely being managed by increasing operational reserves—potentially at the cost of suboptimal economic conditions—in order to maintain the reliability of the system. This chapter looks at the big picture role of forecasting in power systems from generation to consumption and provides a comprehensive review of traditional approaches for forecasting generation and load in various contexts. This chapter then takes a deep dive into the state-of-the-art machine learning and deep learning approaches for power systems forecasting. Furthermore, a case study of multi-time-horizon solar irradiance forecasting using deep learning is discussed in detail. Smart grids form the backbone of the future interdependent networks. For addressing the challenges associated with the operations of smart grid, development and wide adoption of machine learning and deep learning algorithms capable of producing better forecasting accuracies are urgently needed. Along with exploring the implementation and benefits of these approaches, this chapter also considers the strengths and limitations of deep learning algorithms for power systems forecasting applications. This chapter, thus, provides a panoramic view of state-of-the-art of predictive analytics in power systems in the context of future smart grid operations.

1.3.2   Chapter 8: Bilevel Adversary-Operator Cyberattack Framework and Algorithms for Transmission Networks in Smart Grids

Transmission system is one of the most important assets in secure power delivery. Recent advancements toward automation of smart grids and application of supervisory control and data acquisition (SCADA) systems have increased vulnerability of power grids to cyberattacks. Cyberattacks on transmission network, specifically the power transmission lines, are among crucial emerging challenges for the operators. If not identified properly and in a timely fashion, they can cause cascading failures leading to blackouts. This chapter tackles false data injection modeling from the attacker's perspective. It further develops an algorithm for detection of false data injections in transmission lines. To this end, first, a bilevel mixed integer programming problem is introduced to model the attack scenario, where the attacker can target a transmission line in the system and inject false data in load measurements on targeted buses in the system to overflow the targeted line. Second, the problem is analyzed from the operator's viewpoint and a detection algorithm is proposed using l_1 norm minimization approach to identify the bad measurement vector in data readings. In order to evaluate the effectiveness of the proposed attack model, case studies have been conducted on IEEE standard test system.

1.3.3   Chapter 9: Toward Operational Resilience of Smart Energy Networks in Complex Infrastructures

Smart energy systems can mitigate electric interruption costs provoked by manifold disruptive events via making efforts toward proper pre-disturbance preparation and optimal post-disturbance restoration. In this context, effective contingency management in power distribution networks calls for contemplating disparate parameters from interconnected electric and transportation systems. This chapter, while considering transportation issues in power networks' field operations, presents a navigation system for pre-positioning resources such as field crews and reconfiguring the network to acquire a more robust configuration in advance of the imminent catastrophe. Also, after the occurrence of the calamity, this navigator optimally allocates the resources to recover the devastating system. So, providing a coordination framework for manual field operation and automation system, this navigator takes a step from traditionally operated systems accommodation toward smart networks. During the contingency management process, there might be modifications in initial data due to the dynamic and time-varying condition of electric and transportation systems. Therefore, the mentioned navigator copes with a real-time problem of data-driven decision making in which, the decisions need to track online changes to the input data. Decision making by the navigation system in this environment is based on a mixed integer linear programming (MILP) optimization which is described in this chapter in details.

1.3.4 Chapter 10: Control of Cooperative Unmanned Aerial Vehicles: Review of Applications, Challenges, and Algorithms

A system of cooperative unmanned aerial vehicles (UAVs) is a group of agents interacting with each other and the surrounding environment to achieve a specific task. In contrast with a single UAV, UAV swarms are expected to benefit efficiency, flexibility, accuracy, robustness, and reliability. However, the provision of external communications potentially exposes them to an additional layer of faults, failures, uncertainties, and cyberattacks and can contribute to the propagation of error from one component to other components in a network. Also, other challenges such as complex nonlinear dynamic of UAVs, collision avoidance, velocity matching, and cohesion should be addressed adequately. Main applications of cooperative UAVs are border patrol; search and rescue; surveillance; mapping; military. Challenges to be addressed in decision and control in cooperative systems may include the complex nonlinear dynamic of UAVs, collision avoidance, velocity matching, and cohesion. In this chapter, emerging topics in the field of cooperative UAVs control and their associated practical approaches are reviewed.

1.3.5 Chapter 11: An Optimal Approach for Load-Frequency Control of Islanded Microgrids Based on Non-linear Model

Due to the increased environmental and economic challenges, in recent years, renewable based distribution generation has been developed. More penetrations from the side of consumers caused a new concept called microgrids which are able to stand with or without connection to the bulk power system. Control of microgrids in islanded mode is very crucial for decreasing the amplitude of frequency deviations as well as damping speed. This chapter aims to propose an optimal combination of FOPD and fuzzy pre-compensated FOPI approach for load-frequency control of microgrids in islanded mode. The optimization parameter of the control scheme is designed by the differential evolution (DE) algorithm which has been improved by a fuzzy approach. In the optimization, control effort is considered as a constraint. Due to the robustness and flexibility of the proposed method, the simulation results have been improved substantially. Robust performance of the proposed control method is examined through sensitivity analysis.

1.3.6   Chapter 12: PV Design for Smart Cities and Demand Forecasting Using Truncated Conjugate Gradient Algorithm

Concerns over the world the global warming is very important issue. This means that the climate change caused by human activities which affect the environment. The climate change presents a serious threat to nature of the world. This might affect in the future, unless the action taken to avoid such phenomena. In addition, without ambitious mitigation efforts, global temperature rises this century. In the recent years, the countries over the world have their own vision to direct it towards the renewable energy, which is the clean one. Since it is a clean energy, it will help to avoid the results of the global warming. One of these energies is the solar energy. The idea of the solar energy has been raised to improve the sustainability levels of the countries and energy sectors. The decision came to develop the countries with the task of the renewable energy projects. The solar energy plans become important in the recent years. In addition, the integrating variable energy resource (VER) into electric network grid can play main resource from solar photovoltaic (PV). The VER as new resources currently envisioned to be either wind or solar photovoltaic (PV). These types of resources output can be highly variable and depend on weather fluctuations such as wind speed and cloud cover. Since PV generation is highly dependent on weather conditions. This means that the PV generation behavior differs in different regions. In particular, the solar irradiance is affecting the PV generation behavior. This means that the solar power forecasting becomes an important tool for optimal economic dispatch of the electric power network. In this chapter, the artificial intelligence technique is deployed to estimate the calculation of the number of solar power panels required to satisfy given estimated daily electric load for five countries. The artificial intelligence techniques play an important role in modeling and prediction of renewable energy engineering. The main purpose of the present chapter is the design of PV panel which helps to reduce the emission of CO_2 emission, where these panels have a connection to the national electricity grid. This grid feed by the extra generated electricity through the solar power plant. In this case, the power plant becomes more efficient compared with the combined cycle plant. At the same time, the modeling and prediction in renewable energy engineering help the engineers to predict the future estimation of the required estimated load.

References

1. M.H. Amini, H. Arasteh, P. Siano, Sustainable smart cities through the lens of complex interdependent infrastructures: panorama and state-of-the-art, in *Sustainable Interdependent Networks II* (Springer, Cham, 2019), pp. 45–68
2. M.H. Amini, J. Mohammadi, S. Kar, Distributed holistic framework for smart city infrastructures: tale of interdependent electrified transportation network and power grid. IEEE Access **7**, 157535–157554 (2019)
3. Z. Andrea et al., Internet of things for smart cities. IEEE Internet Things J. **1**(1), 22–32 (2014).
4. M.H. Amini, Distributed computational methods for control and optimization of power distribution networks, PhD Dissertation, Carnegie Mellon University, 2019
5. M.H. Amini, A multi-layer physic-based model for electric vehicle energy demand estimation in interdependent transportation networks and power systems, in *Optimization in Large Scale Problems* (Springer, Cham, 2019), pp. 243–253
6. M.H. Amini, A panorama of interdependent power systems and electrified transportation networks, in *Sustainable Interdependent Networks II* (Springer, Cham, 2019), pp. 23–41
7. M.H. Amini, M.P. Moghaddam, O. Karabasoglu, Simultaneous allocation of electric vehicles' parking lots and distributed renewable resources in smart power distribution networks. Sustain. Cities Soc. **28**, 332–342 (2017)
8. M.R. Mozafar, M.H. Amini, M.H. Moradi, Innovative appraisement of smart grid operation considering large-scale integration of electric vehicles enabling V2G and G2V systems. Electr. Pow. Syst. Res. **154**, 245–256(2018)
9. M.H. Amini, O. Karabasoglu, Optimal operation of interdependent power systems and electrified transportation networks. Energies **11**, 196 (2018)
10. M.H. Amini, A. Kargarian, O. Karabasoglu, ARIMA-based decoupled time series forecasting of electric vehicle charging demand for stochastic power system operation. Electr. Pow. Syst. Res. **140**, 378–390 (2016)

Part I
Theoretical Algorithms for Optimization, Learning, and Data Analytics in Interdependent Complex Networks

Chapter 2
Promises of Fully Distributed Optimization for IoT-Based Smart City Infrastructures

M. Hadi Amini, Javad Mohammadi, and Soummya Kar

Abstract Modern wireless communication and sensor technologies enable ubiquitous sensing through distributed agents in various networks. In the context, collaborative interaction of Internet of Things (IoT) connected end-user devices enables achieving certain goals. In this study, we aim at introducing a holistic distributed framework that enhances operational performance of smart cities infrastructures. Then, we explain a fully distributed *consensus + innovations* approach for coordinating among agents. Our proposed *consensus + innovations* sits at the core of the proposed framework. In a nutshell, our distributed algorithm achieves a distributed solution for a collaborative decision making problem through iterative agent-based computations and limited inter-agent communications. This algorithm enables fully distributed coordination of agents, plug-and-play capability, and scalability of the solution algorithm for future network expansion.

Keywords Distributed consensus + innovations · Smart cities · Interdependent infrastructures · Plug-and-play capability

---

Some parts of this book are inspired by our open access article in M.H. Amini, J. Mohammadi, S. Kar, Distributed holistic framework for smart city infrastructures: tale of interdependent electrified transportation network and power grid. IEEE Access 7, 157535–157554 (2019).

M. H. Amini (✉)
School of Computing and Information Sciences, Florida International University, Miami, FL, USA

Sustainability, Optimization, and Learning for InterDependent Networks Laboratory (solid lab), Florida International University, Miami, FL, USA
e-mail: amini@cs.fiu.edu; hadi.amini@ieee.org; www.solidlab.network

J. Mohammadi · S. Kar
Department of Electrical and Computer Engineering, Carnegie Mellon University, Pittsburgh, PA, USA
e-mail: jmohamma@andrew.cmu.edu; soummyak@andrew.cmu.edu

M. H. Amini (ed.), *Optimization, Learning, and Control for Interdependent Complex Networks*, Advances in Intelligent Systems and Computing 1123, https://doi.org/10.1007/978-3-030-34094-0_2

2.1 Introduction

2.1.1 Motivation[1]

In recent years, the Internet of Things (IoT) has been shaping the interaction of heterogeneous agents, including large number of sensing devices, communication technologies, and required services [2, 3]. Internet of Things (IoT) connected sensing and actuation is the key enabler in transitioning urban spaces towards smart sustainable urban environments. Smart city concept lends itself well to address upcoming challenges of conventional cities by providing integrative management and coordination of intelligent infrastructures [4].

A fast increase in the number of urban vehicles originates from the growing size of smart cities, from a network-of-networks perspective, and transportation network expansion, from an independent network-wise viewpoint. This growth imposes several challenges on the traffic management, such as traffic congestion and air pollution [5]. One of the pivotal players and promising solutions in the transition towards smart mobility as a part of smart cities is electric vehicle (EV) [6]. EVs not only can contribute to reducing the air pollution, but also can help managing the traffic congestion more effectively as they can potentially recharge their battery at a wider range of locations as compared with fuel-based vehicles. Further, EVs are acting as coupling agents due to their role in various networks in a smart city [7–9]. First, there are a subset of vehicles in transportation networks. Hence, their mobility pattern and routing decisions affect traffic management. According to [10], improving the quality of urban mobility (e.g., reducing the traffic congestion) is one of the most important goals of smart cities. According to 2018 Global EV Outlook of International Energy Agency (IEA) the total number of EVs on the road exceeded 3 million, which is 50% expansion as compared with 2016. Under two scenarios, referred to as IEA's New Policies Scenario and EV30@30 Scenario, the expected number of EVs on the road reaches 125 million and 220 million by 2030, respectively [11]. We have used the available data regarding the total number of vehicles on the road from 1960 to 2014 to estimate the expected number of vehicles worldwide by 2030 (for more information on the vehicles' historical data please refer to [12–15]). The output of our investigation shows the projected number of vehicles on the road is 3.73 billion by 2030. Consequently, depending on the introduced scenarios by IEA [11] and the projected values based on historical data, 3.34–5.89% of the total vehicles worldwide will consist of EVs. This considerable penetration of EVs not only is worth investigating from the transportation network point of view, but also has a substantial impact on the operation of power distribution networks. Specifically, there is an increasing transition towards deploying more fast charging stations. According to the report by National Renewable Energy Laboratory, 2017, during the past 5 years, Tesla has installed 357 fast charging

---

[1]Some parts of this chapter are inspired by our open access article in [1].

stations with an average of seven charging spots at each station (each fast charging station has capacity of supporting 1–12 electric vehicles). This translates into providing 2478 fast charging spots with a maximum charging rate of 120 kW [16]. As of August 2018, this number has expanded to 1342 T supercharger (fast charging) stations worldwide that translates into 11,013 charging spots [17]. From the transportation perspective, optimal management of EVs and exploiting their flexibility to charge their batteries at different locations can contribute to transportation network congestion management [18, 19]. From the power system perspective, EVs can play a pivotal role in balancing the load and generation by serving as mobile and flexible loads [20–23]. They also can lead to reducing gasoline consumption and air pollution since they consume electricity. Required energy to charge the EVs' batteries can be supplied by distributed, renewable, and less carbon-intensive energy resources [24, 25]. The spatiotemporal nature of EV charge scheduling and EV routing problems can be exploited as an existing platform for enabling fully distributed algorithms. This can not only facilitate peer-to-peer (P2P) energy trading but also allow for local computations instead of centralized extensive computing requirements [26]. Although the interdependent representation of plug-in EV (PEV) charge scheduling problem seems straightforward and clear, it constitutes several layers of complexity and involves multiple networks. The ultimate goal of this paper is to provide a thorough vision towards leveraging the interdependencies of power and transportation networks while taking advantage of PEVs as coupling agents. We outline our motivation for developing distributed solutions and enabling plug-and-play capability for PEV charge scheduling from the IoT lens in three major stages as shown in [1]:

1. From independently operating the networks to fully distributed holistic coordination of interdependent networks.
2. From the interdependent networks towards the Internet-of-Things framework with heterogeneous agents.
3. From the Internet-of-Things framework to smart power grids and electrified transportation networks.

2.1.2 Related Works

In the literature, power systems and electrified transportation networks referred to as coupled, interdependent, joint, and interconnected networks alternatively. We review the literature in a top-down manner. We start from providing a big picture of a notable example of interdependent networks, smart cities. We then review the literature on decentralized/distributed/hierarchical approaches for coordination of plug-in electric vehicles and optimal power flow problem as key players in interdependent power systems and transportation networks.

Transportation electrification has been investigated in the literature mostly from the viewpoint of integrating EVs into smart power grids. EVs can facilitate integration of renewable resources by serving as spatiotemporally flexible demands [20, 27–29]. In [27], simultaneous integration of EV charging stations and renewable resources is leveraged to reduce the loss in power distribution systems. According to [28], coordinated operation of EV charging stations and vehicle-to-grid (V2G) technology facilitate the high penetration of renewable resources. Further, in [29] a decentralized method is proposed for joint planning of EV charging stations and PV generation units.

EVs are not only capable of increasing demand flexibility, but also can inject power in terms of V2G technologies [30, 31]. The required infrastructure (e.g., charging stations) is the key to enabling transportation electrification. In [32] smart charging algorithms for EV charge scheduling have been reviewed. In [33], a smart EV charging framework is introduced to optimize load demand from the view point of power grid as well as meeting the requirements of EV drivers. Further in [34], the required infrastructure of smart charging is introduced, that consists of a mobile application, an optimal charge scheduling method, V2G operation algorithm, and a remote information exchange hardware (e.g., radio-frequency identification). One of the major advantages of our proposed framework as compared with the literature is to upgrade the current assets to enhance their operation by taking their interdependent nature into account; i.e., we do not enforce any network to add new entities for the sake of implementing our distributed approach. Alizadeh et al. [35] introduced charging network operator to address the coordination of multiple charging stations. From the traffic management perspective, some of the previous works developed frameworks that take into account traffic constraints while determining the optimal charge schedule of EVs [19, 36–48]. Traffic flow information is used to develop a supervised predictive energy management approach for plug-in hybrid EVs in [36]. Mobile charging vehicles are introduced in [37] to provide spatially flexible charging stations to EVs. In order to determine the optimal location of multiple types of charging stations (parking lots and mobile charging vehicles), Cui et al. [37] proposed to solve optimal charging station location and EV routing simultaneously; i.e., the proposed formulation minimizes the total traveled distance while satisfying the charging stations' constraints, traffic flow constraints, time constraints, and electricity constraints. Interdependent nature of electric power grids and electrified transportation networks have been modeled in terms of a novel routing strategy in [19], referred to as Charging Station Strategy-Vehicle Powertrain Connected Routing Optimization (CSS-VPCRO). In order to capture the coupling between these two networks, optimal power flow problem and CSS-VPCRO are conducted iteratively. At each iteration, EVs find their optimal route based on their desired charge demand, traffic conditions, and electricity price. They broadcast their expected charging demand as well as optimal charging station location with power system operator. Power system operator updates price signals based on the updated electricity demand and share these signals with EV routing database. EVs update their routing decision accordingly and this closes the loop in CSS-VPCRO [19]. A hybrid dynamic system assessment framework is introduced in [38] for multi-modal transportation electrification. This framework considered EVs as the coupling point of transportation system, electric power grid, and their corresponding information infrastructures which formed a *transportation-electricity nexus*. Vehicular energy networks (VEN) concept is recently introduced by Lam et al. [39, 40]. VEN covers a wide range of systems and decision parameters from both power and transportation networks, including EVs, renewable energy, road network, vehicular traffic, energy path, charging cycles, vehicular ad hoc networks, and wireless power transfer [39]. A hybrid dynamic model is proposed in [49] for transportation electrification that also includes next generation traffic simulation. A hierarchical control architecture is developed in [42] that utilized traffic light status, received through vehicle to infrastructure (V2I) communication platform, and neighboring vehicles' status broadcast via vehicle-to-vehicle (V2V) communication platform for optimal energy management of hybrid EVs. In [50], a thorough investigation is conducted to study open source simulation tools for transportation electrification. Alizadeh et al. tackled these coupled networks from a market design perspective [46]. To this end, two non-profit entities, the independent power system operator and independent transportation system operator collaborate to determine jointly optimal price signals, charging station strategies, and road tolls [46]. Alizadeh et al. further investigate the routing problem of an EV driver considering its charging station choices, dynamic traffic conditions, and locational electricity price values [47].

2.1.3 Contribution

We provide the list of contributions of our proposed framework, as well as advantages of using fully distributed algorithm for enabling plug-and-play capability in two major categories: application-wise and theoretical contributions. Note that this study has two major contributions as compared with [51]: (1) we introduce a holistic framework to identify interdependence among power systems and electrified transportation networks, and the required information exchange to improve the optimal operation of these networks, (2) we investigate plug-and-play capability of our distributed solution, evaluate the effect of mobility on the feasibility of the proposed cooperative charging algorithm, and improve the proposed model by taking into account the power constraint enforced by power system, i.e., we model both mobility patterns from transportation network and power constraint from power system perspective.

Application-Wise Advantages: Making the Case for Distributed Solution Based on Real-World Scenarios and Requirements

1. Our proposed distributed framework allows for optimizing the internal goals of each network operator while taking into account the exogenous information from other influential networks.
2. In the proposed framework, PEV charging demand can be used as mobile dispatchable load to reduce the congestion in power distribution networks.

3. Previous studies on coupled power and transportation networks [19, 38, 52–54] have mainly developed centralized approaches which allow them to solve the entire problem in a synchronized fashion. Although this reduces the complexity of their frameworks and corresponding solutions, there is a missing piece in this puzzle: different agents at each network have their operational specifications; i.e., it is more realistic to solve the problem in distributed fashion. This allows for local decision making of heterogeneous agents from various networks.

Algorithm-Wise Advantages

1. Distributed algorithms enable plug-and-play feature in the expanding networks with emerging elements.
2. Computational complexity and run-time are reduced by decomposing a large-scale problem into several small problems. Further, we use a projection operator to enforce the power constraint at each iteration. This also reduces the complexity of our proposed update rule at each iteration.
3. The robustness to communication failure between the agents is enhanced by ensuring a feasible solution at each iteration.

2.1.4 Organization

The rest of this chapter is structured as follows: Sect. 2.2 provides more details of the proposed framework by elaborately explaining different interdependent layers and their interaction. Section 2.4 presents a general formulation and the corresponding optimally conditions. Section 2.5 is devoted to the proposed *consensus + innovations* distributed algorithm for the general optimization problem. Conclusions are provided in Sect. 2.6. Detailed convergence analysis of the distributed algorithm for general optimization problem is provided in Appendix 1.

2.2 A Novel Holistic Framework for Interdependent Operation of Power Systems and Electrified Transportation networks

The proposed holistic distributed framework covers three networks with four major layers. Note that these networks are considered for the sake of representation the growing interdependencies.

- Power distribution network operated by distribution network operator (DNO).
- Charging coordination network operated by PEV aggregators who are managing charging stations.

- Transportation network operated by transportation network operator (TNO) that includes two sub-networks:
  - Traffic management systems. [5]
  - Parking management systems.

In the rest of this article, we elaborately define the objectives and constraints of PEV charge scheduling problem. Then, we formulate the general optimization problem, to be justified based on objective and constraints of each agent at any arbitrary network. We further provide a *consensus + innovations*-based distributed algorithm with performance guarantees to solve the optimization problem of various agents in a fully distributed fashion.

Consensus + innovations methods have been widely applied to emerging applications, providing a smorgasbord of distributed solutions for networks, such as distributed energy management in smart power grids [55–61], secured distributed inference for the Internet-of-Things (IoT) [62], cooperation and sensing in networked systems [63], cooperative charging of electric vehicles [64], robust economic dispatch in power systems [65], and multi-agent coordination of microgrids [66].

Although we have listed motivations for distributed optimization in interdependent networks, we argue that most of the previous works on distributed optimization for a subset of layers/agents in the interdependent networks could be efficiently conducted using available centralized solutions. In fact, in some cases centralized solution is providing acceptable computational efficiency. To this end, we have categorized the agents/elements of interdependent power and transportation networks into four groups. Our ultimate goal is to motivate application of centralized solutions for some subsets, and distributed solutions to the other subsets. Ultimately, we conclude that a holistic framework, that covers all these four aspects, needs to be operated in a distributed fashion.

Previous works focused on deploying cooperative and non-cooperative solutions to solve charge coordination of PEVs. To this end, several approaches have been proposed, including [67], fully distributed *consensus + innovations* algorithm [51, 68], consensus-based distributed charging control [69], distributed consensus-based charge scheduling [70], hierarchical charge scheduling using Danzig–Wolfe decomposition [71], the alternating direction method of multipliers [72, 73], mean field game theory [74, 75].

Some studies considered the coupling among these networks, such as [1, 15, 34, 45, 49, 51]. This class, however, deployed centralized solutions which increases the complexity of the model from optimization and computational burden perspectives, as well as policy making perspective. Such models have to take the decision variables and constraints of all stakeholders from both networks in a single model. This raises two major challenges: (1) increasing the number of decision variables and constraint directly enlarges the size of optimization problems which makes it hard-to-solve as compared with the independent optimization problems of each network; (2) complicating the operators' roles at each network due to combining various objectives from different stakeholders and various networks, i.e., the questions of *Who is solving this large-scale optimization problem?* and *Who is gaining benefit from solving this problem?* will be more crucial after merging the optimization problems of multiple stakeholders from multiple networks, including but not limited to power system operator, transportation network operator, and EV drivers.

Our Solution: Holistic Agent-Based Distributed Optimization of Interdependent Networks Through the IoT Lens

Our framework provides a solution that takes into account the increasing intelligence and emerging widespread M2M communications enabled by 5G technology. We model each element at any network as a heterogeneous agent with communication capabilities. Depending on their goals and decision making criteria, a cluster of agents may choose to cooperate with each other to reach a consensus towards a common goal. This is the pivotal contribution of our framework that enables spatiotemporal plug-and-play capability, i.e., at any time and at any location, an agent can decide which other agents and entities to communicate/cooperate with. In our knowledge, this study is the first of its kind that provides a holistic model for interdependent power and electrified transportation networks while enabling distributed decision making. For instance, a PEV driver plans to find the optimal route from current location, loc_1 to the destination loc_2. This trip lasts from time t_1 to t_2 and the battery state-of-charge of PEV reduces from SOC_1 to SOC_2. The PEV driver may decide to charge at different location and different time to optimize her/his objective leveraging the spatiotemporal flexibility.

2.3 Definition of Agents and Their Corresponding Features

In this section we identify various agents in the proposed framework, their objectives and constraints, and the time-scale at which each agent is operating. To this end we investigate the agents in three major categories: power system-specific agents, transportation network-specific agents, and coupling agents. These agents can count as one of the following categories: *passive* such as traffic lights, *active decision maker* such as EV charging station aggregators, and *active sensor* such as EV charging stations, based on their functionality. A passive agent only receives commands, which can be basically the output decisions of active agents based on the local optimization at each iteration, and change their state based on the received command, e.g., traffic lights are basically passive agents that are responsible to switch between two status (red and green) to control the traffic. Another set of passive agents are the ones who are responsible for recording data and communicating raw data to the intelligent agents, such as conventional traffic cameras, here referred to as *passive* traffic cameras. Active decision maker agents are the ones who not only receive or fetch the data from other sources and agents, but also use the received data to solve an optimization problem and send the proper command signals to other agents. Sensor agents are the agents that are collecting/receiving

Table 2.1 Agents and their features

Agent	Feature		
	Active/passive	Physical/virtual	Power/transportation/coupling
Electric vehicle (EV)	Active	Physical	Coupling
EV charging station	Active	Physical	Coupling
Demand response aggregator	Active	Virtual	Power
Distributed energy resources	Active	Physical	Power
Transformer agents	Active	Physical	Power
Distribution system operator agent	Active	Physical	Power
Intelligent traffic cameras	Active	Physical	Transportation
Passive traffic cameras	Passive	Physical	Transportation
Road side units	Passive	Physical	Transportation
Traffic lights	Passive	Physical	Transportation
Toll road pricing agent	Active	Virtual	Transportation

data at one layer, and communicate it to other agents at another layer. These agents help us using the current infrastructures with minimum hardware requirements and communication platform. EV charging stations are consummate examples of sensor agents that indirectly enable communication between power distribution network layer and charge coordination layer. Table 2.1 summarized agents from various networks and their features.

2.3.1 Power System-Specific Agents

1. Distribution System Operator: This agent is responsible for maintaining reliable operation of power distribution networks by optimizing the available resources and satisfying the physical constraints of the grid. It mainly manages the power delivery from transmission networks to the customers.
2. Demand Response Aggregator: This agent offers demand response services with two main objectives: reducing the electric load demand of the customers and maximizing its benefit by saving energy. As a commercial entity, it offers the load reductions to the wholesale energy market. The interaction of demand response aggregator and utilities can be modeled as a non-cooperative game (see for example [92]).
3. Distributed Energy Resources (DERs) Agent: This agent potentially covers a wide range of technologies and resources, including energy storage units and renewable energy resources (e.g., PV panels). Its main task is to optimize the internal operation of the corresponding resource and to maximize the benefit of the DER owner.
4. Transformer Agents: This agent is responsible for communicating the transformer's situation to other entities. In the intelligent distribution system, a smart

transformer agent at the main substation of each feeder is capable of conducting optimal power flow with respect to the transformer loading constraint as well as the expected load demand.

2.3.2   Transportation Network-Specific Agents

1. Intelligent Traffic Cameras: These cameras are capable of monitoring the vehicles, local decision making, sharing traffic situation with other agents, and broadcasting command signals to traffic lights to manage congestion.
2. Passive Traffic Cameras: These cameras are only capable of monitoring the traffic situation and sharing it with decision making entities. The main difference of these cameras and intelligent traffic cameras is lack of decision making capability.
3. Road Side Units: Road side units (RSUs) are equipped with communication capability. They communicate with on-board units (installed on the vehicles) to monitor traffic situation, such as location and speed of the vehicles.
4. Traffic Lights: Traffic lights are mainly scheduled to follow a certain schedule. They are equipped with a remotely controllable device which can be managed through the control signals from active/decision maker agents, such as intelligent traffic cameras.
5. Toll Road Pricing Agents.

2.3.3   Coupling Agents

These agents interact with both power system-specific and transportation network-specific agents.

1. Electric vehicle (EV): EVs are one of the major coupling agents. The coupling is caused by their optimal routing decisions that affect the congestion in transportation networks, as well as their spatiotemporal charging decisions that affect both the load demand in power systems and traffic condition of transportation networks. Potential objectives, internal constraints, and external limitation of EVs are explained in the following paragraph.

   Electric vehicles goals may include: (1) finding optimal route, (2) reducing charging cost, and (3) leveraging the flexibility of their load demand in terms of time and location (spatiotemporal flexibility) to reduce energy cost. Constraints of the EVs can be categorized into: (1) internal constraints which are enforced by the driver or technical specifications of the EV, such as charging rate limit, minimum state-of-charge, time limits to arrive destination, and duration of stay at charging station; (2) external constraints which are mainly caused by exogenous inputs from other agents, such as limits enforced by power network agents (e.g., hourly demand limit and locational marginal price variations due to line congestion), and limits enforced by transportation network agents (e.g., traffic conditions and traffic congestion pricing).

Goals

- Finding their optimal route
- Reducing their charging cost
- Leverage the flexibility of their load demand in terms of time and location (spatiotemporal flexibility) to reduce energy cost

Constraints

Internal Constraints

- Charging rate
- Minimum state-of-charge
- Time limits to arrive destination
- Duration of stay at charging station

External Constraints

- Limits enforced by power network agents (e.g., hourly demand limit and locational marginal price variations due to line congestion)
- Limits enforced by transportation network agents (e.g., traffic conditions and traffic congestion pricing)

2. Charging station agent[2] : Charging stations play a pivotal role in modeling the interdependency among power systems and transportation networks. First,

---

[2]There are various types of charging stations that can enable communication. For instance, Eaton offers the following four models with different functionality:

A series: Single phase, no communication, proper for residential uncontrolled applications, up to 7.4 kW charging capacity.

X series: Single/three phase, communication and building management system integration capability, proper for controlled charging at the residential level, up to 7.4 kW charging capacity

S series: Single/three phase, intelligent load management and communication capability; offers the features of X series as well as enabling UDP (the standard protocol for integrating a device into other operating systems, such as a smart home system) and OCPP (the standard protocol that is used if several charging stations are networked together), two options for the charging capacity: 7.4 kW or 22 kW.

xChargeIn M series: integrator (master) for networking a number of **S series** equipped with online communication

The **M series** serves as a master device in online or offline charging systems and manages the connected vehicles via individual charging stations of the **S series**. A charging system can consist of one **M series** master station and up to 15 **S series** charging stations.

Source: http://www.eaton.eu/Europe/Electrical/ProductsServices/Residential/xChargeIn/index.htm.

they communicate with power distribution system operator, demand response aggregator, and other entities to optimize their cost of energy. Second, they try to find the optimal strategy to attract more EVs. After the EVs plugged in their batteries, the charging station agent needs to make sure to satisfy all EVs energy needs while maximizing its own benefit. Note that the charging station agent can also participate in demand side management programs leveraging its flexibility based on the plugged in EVs, i.e., the optimal decision to charge or not to charge EVs at each timestep can vary based on the market signals, load demand, and availability of other distributed energy resources. Potential objectives, internal constraints, and external limitation of charging station agents are explained below.

EV charging station goals may include: (1) maximizing their profit by offering optimal price signals to EV agents, (2) meeting all EVs' charging demand expectations, (3) leveraging the flexibility of EV load demand to increase their profit, and (4) providing ancillary services/demand side management to power systems. Similar to EV agent, charging station agent may also enforce internal constraints caused by the decision making parameters of the aggregator or technical specification, such as maximum capacity in terms of charging spots and maximum capacity in terms of total hourly power demand. There are also externally constraints that are enforced by exogenous agents, such as limits enforced by power network agents (e.g., line congestion limits) as well as limits of transportation network agents (e.g., traffic conditions that affect the time for EVs to arrive charging stations).

Goals

- Maximizing their profit by offering optimal price signals to EV agents
- Meeting all EVs' charging demand expectations
- Leveraging the flexibility of EV load demand to increase their profit
- Providing ancillary services/demand side management to power systems
  Constraints

Internal Constraints

- Maximum capacity in terms of charging spots
- Maximum capacity in terms of total hourly power demand

External Constraints

- Limits enforced by power network agents (e.g., line congestion limits)
- Limits of transportation network agents (e.g., traffic conditions that affect the time for EVs to arrive charging stations)

2.4   General Optimization Problem

2.4.1   Problem Formulation

Formulate the centralized optimization problem, shown in (2.1). Let Ω_{agents}, Ω_{ineq}, and Ω_{eq} denote sets of all agents, inequality constraints, and equality constraint, respectively.

$$\underset{x_k}{\text{minimize}} \quad \sum_{k \in \Omega_{\text{agents}}} f_k(x_k) \tag{2.1a}$$

$$\text{s.t.} \quad g_j(x) \leq 0; \quad (: \mu_j) \quad j \in \Omega_{ineq} \tag{2.1b}$$

$$h_j(x) = 0; \quad (: \lambda_j) \quad j \in \Omega_{eq} \tag{2.1c}$$

where $f_k(\cdot)$ and x_k denote the objective function and variable(s) of agent k in the network, respectively. Depending on the problem definition, agent can represent a wide range of physical or virtual entities, including power distribution network bus, microgrid operator, demand response aggregator, and charging station aggregator. Functions $g_j(\cdot)$ and $h_j(\cdot)$ denote corresponding functions of inequality and equality constraints, respectively.

2.4.2   Optimality Conditions

Formulate the *Lagrangian* for the optimization problem in (2.1), as shown in (2.2).

$$\mathcal{L} = \sum_{k \in \Omega_{\text{agents}}} f_k(x_k) + \sum_{j \in \Omega_{ineq}} \mu_j g_j(x) + \sum_{j \in \Omega_{eq}} \lambda_j h_j(x).$$

Derive the first-order optimality conditions, as provided in (2.2).

$$\begin{cases} \frac{\partial \mathcal{L}}{\partial x_k} = 0, & \forall k \in \Omega_{\text{agents}} \\ \frac{\partial \mathcal{L}}{\partial \mu_j} \leq 0, & \forall j \in \Omega_{ineq} \\ \frac{\partial \mathcal{L}}{\partial \lambda_j} = 0, & \forall j \in \Omega_{eq}. \end{cases} \tag{2.2}$$

2.5   *Consensus + Innovations* **Based Distributed Algorithm**

2.5.1   *Distributed Decision Making: General Distributed Update Rule*

Distributed iterative approach for a generic optimization problem is followed to solve the first-order optimality conditions in (2.2). The iterative model only needs information exchange between physically connected agents at each iteration. Let Ω_i denote the neighboring set of agent i. Let $y_i(k) = [x_i(k), \mu_j(k), \lambda_j(k)], j \in \Omega_i$ denote the variable associated with agent i at iteration k. The general format of the local updates which is performed by all agents at each iteration is shown in (2.3).

$$y_i(k+1) = \mathbb{P}[y_i(k) + \rho_i s_i(y_j(k))]_{\mathcal{F}}, \quad j \in \Omega_i \tag{2.3}$$

where $s_i(\cdot)$ reflects the first order optimality constraints related to agent i, and ρ_i denotes the vector of tuning parameters. Further, $\mathbb{P}$ is the projection operator to project x_i onto its determined feasible space, denoted by $\mathcal{F}$.

Note that, $s_i(y_j(k))$ only depends on the iterates $y_j(k)$ of neighboring nodes j in the physical neighborhood of i. Hence, a distributed implementation of (2.3) is possible.

The projection operator settings, tuning parameters, and corresponding constraints vary based on the network objectives, constraints, and decision making variables. We later elaborate on each network's optimization problem as well as network-oriented distributed algorithm that is tailored for each network.

2.5.2   *Agent-Based Distributed Algorithm*

Here we present a more detailed formulation of *consensus + innovation* based distributed algorithm at the intra-network layer. Let inter-agent communication graph to be connected.

Agent i updates its local variables, i.e., variables that are directly corresponding to this agent, i.e., y_i. Let k represent the iteration counter. The corresponding variables of agent i are updated using (2.4).

$$y_i(k+1) = \mathbb{P}[y_i(k) + \rho_i^{\mathcal{C}} \overbrace{s_i(y_j(k))}^{\text{neighborhood consensus}}$$
$$+ \rho_i^{\mathcal{I}} \underbrace{s_i(y_i(k))}_{\text{local innovation}}]_{\mathcal{F}}, \quad j \in \Omega_i \tag{2.4}$$

where $\rho_i^{\mathcal{C}}$ denotes positive tuning parameters corresponding to consensus among agent i and its neighboring agents $j \in \omega_i$. Further, $\rho_i^{\mathcal{I}}$ is the tuning parameter for the local innovation term. In (2.4), the first and second terms represent the neighborhood *consensus* and local *innovation*, respectively.

Consequently, the update rules for all the variables at the intra-network optimization of network $\mathcal{N}$ in a dense form are provided by (2.5)

$$X_{\mathcal{N}}(k+1) = \widetilde{X}_{\mathcal{N}}(k) - A_{\mathcal{N}}\widetilde{X}_{\mathcal{N}}(k) + C_{\mathcal{N}}$$
$$\widetilde{X}_{\mathcal{N}}(k+1) = \mathbb{P}\,[X_{\mathcal{N}}(k+1)]_{\mathcal{F}} \qquad (2.5)$$

where $X_{\mathcal{N}}$ is the vector of the stacked variables, i.e., y_i, for all agents, and $\mathbb{P}$ is the projection operator which ensures that the Lagrange multipliers for the inequality constraints stay positive and the box constrained variables stay within the given bound. Further, $\mathcal{F}$ represents the feasible space spanned by positiveness and box constraints. Hence, $\widetilde{X}$ is the vector of the stacked projected variables.

The detailed convergence analysis of the proposed distributed algorithm is provided in Appendix 1.

2.6 Conclusions

We develop a holistic agent-based distributed algorithm and framework for the IoT-based interdependent networks, with a major focus on interdependent power systems and electrified transportation networks. Our solution enables distributed coordination of agents in the network-of-networks, such as smart city infrastructures. To this end, we propose a fully distributed *consensus + innovations* approach. Our distributed iterative algorithm achieves a distributed solution of the decision making for each agent through local computations and limited communication with other neighboring agents that are influential in that specific decision. For instance, the optimal routing decision of a PEV involves a different set of agents as compared with the optimal charging strategy of the same PEV. The exogenous information from an external network/agent can affect internal operation of the other agents. For instance, having some information about traffic congestion at the transportation networks changes the decision of electric vehicles (EVs) to charge their battery at another location. Our approach constitutes solving an iterative problem, which utilizes communication at the smart city layer, as a network of different infrastructures that enables fully distributed coordination of agents, plug-and-play capability, and scalability of the solution algorithm for future expansion of each network.

Appendix 1: Convergence Analysis[3]

This section presents a formal proof that any limit point of the proposed algorithm in (2.5) is the optimal solution of the optimization problem in (2.1). Moreover, it introduces a sufficient condition for the convergence of the proposed algorithm.

In the following Theorem 2.1, we first show that a fixed point of the proposed iterative scheme necessarily satisfies the optimality conditions (2.2) of the original optimization problem.

Theorem 2.1 *Let X^* be a fixed point of the proposed algorithm defined by (2.5). Then, X^* satisfies all of the optimality conditions of the original problem (2.2).*

Proof To prove this theorem, we verify the claim that X^* fulfills all of the first order optimality conditions. Note that X^* is the vector of stacked variables.

Claim 1 X^* fulfills the optimality conditions which enforce the positivity of the Lagrangian multipliers associated with the inequality constraints, i.e., $\mu_j^* \geq 0$.

Verification by contradiction: Let us assume on the contrary that in X^* one of the multiplier variables, say μ_j^*, is negative. Now, note that, evaluating (2.4) at X^* results in a non-negative value for μ_j due to the projection of μ_j into the set of positive reals. This contradicts the fact that X^* is a fixed point of (2.2).

Claim 2 X^* satisfies the optimality conditions associated with the inequality constraints, $\frac{\partial \mathcal{L}}{\partial \mu_j} \leq 0$.

Verification by contradiction: Let us assume that X^* does not fulfill $\frac{\partial \mathcal{L}}{\partial \mu_j} \leq 0$ for all j, i.e., there exists j such that $\frac{\partial \mathcal{L}}{\partial \mu_j}(X^*) > 0$. This implies that the value of the innovation term in (2.4) is negative when evaluated at X^*. Also, note that, based on the claim 1, $\mu_j^* \geq 0$. Therefore, evaluating (2.4) for the inequality constraints at X^* results in a value greater than μ_j^* which contradicts the fact that X^* is a fixed point of (2.4). Similar arguments can be used to prove that X^* fulfills the KKT conditions corresponding to the equality constraints, $\frac{\partial \mathcal{L}}{\partial \lambda_j} = 0, \quad \forall j \in \Omega_{eq}$.

Claim 3 X^* satisfies the optimality conditions associated with the complementary slackness condition, i.e., for all $j \in \Omega_{ineq}$, we have $\mu_j^* \cdot \left(g_j(x^*) \right) = 0$.

Verification by contradiction: Let us assume on the contrary that X^* does not satisfy the above complementary slackness condition, i.e., there exists a value for j such that both μ_j^* and $g_j(x^*)$ are non-zero. Hence, according to the claims 1 and 2, we must have, $\mu_j^* > 0$ and $g_j(x^*) < 0$, respectively. Now, note that evaluating (2.4) at X^*, results in a value less than μ_j^*, which clearly contradicts the fact that X^* is a fixed point of (2.4).

---

[3]This appendix is inspired by our work in [1].

We now discuss the consequences of Theorem 2.1. To this end, note that, since the proposed iterative scheme (2.5) involves continuous transformations of the updates, it follows that, if (2.5) converges, the limit point is necessarily a fixed point of the iterative mapping. Since, by Theorem 2.1, any fixed point of (2.5) solves the first order optimality conditions (2.2), we may conclude that, if (2.5) converges, it necessarily converges to a solution of the first order optimality conditions (2.2). This immediately leads to the following optimality of limit points of the proposed scheme.

Theorem 2.2 *Assume that the original optimization problem (2.1) has a feasible solution that lies in the interior of the corresponding constraint set. Further, suppose the proposed algorithm introduced by (2.5) converges to a point $X^\star$. Then X^* constitutes an optimal solution of the original problem (2.1).*

Proof By Theorem 2.1 and the above remarks, $X^\star$ satisfies the optimality conditions (2.2). Since the original optimization problem is a convex problem and, by assumption, is strictly feasible, it follows that the primal variables $(x_i^\star)$ in X^* constitute an optimal solution to the original problem (2.1).

Consequently, we note that Theorems 2.1 and 2.2 guarantee that any fixed point of the proposed distributed algorithm constitutes an optimal solution to the original problem, and, if the scheme achieves convergence, the limit point is necessarily an optimal solution of the original problem. Finally, we note, that whether the scheme converges or not depends on several design factors, in particular, the tuning parameters ρ_i^C and ρ_i^T.

References

1. M.H. Amini, J. Mohammadi, S. Kar, Distributed holistic framework for smart city infrastructures: tale of interdependent electrified transportation network and power grid. IEEE Access **7**, 157535–157554 (2019)
2. A. Zanella, N. Bui, A. Castellani, L. Vangelista, M. Zorzi, Internet of things for smart cities. IEEE Internet Things J. **1**(1), 22–32 (2014)
3. F. Montori, L. Bedogni, L. Bononi, A collaborative internet of things architecture for smart cities and environmental monitoring. IEEE Internet Things J. **5**(2), 592–605 (2018)
4. S.E. Bibri, J. Krogstie, Smart sustainable cities of the future: an extensive interdisciplinary literature review. Sustain. Cities Soc. **31**, 183–212 (2017)
5. S. Djahel, R. Doolan, G.-M. Muntean, J. Murphy, A communications-oriented perspective on traffic management systems for smart cities: challenges and innovative approaches. IEEE Commun. Surv. Tutor. **17**, 125–151 (2015)
6. L. Grackova, I. Oleinikova, G. Klavs, Electric vehicles in the concept of smart cities, in *2015 IEEE 5th International Conference on Power Engineering, Energy and Electrical Drives (POWERENG)* (IEEE, New York, 2015), pp. 543–547
7. M.H. Amini, H. Arasteh, P. Siano, Sustainable smart cities through the lens of complex interdependent infrastructures: panorama and state-of-the-art, in *Sustainable Interdependent Networks II* (Springer, Cham, 2019), pp. 45–68
8. M.H. Amini, A. Islam, Allocation of electric vehicles' parking lots in distribution network, in *ISGT 2014* (IEEE, New York, 2014)

9. M.H. Amini, A panorama of interdependent power systems and electrified transportation networks, in *Sustainable Interdependent Networks II* (Springer, Cham, 2019), pp. 23–41
10. A.M. Annaswamy, Y. Guan, H.E. Tseng, H. Zhou, T. Phan, D. Yanakiev, Transactive control in smart cities. Proc. IEEE **106**(4), 518–537 (2018)
11. Global EV outlook 2018. International Energy Agency (IEA). https://www.iea.org/gevo2018/. Accessed 16 July 2018
12. S.C. Davis, S.E. Williams, R.G. Boudy, Transportation energy data book: Edition 35 (pdf). vehicle technologies office, office of energy efficiency and renewable energy, U.S. Department of Energy. https://cta.ornl.gov/data/editions/Edition35_Full_Doc.pdf, July 2016
13. S.C. Davis, S.E. Williams, R.G. Boudy, Transportation energy data book: Edition 33 (pdf). office of energy efficiency and renewable energy, U.S. Department of Energy. https://info.ornl.gov/sites/publications/files/Pub50854.pdf, July 2014
14. S.C. Davis, S.E. Williams, R.G. Boudy, Transportation energy data book: Edition 31 (pdf). Office of energy efficiency and renewable energy, U.S. Department of Energy. https://cta.ornl.gov/data/editions/Edition31_Full_Doc.pdf, July 2012
15. S.C. Davis, S.E. Williams, R.G. Boudy, Transportation energy data book: Edition 30 (pdf). Office of energy efficiency and renewable energy, U.S. Department of Energy. https://info.ornl.gov/sites/publications/files/Pub31202.pdf, June 2011
16. E.W. Wood, C.L. Rames, M. Muratori, S. Srinivasa Raghavan, M.W. Melaina, "National plug-in electric vehicle infrastructure analysis," National Renewable Energy Lab.(NREL), Golden, CO (United States), Tech. rep., 2017
17. Charge on the road, Tesla supercharger network map. https://www.tesla.com/supercharger, August 2018
18. B. Bilgin, P. Magne, P. Malysz, Y. Yang, V. Pantelic, M. Preindl, A. Korobkine, W. Jiang, M. Lawford, A. Emadi, Making the case for electrified transportation. IEEE Trans. Transp. Electr. **1**(1), 4–17 (2015)
19. M.H. Amini, O. Karabasoglu, Optimal operation of interdependent power systems and electrified transportation networks. Energies **11**(1), 196 (2018)
20. F. Mwasilu, J.J. Justo, E. Kim, T.D. Do, J. Jung, Electric vehicles and smart grid interaction: a review on vehicle to grid and renewable energy sources integration. Renew. Sustain. Energy Rev. **34**, 501–516 (2014)
21. W. Su, H. Eichi, W. Zeng, M.-Y. Chow, A survey on the electrification of transportation in a smart grid environment. IEEE Trans. Ind. Inf. **8**(1), 1–10 (2012)
22. C. Li, C. Liu, K. Deng, X. Yu, T. Huang, Data-driven charging strategy of PEVs under transformer aging risk. IEEE Trans. Control Syst. Technol. **26**(4), 1386–1399 (2018)
23. J. Xiong, K. Zhang, Y. Guo, W. Su, Investigate the impacts of PEV charging facilities on integrated electric distribution system and electrified transportation system. IEEE Trans. Transp. Electr. **1**(2), 178–187 (2015)
24. T.H. Bradley, A.A. Frank, Design, demonstrations and sustainability impact assessments for plug-in hybrid electric vehicles. Renew. Sustain. Energy Rev. **13**(1), 115–128 (2009)
25. C. Samaras, K. Meisterling, Life cycle assessment of greenhouse gas emissions from plug-in hybrid vehicles: implications for policy (2008)
26. W. Tushar, C. Yuen, H. Mohsenian-Rad, T. Saha, H.V. Poor, K.L. Wood, Transforming energy networks via peer to peer energy trading: potential of game theoretic approaches. Preprint. arXiv:1804.00962 (2018)
27. M.H. Amini, M.P. Moghaddam, O. Karabasoglu, Simultaneous allocation of electric vehicles parking lots and distributed renewable resources in smart power distribution networks. Sustain. Cities Soc. **28**, 332–342 (2017)
28. H.N. Nguyen, C. Zhang, J. Zhang, Dynamic demand control of electric vehicles to support power grid with high penetration level of renewable energy. IEEE Trans. Transp. Electr. **2**(1), 66–75 (2016)
29. H. Zhang, S.J. Moura, Z. Hu, W. Qi, Y. Song, Joint PEV charging network and distributed PV generation planning based on accelerated generalized benders decomposition. IEEE Trans. Transp. Electr. **4**(3), 789–803 (2018)

30. K.M. Tan, V.K. Ramachandaramurthy, J.Y. Yong, Integration of electric vehicles in smart grid: a review on vehicle to grid technologies and optimization techniques. Renew. Sustain. Energy Rev. **53**, 720–732 (2016)
31. E. Sortomme, M.A. El-Sharkawi (2011). Optimal charging strategies for unidirectional vehicle-to-grid. IEEE Trans. Smart Grid **2**(1), 131–138
32. Q. Wang, X. Liu, J. Du, F. Kong, Smart charging for electric vehicles: a survey from the algorithmic perspective. IEEE Commun. Surv. Tutor. **18**(2), 1500–1517 (2016)
33. S. Mal, A. Chattopadhyay, A. Yang, R. Gadh, Electric vehicle smart charging and vehicle-to-grid operation. Int. J. Parallel Emergent Distrib. Syst. **28**(3), 249–265 (2013)
34. C.-Y. Chung, J. Chynoweth, C.-C. Chu, R. Gadh, Master-slave control scheme in electric vehicle smart charging infrastructure. Sci. World J. (2014). https://doi.org/10.1155/2014/462312
35. M. Alizadeh, H.-T. Wai, A. Goldsmith, A. Scaglione, Retail and wholesale electricity pricing considering electric vehicle mobility. IEEE Trans. Control Netw. Syst. **6**(1), 249–260 (2018)
36. C. Sun, S.J. Moura, X. Hu, J.K. Hedrick, F. Sun, Dynamic traffic feedback data enabled energy management in plug-in hybrid electric vehicles. IEEE Trans. Control Syst. Technol. **23**(3), 1075–1086 (2015)
37. S. Cui, H. Zhao, C. Zhang, Multiple types of plug-in charging facilities location-routing problem with time windows for mobile charging vehicles. Sustainability **10**(8), 1–27 (2018)
38. T.J. van der Wardt, A.M. Farid, A hybrid dynamic system assessment methodology for multimodal transportation-electrification. Energies **10**(5), 653 (2017)
39. A.Y. Lam, K.-C. Leung, V. O. Li, Vehicular energy network, IEEE Trans. Transp. Electr. **3**(2), 392–404 (2017)
40. A.Y. Lam, V.O. Li, Opportunistic routing for vehicular energy network. IEEE Internet Things J. **5**(2), 533–545 (2018)
41. J. James, A.Y. Lam, Autonomous vehicle logistic system: joint routing and charging strategy. IEEE Trans. Intell. Transp. Syst. (2017). https://doi.org/10.1109/TITS.2017.2766682
42. B. HomChaudhuri, R. Lin, P. Pisu, Hierarchical control strategies for energy management of connected hybrid electric vehicles in urban roads. Transp. Res. Part C: Emerg. Technol. **62**, 70–86 (2016)
43. D.F. Opila, Uncertain route, destination, and traffic predictions in energy management for hybrid, plug-in, and fuel-cell vehicles, in *American Control Conference (ACC), 2016* (IEEE, New York, 2016), pp. 1685–1692
44. A. Sarker, H. Shen, J.A. Stankovic, MORP: data-driven multi-objective route planning and optimization for electric vehicles, in *Proceedings of the ACM on Interactive, Mobile, Wearable and Ubiquitous Technologies*, vol. 1(4), p. 162 (2018)
45. X. Zeng, J. Wang, Optimizing the energy management strategy for plug-in hybrid electric vehicles with multiple frequent routes. IEEE Trans. Control Syst. Technol. **27**(99), pp. 1–7 (2017)
46. M. Alizadeh, H.-T. Wai, M. Chowdhury, A. Goldsmith, A. Scaglione, T. Javidi, Optimal pricing to manage electric vehicles in coupled power and transportation networks. IEEE Trans. Control Netw. Syst. **4**, 863–875 (2016)
47. M. Alizadeh, H.-T. Wai, A. Scaglione, A. Goldsmith, Y. Y. Fan, T. Javidi, Optimized path planning for electric vehicle routing and charging, in *2014 52nd Annual Allerton Conference on Communication, Control, and Computing (Allerton)* (IEEE, New York, 2014), pp. 25–32
48. J. James, A.Y. Lam, S.-C. Tan, Energy exchange coordination of off-grid charging stations with vehicular energy network, in *2017 IEEE International Conference on Smart Grid Communications (SmartGridComm)* (IEEE, New York, 2017), pp. 375–380
49. A.M. Farid, A hybrid dynamic system model for multimodal transportation electrification. IEEE Trans. Control Syst. Technol. **25**(3), 940–951 (2017)
50. D.F. Allan, A.M. Farid, A benchmark analysis of open source transportation-electrification simulation tools, in *2015 IEEE 18th International Conference on Intelligent Transportation Systems (ITSC)* (IEEE, New York, 2015), pp. 1202–1208

51. J. Mohammadi, G. Hug, S. Kar, A fully distributed cooperative charging approach for plug-in electric vehicles. IEEE Trans. Smart Grid **9**(4), 3507–3518 (2016)
52. W. Wei, L. Wu, J. Wang, S. Mei, Network equilibrium of coupled transportation and power distribution systems. IEEE Trans. Smart Grid **9**, 6764–6779 (2017)
53. W. Wei, L. Wu, J. Wang, S. Mei, Expansion planning of urban electrified transportation networks: a mixed-integer convex programming approach. IEEE Trans. Transp. Electr. **3**(1), 210–224 (2017)
54. A.M. Farid, Symmetrica: test case for transportation electrification research. Infrastruct. Complexity **2**(1), 9 (2015)
55. S. Kar, G. Hug, J. Mohammadi, J.M. Moura, Distributed state estimation and energy management in smart grids: a consensus + innovations approach. IEEE J. Select. Top. Signal Process. **8**(6), 1022–1038 (2014)
56. A. Kargarian, J. Mohammadi, J. Guo, S. Chakrabarti, M. Barati, G. Hug, S. Kar, R. Baldick, Toward distributed/decentralized DC optimal power flow implementation in future electric power systems. IEEE Trans. Smart Grid **9**(4), 2574–2594 (2018)
57. D.K. Molzahn, F. Dörfler, H. Sandberg, S.H. Low, S. Chakrabarti, R. Baldick, J. Lavaei, A survey of distributed optimization and control algorithms for electric power systems. IEEE Trans. Smart Grid **8**(6), 2941–2962 (2017)
58. M.H. Amini, B. Nabi, M.R. Haghifam, Load management using multi-agent systems in smart distribution network, in *2013 IEEE Power and Energy Society General Meeting* (IEEE, New York, 2013), pp. 1–5
59. A. Imteaj, M.H. Amini, J. Mohammadi, Leveraging decentralized artificial intelligence to enhance resilience of energy networks. arXiv preprint arXiv:1911.07690 (2019)
60. M.H. Amini, S. Bahrami, F. Kamyab, S. Mishra, R. Jaddivada, K. Boroojeni, P. Weng, Y. Xu, Decomposition methods for distributed optimal power flow: panorama and case studies of the DC model, in *Classical and recent aspects of power system optimization* (Academic Press, Cambridge, 2018), pp. 137–155
61. M. Ali et al., A system of systems engineering framework for modern power system, in *Sustainable Interdependent Networks II: From Smart Power Grids to Intelligent Transportation Networks* (2019), p. 217
62. Y. Chen, S. Kar, J.M. Moura, The Internet of Things: secure distributed inference. IEEE Signal Process. Mag. **35**, 64–75 (2018)
63. S. Kar, J.M. Moura, Consensus + innovations distributed inference over networks: cooperation and sensing in networked systems. IEEE Signal Process. Mag. **30**(3), 99–109 (2013)
64. J. Mohammadi, S. Kar, G. Hug, Distributed cooperative charging for plug-in electric vehicles: a consensus + innovations approach, in *2016 IEEE Global Conference on Signal and Information Processing (GlobalSIP)* (IEEE, New York, 2016), pp. 896–900
65. S. Kar, G. Hug, Distributed robust economic dispatch in power systems: a consensus + innovations approach, in *Power and Energy Society General Meeting, 2012 IEEE* (IEEE, New York, 2012), pp. 1–8
66. G. Hug, S. Kar, C. Wu, Consensus + innovations approach for distributed multiagent coordination in a microgrid. IEEE Trans. Smart Grid **6**(4), 1893–1903 (2015)
67. P. Bucić, V. Lešić, M. Vašak, Distributed optimal batteries charging control for heterogenous electric vehicles fleet, in *2018 26th Mediterranean Conference on Control and Automation (MED)* (IEEE, New York, 2018), pp. 837–842
68. J. Mohammadi, M. González Vayá, S. Kar, G. Hug, A fully distributed approach for plug-in electric vehicle charging, in *2016 Power Systems Computation Conference (PSCC)* (IEEE, New York, 2016), pp. 1–7
69. Y. Xu, Optimal distributed charging rate control of plug-in electric vehicles for demand management. IEEE Trans. Power Syst. **30**(3), 1536–1545 (2015)
70. N. Rahbari-Asr, M.-Y. Chow, Cooperative distributed demand management for community charging of PHEV/PEVs based on KKT conditions and consensus networks. IEEE Trans. Ind. Inf. **10**(3), 1907–1916 (2014)

71. M.H. Amini, P. McNamara, P. Weng, O. Karabasoglu, Y. Xu, Hierarchical electric vehicle charging aggregator strategy using Dantzig-Wolfe decomposition, in *IEEE Design & Test*, 2017

72. J. Rivera, P. Wolfrum, S. Hirche, C. Goebel, H.A. Jacobsen, Alternating direction method of multipliers for decentralized electric vehicle charging control, in *52nd Annual Conference on Decision and Control (CDC)* (2013), pp. 6960–6965

73. M. González Vayá, G. Andersson, S. Boyd, Decentralized control of plug-in electric vehicles under driving uncertainty, in *IEEE PES Innovative Smart Grid Technologies Conference*, Istanbul, 2014

74. Z. Ma, D.S. Callaway, I.A. Hiskens, Decentralized charging control of large populations of plug-in electric vehicles. IEEE Trans. Control Syst. Technol. **21**(1), 67–78 (2013)

75. F. Parise, M. Colombino, S. Grammatico, J. Lygeros, Mean field constrained charging policy for large populations of plug-in electric vehicles, in *53rd Annual Conference on Decision and Control (CDC)* (2014), pp. 5101–5106

76. W.-J. Ma, V. Gupta, U. Topcu, Distributed charging control of electric vehicles using online learning. IEEE Trans. Autom. Control 62(10), 5289–5295 (2017)

77. J. Mohammadi, M.G. Vayá, S. Kar, G. Hug, A fully distributed approach for plug-in electric vehicle charging, in *Power Systems Computation Conference (PSCC), 2016* (IEEE, New York, 2016), pp. 1–7

78. R. Carli, M. Dotoli, A distributed control algorithm for optimal charging of electric vehicle fleets with congestion management. IFAC-PapersOnLine **51**(9), 373–378 (2018)

79. Z. Fan, A distributed demand response algorithm and its application to PHEV charging in smart grids. IEEE Trans. Smart Grid **3**(3), 1280–1290 (2012)

80. S. Zou, I. Hiskens, Z. Ma, Consensus-based coordination of electric vehicle charging considering transformer hierarchy. Control Eng. Pract. **80**, 138–145 (2018)

81. L. Gan, U. Topcu, S.H. Low, Stochastic distributed protocol for electric vehicle charging with discrete charging rate, in *Power and Energy Society General Meeting, 2012 IEEE* (IEEE, New York, 2012), pp. 1–8

82. J. Li, C. Li, Y. Xu, Z. Dong, K. Wong, T. Huang, Noncooperative game-based distributed charging control for plug-in electric vehicles in distribution networks, in *IEEE Transactions on Industrial Informatics*, 2016

83. J. Li, C. Li, Z. Wu, X. Wang, K.L. Teo, C. Wu, Sparsity-promoting distributed charging control for plug-in electric vehicles over distribution networks. Appl. Math. Modell. **58**, 111–127 (2018)

84. A. Ghavami, K. Kar, A. Gupta, Decentralized charging of plug-in electric vehicles with distribution feeder overload control. IEEE Trans. Autom. Control **61**(11), 3527–3532 (2016)

85. L. Gan, U. Topcu, S.H. Low, Optimal decentralized protocol for electric vehicle charging. IEEE Trans. Power Syst. **28**(2), 940–951 (2013)

86. L. Zhang, V. Kekatos, G.B. Giannakis, Scalable electric vehicle charging protocols. IEEE Trans. Power Syst. **32**(2), 1451–1462 (2017)

87. Z. Ma, N. Yang, S. Zou, Y. Shao, Charging coordination of plug-in electric vehicles in distribution networks with capacity constrained feeder lines, in *IEEE Transactions on Control Systems Technology*, 2017

88. A.D. Hilshey, P.D. Hines, P. Rezaei, J.R. Dowds, Estimating the impact of electric vehicle smart charging on distribution transformer aging. IEEE Trans. Smart Grid **4**(2), 905–913 (2013)

89. S. Shao, M. Pipattanasomporn, S. Rahman, Grid integration of electric vehicles and demand response with customer choice. IEEE Trans. Smart Grid **3**(1), 543–550 (2012)

90. M.E. Khodayar, L. Wu, M. Shahidehpour, et al., Hourly coordination of electric vehicle operation and volatile wind power generation in SCUC. IEEE Trans. Smart Grid **3**(3), 1271–1279 (2012)

91. M.G. Vayá, G. Andersson, S. Boyd, Decentralized control of plug-in electric vehicles under driving uncertainty, in *IEEE PES Innovative Smart Grid Technologies Conference*, 2014

92. B. Chai, J. Chen, Z. Yang, Y. Zhang, Demand response management with multiple utility companies: a two-level game approach. IEEE Trans. Smart Grid **5**(2), 722–731 (2014)

Chapter 3
Evolutionary Computation, Optimization, and Learning Algorithms for Data Science

Farid Ghareh Mohammadi, M. Hadi Amini, and Hamid R. Arabnia

Abstract A large number of engineering, science, and computational problems have yet to be solved in a computationally efficient way. One of the emerging challenges is how evolving technologies grow towards autonomy and intelligent decision making. This leads to collection of large amounts of data from various sensing and measurement technologies, e.g., cameras, smart phones, health sensors, smart electricity meters, and environment sensors. Hence, it is imperative to develop efficient algorithms for generation, analysis, classification, and illustration of data. Meanwhile, data is structured purposefully through different representations, such as large-scale networks and graphs. Therefore, data plays a pivotal role in technologies by introducing several challenges: *how to present*, *what to present*, *why to present*. Researchers explored various approaches to implement a comprehensive solution to express their results in every particular domain, such that the solution enhances the performance and minimizes cost, especially time complexity. In this chapter, we focus on data science as a crucial area, specifically focusing on a curse of dimensionality (CoD) which is due to the large amount of generated/sensed/collected data, especially large sets of extracted features for a particular purpose. This motivates researchers to think about optimization and apply nature-inspired algorithms, such as meta-heuristic and evolutionary algorithms (EAs) to solve large-scale optimization problems. Building on the strategies of these algorithms, researchers solve large-scale engineering and computational problems

F. G. Mohammadi · H. R. Arabnia
Department of Computer Science, Franklin College of Arts and Sciences, University of Georgia, Athens, GA, USA
e-mail: farid.ghm@uga.edu; hadi.amini@ieee.org

M. H. Amini (✉)
School of Computing and Information Sciences, Florida International University, Miami, FL, USA

Sustainability, Optimization, and Learning for InterDependent Networks Laboratory (solid lab), Florida International University, Miami, FL, USA
e-mail: amini@cs.fiu.edu; hadi.amini@ieee.org; www.solidlab.network

© Springer Nature Switzerland AG 2020
M. H. Amini (ed.), *Optimization, Learning, and Control for Interdependent Complex Networks*, Advances in Intelligent Systems and Computing 1123, https://doi.org/10.1007/978-3-030-34094-0_3

37

with innovative solutions. Although these algorithms look un-deterministic, they are robust enough to reach an optimal solution. To that end, researchers try to run their algorithms more than usually suggested, around 20 or 30 times, then they compute the mean of result and report only the average of 20/30 runs' result. This high number of runs becomes necessary because EAs, based on their randomness initialization, converge the best result, which would not be correct if only relying on one specific run. Certainly, researchers do not adopt evolutionary algorithms unless they face a problem which is suffering from placement in local optimal solution, rather than global optimal solution. In this chapter, we first develop a clear and formal definition of the CoD problem, next we focus on feature extraction techniques and categories, then we provide a general overview of meta-heuristic algorithms, its terminology, and desirable properties of evolutionary algorithms.

Keywords Evolutionary algorithms · Dimension reduction (auto-encoder) · Data science · Heuristic optimization · Curse of dimensionality (CoD) · Supervised learning · Data analytic · Feature extraction · Optimal feature selection · Big data

3.1 Introduction

3.1.1 Overview

A large number of engineering, science, and computational problems have yet to be solved in a more computationally efficient way. One of the emerging challenges is the evolving technologies and how they enhance towards autonomy. This leads to collection of large amount of data from various sensing and measurement technologies, such as cameras, smart phones, health sensors, and environment sensors. Hence, generation, manipulation, and illustration of data grow significantly. Meanwhile, data is structured purposefully through different representations, such as large-scale networks and graphs. Therefore, data plays a pivotal role in technologies by introducing several challenges: *how to present, what to present, why to present*. Researchers explored various approaches to implement a comprehensive solution to express their results in every particular domain, such that the solution enhances the performance and minimizes cost, especially time complexity. In this chapter, we focus on data science as a crucial area; specifically focusing on curse of dimensionality (CoD) which is due to the large amount of generated/sensed/collected data, especially large sets of extracted features for a particular purpose. This motivates researchers to think about optimization and apply nature-inspired algorithms, such as meta-heuristic and evolutionary algorithms (EAs) to solve large-scale optimization problems. Building on the strategies of these algorithms, researchers solve large-scale engineering and computational problems with innovative solutions. Although these algorithms look un-deterministic, they are robust enough to reach an optimal solution. To that end, researchers try to run their algorithms more than usually suggested, around 20 or 30 times, then they compute the mean of result and report only the average of 20/30 runs' result. This high number of runs becomes necessary because EAs, based on their randomness initialization, converge the best result, which would not be correct if only relying on one specific run. Certainly, researchers do not adopt evolutionary algorithms unless they face a problem which is suffering from placement in local optimal solution, rather than global optimal solution. In this chapter, we first develop a clear and formal definition of the CoD problem, next we focus on feature extraction techniques and categories, then we provide a general overview of meta-heuristic algorithms, its terminology, and desirable properties of evolutionary algorithms.

3.1.2 Motivation

In the last twenty years, computer usage has proliferated significantly, and it is most likely that you could find technologies and computers almost anywhere you want to work and live. A large amount of data is being generated, extracted, and presented through a wide variety of domains, such as business, finance, medicine, social medias, multimedia, all kinds of networks, and many other sources due to this spectacular growth. This increasingly large amount of data is often referred to as big data. In addition, distributed systems and networks are not performing as well as they did as in the past [1]. Hence, it is imperative to leverage new approaches which optimize and learn to use these devices. Further, distributed optimization and learning algorithms lend themselves as promising solutions to deal with information privacy, scalability, as well as (near) real-time decision making capability; applications of such algorithms include optimal operation of smart city infrastructures, interdependent power and transportation networks [2–4], artificial intelligence for energy system resilience [5], energy management and optimal power flow problem [6, 7], and learning at the IoT device level [8]. Moreover, big data also requires that scientists propose new methods to analyze the data. Obtaining a proper result, thus, requires an unmanageable amount of time and resources. This problem is known as the curse of dimensionality (CoD) which is discussed in the next sub-section in detail. Ghareh Mohammadi and Arabnia have discussed application of evolutionary algorithms on images, specifically focused on image steganalysis [9]. But, in this study we expanded our investigation and consider large-scale engineering and science problems carefully.

In machine learning, the majority of problems require a fitness function which optimizes a gradient value to lead a global optimum accurately [10]. This function is also known as an objective function and may have different structures for different problems. In machine learning, we work with three categories of data: one supervised, one semi-supervised, and one unsupervised. These categories also have different learning processes based on their types. Supervised datasets are the most common dataset and are characterized by having a ground truth with which to compare results. Supervised learning algorithms normally take a supervised dataset and then divide them into two parts: train and test. After that, one of the supervised learning algorithms learns from train data, predicts test data, and compares the result with the ground truth to ascertain the accuracy of the algorithm performance. The most common types of supervised learning algorithms are classification and regression. It is noteworthy that regression has different algorithms which mainly focus on time series problems. The only exception is that regression algorithms have a particular algorithm, logistic regression, which is considered as a classification, rather than regression, algorithm [11]. In this chapter, we focus on supervised datasets and supervised learning algorithms.

On the other hand, unsupervised learning algorithms follow the process of using unsupervised datasets which do not have any ground truth to compare their result, which makes classifying and evaluating the performance of the algorithm problematic. The absence of a ground truth is increasingly common through all domains such as web-based, engineering, etc. data and it is necessary to address this problem. Unsupervised learning takes more steps to analyze features and find the most relevant features with the best possible positive relation. Clustering and representation learning (RL) algorithms are the most common algorithms in unsupervised learning category. K-means is an important clustering algorithm that attempts to find k clusters located close to each other. The main problem of k-means is its bias-k towards the problem. In other words, k-means needs to have k number set in advance before running the algorithms. RL also works for supervised datasets, although its nature behaves in an independent way per task [12].

Semi-supervised datasets fall somewhere between supervised and unsupervised datasets in terms of characteristics. This means that semi-supervised learning algorithms take a dataset which provides ground truth value for some instances but not for others. Expectation maximization (EM) is the most important and robust technique for working with these datasets [13]. Moreover, EM is also able to handle missing values of a given dataset properly. Real data always involves missing values, and researchers struggle with this problem.

Feature extractor (FE) which is discussed in details in the next section is almost a universal technique which is capable of applying on these three types of problems to aim for dimension reduction. Meanwhile, the majority of problems and dataset have been so far used are supervised datasets. But it does not mean that FE does not apply on unsupervised or semi-supervised datasets. For instance, for unsupervised dataset, it is normal to use dimension reduction or auto-encoder techniques for that.

There has been numerous challenges in the literature regarding the deployment of evolutionary algorithms for computation, optimization, and learning. These studies can be reviewed in the following major aspects: curse of dimensionality [14, 15], nature-inspired computation [1, 16], nature-inspired meta-heuristic computation [17–19], and nature-inspired evolutionary computation [20–24]. These studies are elaborately reviewed in the following.

3.1.3   Curse of Dimensionality

Curse of dimensionality is related to the fact that the input data is too huge that no human being can analyze it. In machine learning, recently, researchers work with high-dimensional data. For instances, if we are analyzing three channel images, such as RGB, HSV images, sized 512×512, we are working in a space with $512 \times 512 \times 3$ dimensions. Altman and Krzywinski [14] believe that having more data is much better than having few or nothing. This overabundance of data is called the curse of dimensionality (CoD) which causes problems in big data era such as data sparsity, multiple testing, which researchers [15] proposed a new approach to solve the problem, and most importantly over-fitting which is opposite of under-fitting. Beside these problems, CoD also brings high time complexity problem which makes scientists suffering from waiting too much time to get a result.

The world of information technology CoD not only causes a wide range of problems to scientists, but also has a wide adversely affect other majors, such as engineering [25], medicine [26, 27], cognitive science [28, 29], bioinformatics [30], and even optimization problems [10, 31, 32].

Classification in big data suffers from plenty of problems and issues, one of which is considered very challenging named CoD. Traditional feature extraction techniques also are not able to solve this problem technically any more due to some limitation [32]. According to the research studies have accomplished, scientist proposed a new approach to solve this problem. Researchers introduce nature-inspired computation which enable to simulate traditional feature extraction techniques in a way that improve the performance of classification.

3.1.4   Nature-Inspired Computation

Pure and basic machine learning algorithms are not capable of solving emerging challenging issues in the world of technologies any more. It is needed to adopt a new approach to face these problems and leverage decent machine learning algorithms. Finally, scientists discovered that combining machine learning algorithms in a technical way may solve the problems. This mixture of machine learning techniques is called nature-inspired computation, but it still is considered an advanced machine learning algorithm.

Majority of scientific and technological developments leverage inspiring from the nature towards their goal, especially robotics simulate how the nature works. In world of computer science, each tool or software development process is needed to have strong synchronization, robustness, manageability, parallelization, scalability, distributedness, redundancy, adaptability, cooperation. Indeed, the nature provides the same properties. Therefore, the nature-inspired techniques play an important role in computing environments. Concretely, the nature-inspired techniques are adopted to develop practical algorithms to solve data-driven optimization problems [16].

Researchers in [1, 16] categorized nature-inspired computation. In [16] the authors classified them into six different categories such as swarm intelligence, natural evolution, molecular biology, immune system, and biological cells. But here, we provide another applicable way to express the nature-inspired computation towards solving problems: one meta-heuristic and one evolutionary computation.

3.1.5 Nature-Inspired Meta-Heuristic Computation

A meta-heuristic is an advanced procedure developed to seek and generate a sufficiently tuned solution to data-driven optimization problems [17]. It involves, high level view, two types of computations. The first and foremost one is population based computation which is well-known as evolutionary algorithms, second one is non-population computation such as tabu search (TS), stochastic local search (SLS), iterated local search (ILS), guided local search (GLS). For more information about this classification, please refer to [18]. Further, Razavi and Sajedi [19] proposed a single-based meta-heuristic algorithm, vortex search algorithm (VSA) is inspired by the vortices. In this chapter, we mainly focus on the former classification, evolutionary algorithms which is discussed next sub-section properly.

3.1.6 Nature-Inspired Evolutionary Computation

Evolutionary algorithms (EAs) are invented not more than 28 years and are not pretty old computational algorithm [21]. Research studies have been accomplished new evolutionary algorithms in engineering and computational science [22–24]. EAs are known as population-based algorithm. Their learning process comes from interactions between multiple candidate solutions called food source or population. EAs are particular optimization type of meta-heuristics designed to solve optimization problems [22]. This chapter discusses classical EAs and other popular methods including memetic algorithms (MAs), particle swarm optimization (PSO), and artificial bee colony (ABC), ant colony optimization (ACO), grey wolf optimizer (GWO), and coyote optimization algorithm (COA).

3.1.6.1 Evolutionary-Based Memetic Algorithms

Memetic algorithms (MAs) are one of the particular growing research studies within EA. Based on a population based search and local search, MAs have practically succeeded in a variety of engineering and science problem domains, in particular for NP-hard optimization problems [20, 22]. Memetic algorithms intrinsically exploit all available sources, however, traditional EAs fail to do that. Population based search MAs leverage recombination (or crossover operator) which is an important process within MAs. For the search process, it is essential to have three parameters ready: one neighborhood relation, one guiding function, and a search space which provides borders of the problem.

The search space is also important to provide comprehensive knowledge for guiding function works. The implication of search space is to influence the dynamics of the search algorithm. These dynamics stand for the relationships, which are accessible, among the configurations. Thus, these relationships depend on neighborhood function. For more information about this topic, please refer to [20].

3.1.7 Organization

The rest of this study is organized as follows. In Sect. 3.2, we have discussed the feature extraction techniques and their categories. First, feature extraction from a sample object like image against feature extraction from given datasets is mentioned. Next, the feature extraction from dataset has selected to discover it. It has three types including feature selection, dimension reduction (auto-encoder), and feature generation. Then, we introduce nature-inspired algorithms and their application, together with related pseudocode in solving large-scale engineering and science problems, particularly CoD problem. The summary of evolutionary algorithms discussed in this chapter is as follows: genetic algorithm (GA), artificial bee colony (ABC), ant colony optimization (ACO), grey wolf optimizer (GWO), coyote optimization algorithm (COA), and particle swarm optimization. In general, Fig. 3.1 represents the overall structure of this study.

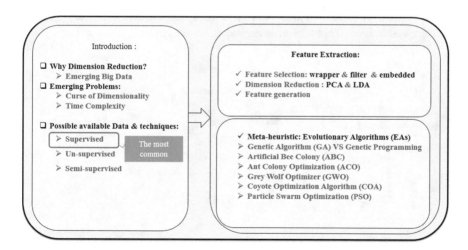

Fig. 3.1 Overall structure of this study

3.2 Feature Extraction Techniques

It is worth mentioning, in the world of science, "feature extraction" is used to refer to two completely separate applications. There are two different processes, one occurring before raw data generation and one taking place after data has generated. The process of feature extraction before having raw data works to extract features using some advancing techniques, to export information from the objects. For example, if we want to extract features from images, we need to adopt advanced image processing techniques, like a feature extractor, for that end. Therefore, based on the generated data, we will have a set of raw data. Then, in pre-processing techniques, a second type of feature extraction is used for dimension reduction. Three major differences separate these two types of feature extraction. The first difference is their input value; the input value of the first algorithm is not particular features, but the second feature extraction accepts only features of any dataset. Second, the first type of feature extraction is domain specific, while the second type is not domain specific. Third, the former does not adopt machine learning algorithms, but the latter type does. Basically, both of them work with data, take values, and generate outputs. The scope of the first algorithm is dynamic and would be any multimedia or social networks, etc. On the other hand, the second one has an almost stationary scope of input data.

General overview of testing and evaluating given dataset is shown in Fig. 3.2. On the top of the figure, it clearly presents that three separate steps are required to be done in advance before generating a proper result. Pre-processing plays a main role in each problem of engineering and optimization problems. Then, a classification algorithm is selected to make a model based on the train data. Finally, the classifier attempts to predict the test data based on the learned data.

Once data is generated and dataset is ready to be evaluated, we call the dataset, raw dataset. This dataset is needed to be converted into a standard dataset

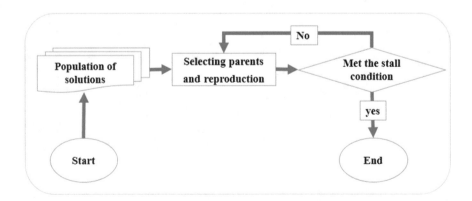

Fig. 3.2 General process of evolutionary algorithms

which enables classifiers to examine in a professional way and obtain a higher performance. The most common problems of raw dataset consist of curse of dimensionality (CoD), heterogeneous features in case of values and type, missing values, outliers. In this chapter, we discuss in detail how evolutionary algorithms (EAs) are adopted to solve the CoD problems, the bottom of the Fig. 3.2 depicts the idea where EAs are explicitly embedded into pre-processing and enhances the classifier's performance. Concretely, EAs attempt to optimize the process of feature extraction in an innovative way.

Feature extraction (FE), which is one of the most popular pre-processing techniques, is the process of shrinking the number of dimension (features) and the capability of having adapted diversity while considering strong mapping between features and target values. FE aims to decrease the feature dimension as minimum possible as it keeps the same performance. A feature extractor is considered as the best one which is capable of decreasing the feature dimension and meanwhile improving the performance. The better result is obtained by the better FE. FE techniques are intrinsically classified into three broad groups: one auto-encoder, one feature selection (FS), and feature generation. The first two are the most common techniques in the scope of dimension reduction. Meanwhile researcher can leverage feature generation (such as [33, 34]) to improve a classifier performance. The former technique is also known as dimension reduction (DR) which attempts to transform given dimension to a new dimension with strong linear connectivity of original dimension. The most popular auto-encoder algorithm is principal component analysis (PCA).

PCA completely is used to generate a new dimension using a certain formula and convert the given data into new dimension. The idea behind PCA is that it leverages singular value decomposition (SVD) theorem to seek for the most relevant and correlated features and the relationship between each other. Although PCA is used to emphasize variation and bring out strong patterns in a dataset, it may not guarantee to reach an optimal solution in some datasets. PCA fails once your special visualization of instances leads to loss of information. It tries to convert input data into new dimension using a linear function. Circle-based and sine- or cosine-based distribution of instances are the most popular situations that PCA fails. PCA failure means that the FE did not obtain a better performance while decreasing the dimension, not only that, but also it did not yield same performance. If PCA does not yield a better result, it means that features are not correlated or have non-linear relationships. However, researchers often used to enable data easy to explore and visualize, in case for representation learning (RL) [35].

Feature selection (FS), the latter one, which is the process of choosing proper sets of relevant features rather than converting to a new dimension. FS covers the lack of auto-encoders properly by keeping the original values of features during the process; meanwhile, it is most likely to decrease the number of features/dimension. Feature selection mainly provides three kinds of categories: filter-based, wrapper-based, and embedded FS. Filter-based FS is the easy technique to implement and can be adapted to each engineering problem independently. It tries to examine given dataset features separately non-dependently with respect to their target. It attempts to calculate the goodness of each feature separately. However, the wrapper-based feature selection relies on a set of selected features and calculated their goodness using classifiers. Wrapper-based FS is a special kind of filter-based FS such that wrapper-based FS has capability of using some hyper-parameter function for evaluation. Therefore, the pace of running filter-based is high in comparison with wrapper-based. So, it is recommended for real-time systems because of low time complexity. Furthermore, filter-based is cheaper than wrapper-based. But the wrapper-based feature selection [23, 36, 37] yields a better result than filter-based feature selection. By advancing technologies, wrapper-based FS also can be adopted in every system, even real-time decision making system [36]. The third one, embedded feature selection which is similar to the wrapper-based feature selection to select the best subsets of features. However, it has an important drawback, which is time complexity in comparison with earlier feature selection, when it tries to train the model. One of the popular embedded feature selections is regularization which provides both training and making model section, together with automatic feature selection at the same time. Furthermore, researchers [38, 39], proposed another type of feature selection, combined (hybrid) methods, which mixes evolutionary algorithms together with filter-based or wrapper-based algorithms.

Feature generation is considered the third type of feature extractor techniques. Feature generation is a technique between feature selection and dimension reduction. It starts to examine the features and generates new features using the original given features. In this case, you first increase the feature dimension then remove irrelevant features. Unlike dimension reduction, no new dimension is generated. Feature generation keeps the original features for generating new features. Then, feature generation can do feature selection based on the generated features [33].

3.3   Bio-Inspired Evolutionary Computation

Engineering problems and other sensitive optimization need to reach the global optimum. However, machine learning algorithms are not useful anymore. So, it is required scientists adopt new kind of algorithms have been proved completely in nature for years. In this section, we provide general overview of nature-inspired algorithms and their terminology. Table 3.1 provides complete definitions for abbreviation which are used in this chapter.

3.3.1   Overview of Evolutionary Algorithms

Everything in EA starts to explain the problem and proper solutions. The first important step in evolutionary algorithm is representation. After that, in each step,

Table 3.1 List of abbreviations

Abb	Definition
ABC	Artificial bee colony
ACOAR	Ant colony optimization attribute reduction
BA	Bee algorithm
BCO	Bee colony optimization
BOA	Butterfly optimization algorithm
CNN	Convolutional neural network
COA	Coyote optimization algorithm
CoD	Curse of dimensionality
CSO	Chicken swarm optimization
CCSO	chaotic chicken swarm optimization
CRO	Coral reefs optimization
DA	Dragonfly algorithm
DR	Dimension reduction
EAs	Evolutionary algorithms
FE	Feature extraction
EM	Expectation maximization
EP	Evolutionary programming
FS	Feature selection
FSA	Fish swarm algorithm
GA	Genetic algorithm
GANs	Generative adversarial networks
GGA	Generational genetic algorithm
GLS	Guided local search
GP	Genetic programming
GWO	Grey wolf optimizer
HBMO	Honey bee mating optimization
IFAB	Image steganalysis using FS based on ABC
IoT	Internet of things
ILS	Iterated local search
IWOA	Improved whale optimization algorithm
MAs	Memetic algorithms
ML	Machine learning
PCA	Principal component analysis
PEAs	Parallel evolutionary algorithms
RFPSO	RelieF and PSO algorithms
RL	Representation learning
RNN	Recurrent neural network
SLS	Stochastic local search
SSGA	Steady state genetic algorithm
SVD	Singular value dimension

(continued)

Table 3.1 (continued)

Abb	Definition
SVM	Support vector machine
TMABC-FS	Two-archive multi-objective ABC algorithm for FS
TS	Tabu search
VSA	Vortex search algorithm
WOA	Whale optimization algorithm
WANFIS	Whale adaptive neuro-fuzzy inference system

EA works based on this representation. Figure 3.2 depicts a general overview of each evolutionary algorithm's procedure. It is extremely necessary how to present your sample solutions. Two approaches are given: a one-hot representation and an integer representation. The former one is also known as binary representation. The number of "1" in the solution shows the number of parameters have to be involved to yield a result. "1" represents that which specific features are selected and "0" stands for the features which are not considered in a specific solution. In this case, your solution's length would be as same as the input feature dimension. If feature dimension become too big, handling the food source are going to be a challenging issues which waste resources and yields high time complexity. However, the integer representation works good even with high feature dimension. But it still has a big disadvantage which you need to set the reduced length of your feature vector in initialization step.

Second important step is generating a population based on the descriptive model of representation. This population mostly is generated randomly with considering the representation limitation. Third step is fitness function and evaluation process. It is important to provide a tuned fitness function (objective function) towards their application of the evolutionary algorithms.

The next step is to select two possible solutions as parents of new generations. Selection strategy has two broad categories: one uniform parent selection and one ununiform parent selection. In the former one, each solution has the same chance to be selected. However, the latter one has different structures and criteria, and parents are selected based on those. The ununiform parent selection has different strategies, the most important strategies are proportional selection which is also known as roulette wheel, ranked based selection, and tournament selection.

Roulette wheel and tournament are the most widely used selection methods in GA. Roulette considers the fitness value for each chromosomes with respect to their probabilities, using the Eq. (3.1) where $p[i]$ stands for the probability of selecting a specific chromosome i, $f[i]$ goes for the fitness value of each chromosome of index i.

$$p[i] = \frac{f[i]}{\sum f[i]} \qquad (3.1)$$

Moreover, the tournament selection is pretty simpler than Roulette wheel. The idea is that it takes k chromosomes and selects based on the fitness value of each chromosome. The best fitness value goes for the lucky chromosome to be selected.

After that, EA tries to reproduce new generation and update the population. EA takes two parents and regenerates new offspring based on crossover operator. The crossover or recombination, which is one of the genetic operators used to recombine two chromosomes to generate new offspring. The crossover operator includes uniform crossover, arithmetic crossover, and k-point crossover which is a classical one. Once crossover step is done, mutation should be done with a specific rate. The mutation may change one or more components.

Finally, the stall condition is set to check once new generation is produced. If the new generation met the condition, EA stops running and returns the best solution which satisfied the condition.

3.3.2   Genetic Algorithm vs. Genetic Programming

It is a common mistake that to think genetic algorithm (GA) is the same genetic programming (GP). Generally speaking, researchers have used these two algorithms interchangeably. But, from a technical point of view they are completely different techniques. In this sub-section, we provide a clear definition of each of them.

3.3.2.1   Genetic Algorithm

Genetic algorithm is one of the basic but important evolutionary algorithm. It has been applied on majority of problems such as engineering, medicine, finance, etc. GA provides two kinds of approaches towards solving problems [40]: one steady state genetic algorithm (SSGA) and one generational genetic algorithm (GGA). They are different based on their procedure and updating mechanism function of whole process, but they do the same process of parent selection, reproduction, and population update. In the literature, some studies deployed GA as an effective tool for solving large-scale optimization problems, including optimal allocation of electric vehicle charging station and distributed renewable resource in power distribution networks [41], resource optimization in construction projects [42], and allocation of electric vehicle parking lots in smart grids [43]. Algorithm 1 illustrates a pseudocode of basic GA in detail.

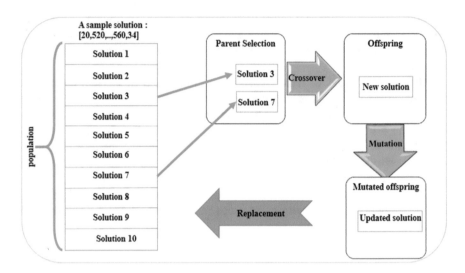

Fig. 3.3 SSGA (steady state genetic algorithm): process of updating the population

---

Algorithm 1 Implementation of GA algorithm for feature selection

---

Input: $S = \{x_0, x_1, x_2, \ldots, x_n\}$, $max_{iteration} \geq 0$, t=0, $\alpha_M \in [0, 1]$, $random_{number} \in [0, 1]$,
     $Best_{solution} = \emptyset$.
Output: $Best_{solution}$: $An optimal subset of features$(F) , $F = \{x_0, x_1, x_2, \ldots, x_m\}$, m$\leq$ n ,
     $(\forall f_i \in F) \in S$, $F_{length} \leq S_{length}$.
 1: **for** t=0 $\cdots$ $max_{iteration}$ **do**
 2:      Call parent selection function
 3:      Call crossover method to generate offspring
 4:      **if** $random_{number} \leq \alpha_M$ **then**
 5:            Call mutation function
 6:            Return offspring
 7:      **end if**
 8:      Call fitness function to evaluate the chromosome
 9:      **if** any chromosome obtained the best score **then**
10:            Update the $Best_{solution}$
11:      **end if**
12: **end for**

In SSGA, GA works with a stationary population which the size of that will be the same and just it's solutions get updated each iteration. Moreover, SSGA is an in-place algorithm, in which their population does not need another space to update. Like normal process, SSGA also starts with a problem representation and fitness function, then initialize the selection strategy, crossover and mutation operators. After that, SSGA takes another step to update the population with replacement strategy. Figure 3.3 depicts that how two solutions are selected, crossover and mutation operators are applied, and then new solution is replaced with the worse solution.

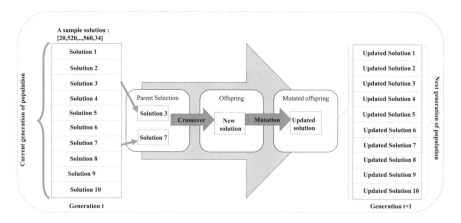

Fig. 3.4 GGA (generational genetic algorithm): the process of generating a new population (generation $t + 1$)

Further, GGA produces a new population each iteration. So, GGA is not an in-place algorithm since it generates a new population each iteration. GGA follows the same structure of EA except the last step which is replacement. GGA skips this step since it generates a new population in each iteration. Therefore, the replacement step is not required. Figure 3.4 shows complete process of generating a new population (generation $t + 1$) from current population (generation t).

From technical point of view, scientists can apply either GGA or SSGA based on the problem model and strategies. However, SSGA converges faster than GGA since parents always are selected through the same population and then replaced the worse solution with the another best solution. Hence, most of the research studies are accomplished using SSA. Moreover, most evolutionary algorithms are discussed here also use the same strategies to converge faster towards global optimum. But, SSGA still has a disadvantage that may stuck in a local optimum.

3.3.2.2 Genetic Programming

Genetic programming (GP) is proposed by Koza in 1992 [44]. It is noteworthy that this idea is introduced date back to 50s. GP evolves computer programs which are represented as trees. Each tree consists of two sections: a function set and second is terminal set. Both of them provide constant sets of symbols. The former one always plays non-leaf nodes role and the latter one plays leaf nodes role. Figure 3.5 shows an example of presenting a problem $4 \times \tan(x) + y^2$.

Similar to GA that crossover is conducted on vectors, in GP crossover is done through a tree and only needs to choose two sub-trees. Figure 3.6 expresses that the first two trees have two subsets which are selected as a parent. Second tree the below are the new offsprings which are generated based on parents. GP is mostly

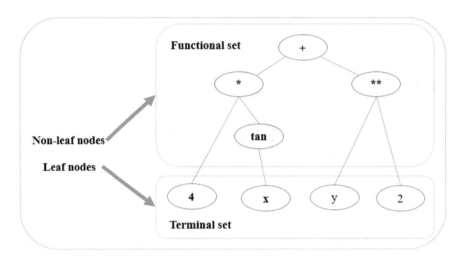

Fig. 3.5 Tree presentation of a problem

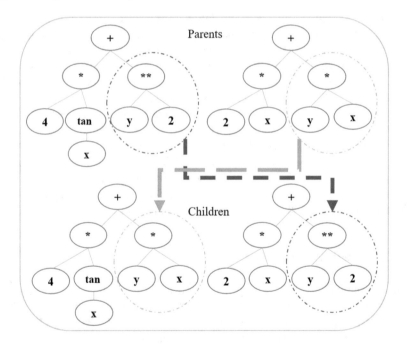

Fig. 3.6 Crossover operator in genetic programming

generational genetic algorithm. Thus, GP is not an in-place algorithm. GP is useful for solving engineering and computational problems (e.g., [45]).

Genetic programming has specific advantages over genetic algorithms. Here, we address the most important characteristics of GP. Genetic programming has a wide variety of representation models which make it pretty flexible against genetic algorithm. This flexibility of GP comes from its tree-based properties. Another important feature of GP is its application over GA. GP has greater applications in comparison with GA. In spite of considering positive features of GP, it also has disadvantages which should bear in mind. The most disadvantages of GP are its speed which is extraordinary slow. Another point is its lack of handling a large number of input data which makes also hard to handle required related population.

There is still another algorithm that attracts researcher's attention called evolutionary programming (EP). Fogel et al. [46] originally introduced evolutionary programming. It is classified as one of the major evolutionary algorithms. It resembles genetic programming, but it does have a non-variable structure of the program to be optimized. Classical EP develops gradually finite state machine or every structure similar to it. EO always works with mutation only and does not consider crossover at all. It worth mentioning that EP uses a fitness function based on the training sequences. This feature enables EP yields a better result for prediction in time series problem and sequence problems like DNA and RNA.

3.3.3 Artificial Bee Colony Algorithm

In the bees population, the process of mating and generating new offspring, finding new food sources and gathering the nectar, sharing information in hive, allocating tasks, onlooker and scout bees, all of these have been inspired properly and nature-based evolutionary algorithms have been presented. To be specific about the algorithms, honey bee mating optimization (HBMO), bee colony optimization, bee algorithm (BA), and artificial bee colony (ABC) are the most popular research studies accomplished based on these algorithms [47]. Karaboga et al. [47] present statistical overview of using these algorithms in scientific papers. It is worth mentioning that ABC has received the highest amount of usage with respect to its application in engineering and science problems. Among all research studies had been done, according to the [47] ABC, BA, BCO, and HBMO are found the most useful application, from the highest number to the lowest number, in large-scale engineering problems. ABC has been considered as the most useful algorithm in several different fields and majority of research studies leverage ABC in their problems, such as training neural network (NN), solving electrical, mechanical, software, control, and civil engineering problems, facing wireless sensor networks issues, optimizing protein structure, and most importantly solving image processing problems. In this chapter, we address emerging challenges like CoD problem in big data and provide practical engineering solutions using ABC and other related algorithms.

Here, we will discuss artificial bee colony (ABC) which is inspired by a set of sequential processes such as the process of seeking for a bunch of flowers, sharing information in the hive regarding that, and allocating employed, onlooker and scout bees. Karaboga introduced ABC [48] which is compatible with continuous problems

---

Algorithm 2 Implementation of ABC algorithm for feature selection

---

Input: $S = \{x_0, x_1, x_2, \ldots, x_n\}$, $P_{size}=2*n$, $limit \geq 0$, $0 \leq lower_{Bound} \leq n/2$, $lower_{Bound} \leq upper_{Bound} \leq n$, $max_{iteration} \geq 0$, $t=0$, $v = random_{number} \in [0, 1]$, $v' = random_{number} \in [0, 1]$, $Best_{solution} = \emptyset$.

Output: $Best_{solution}$: *An optimal subset of features* (F) , $F = \{x_0, x_1, x_2, \ldots, x_m\}$, $m \leq n$, $(\forall f_i \in F) \in S$, $F_{length} \leq S_{length}$.

1:   Call fitness function to evaluate the whole food source (S) (primary evaluation of each food)
2: **for** { **dot**=0 $\cdots$ $max_{iteration}$ }
3:     Call Employed bees to update the food source regarding their evaluation
4:     Call Onlooker bees to exploit the local foods to generate new food (solution)
5:     Choose parents and generate a new food (solution) based on $V_i= f_i+v*(f_i - f_j)$
6:     **if** limit is met **then** :
7:       { Call scout bee to explore new (unseen) food source to prevent from local optimum using
8:        $X_i= X_{upperBound}+v' * \{ (X_{upperBound}-X_{lowerBound})\} \}$
        return $NewSolution$
9:     **end if**
10:    Call fitness function to evaluate the Solution
11:    **if** any Solution obtained the best score **then**
12:       {Update the $Best_{solution}$}
13:    **end if**
14: **end for**

---

in 2005. Algorithm 2 presents a general procedure of given ABC. A large number of research studies have accomplished using this algorithm [49, 50] and even convert that into a way that it also works with discrete problems [23, 27, 37]. Not only those, but also ABC is applied on optimization problems as an optimizer [50–53].

Artificial bee colony interacts with three groups of bees to have work done. The first group is employed bees, second is onlooker, and last one is scout. In initialization step the number of these groups is set. The employed bees together with onlooker bees create a population which has an equal amount of two groups. ABC starts with initialization step which has positive impact on converging in ABC. Among initialization variables, limit is important criteria and provides a condition when an employed bee converts into scout bee; at a time, we only have one scout bee.

3.3.4 Particle Swarm Optimization Algorithm

The particle swarm optimization (PSO) is one of the population-based meta-heuristic algorithms and optimization techniques. PSO is inspired from social-psychological principles [54]. In 1995 particle swarm optimization first introduced by Kennedy and Eberhart [55]. The PSO is based on the simulation of common animal social behaviors, for instances: fish schooling, bird flocking. PSO like other evolutionary algorithms searches for the global optimum rather than local optimum. However, the particle swarm trapped into local optimum easily when feature

---

Algorithm 3 Implementation of PSO algorithm for feature selection [48]

Input: $S = \{x_0, x_1, x_2, \ldots, x_n\}$, $particles_{number} \geq 1$, $acceleration_{coefficient}(c_1, c_2) \in [0, 1]$,
 $max_{velocity}$, t=0, $min_weight, max_weight = random_{number} \in [0, 1]$, $Best_{solution} = \emptyset$.

Output: $Best_{solution}$: $An optimal subset of features$(F) , $F = \{x_0, x_1, x_2, \ldots, x_m\}$, m$\leq$ n ,
 $(\forall f_i \in F) \in S$, $F_{length} \leq S_{length}$.

1: **for** t=0 $\cdots$ $max_{iteration}$ **do**
2:    **for** i=0 $\cdots$ $particles_{number}$ **do**
3:        Call fitness (= objective) function to for the current particle
4:        Save the best personal location
5:        Save the best global location
6:    **end for**
7:    Update the $inertia_{weight}$
8:    **for** i=0 $\cdots$ $particles_{number}$ **do**
9:        Update the velocity
10:        Update the position
11:    **end for**
12:    **if** condition met **then**
13:        Return the best global location as the global optimum
14:    **end if**
15: **end for**

---

dimension grows significantly. Algorithm 3 presents a pseudocode for a standard PSO for solving high dimensionality problem. The whole process of PSO usually initializes groups of random particles and computes fitness for each particle within iterations in order to converge into global optimum. Each particle is considered as a single solution to our problem.

PSO follows two simple yet essential steps to have completed optimization process to find the minimum optimum or maximum optimum. The first step is communication among particles. Each particle shares their information with other particles after moving in their direction. This process makes them find a proper way toward the goal. Each time, based on the problem (maximum/minimum optimization), particles follows the particle and consider the particle that match the problem goal. For instance, each iteration particles call fitness function to get fitness of their location. Then, among the particles, one has the best value which is set to best personal location. The best value is examined based on the problem, if it is minimum optimization then the best value goes for the particle that has the minimum value. Moreover, if the problem is maximum optimization the best value goes for the particle that has the maximum value. When this value is set, each particle updates their direction and moves toward this values. It is obvious that the one has the best value does not move unless other particles find the best value. The second step which each particle does is to learn. They can learn how to update their direction after each iteration and tune the parameters.

The PSO does not have parent selection, recombination, and mutation steps [56]; thus, this enables PSO to behave in a particular way in comparison with other evolutionary algorithms. Concretely, each member within the population do not get updated nor removed. Hao et al. [57] introduced a new PSO with added crossover operator. Zhang et al. [58] proposed a binary PSO with mutation operator to address CoD problem using feature selection techniques to solve it. The crossover enables the particles do not stop in the local optimum by sharing the other particles' information. In [32] PSO is classified into three different versions: classical PSO, scale-free PSO, and binary PSO.

Few parameters are required to adjust, and enable PSO easy to implement, make popular stochastic and yet powerful swarm-based algorithm. Inertia weight becomes more important than other due to its ability of having a trade-off between the exploration and exploitation process within a search space. In addition, inertia weight has positive affect convergence rate in PSO [59].

In the literature, some studies deployed PSO as an effective tool for solving large-scale optimization problems, including optimal allocation of electric vehicle charging station and distributed renewable resource in power distribution networks [41], designing power system stabilizers [60], distribution state estimation [61], and reactive power control [62].

3.3.5 Ant Colony Optimization (ACO)

Ant colony optimization is another popular evolutionary algorithm which is presented in 1999 by Dorigo, Marco, and Di Caro [63], and Socha and Dorigo introduced continuous domain of it [64]. Basically, ACO is one of the stochastic search processes. Once ants explore a new food source, they try to lay some pheromone to mark the way which leads to the food. The pheromone is a chemical odorous material which is produced and used by ants to communicate with other ants in an indirect way. Each ant tries to produce it and lays it on their way. So others can follow the odor to seek for the food; meanwhile, they also produce the same amount of pheromone. On the other hand, further, as we inspired natural behavior, this chemical material is susceptible to evaporate. Thus, the amount of pheromone on specific path will increase by keeping ants on the same path, However, each iteration we have a pheromone reduction process which has negative affect the total amount of pheromone on a particular path. In other words, if any ants do not select the path used to be chosen, then the path would disappear. Algorithm 4 presents an overall procedure of ACO for feature selection.

Algorithm 4 Implementation of ACO algorithm for feature selection [65]

Input: S=$\{x_0, x_1, x_2, \ldots, x_n\}$, K$\} \geq 1$, η and τ, $t = 0$, $best_{solution}$ =Ø.

Output: $Best_{solution}$: *An optimal subset of features* (F), $F = \{x_0, x_1, x_2, \ldots, x_m\}$, m$\leq$ n, ($\forall f_i \in F) \in S$, $F_{length} \leq S_{length}$.

1: Call fitness function to calculate the fitness of each feature
2: **for** t=0 $\cdots$ $max_{Iteration}$ **do**
3:      Generate K ants
4:      **for** each ant $\in Ants(K)$ **do**
5:          Generate a subset of features
6:          call fitness function to evaluate the generated subset
7:          Update the best local and global optimum
8:          **if** condition met **then**
9:              Return the best global location as the global optimum
10:          **else**
11:              Update the η and τ
12:          **end if**
13:      **end for**
14: **end for**

3.3.6 Grey Wolf Optimizer (GWO)

Grey wolf optimizer (GWO), which is a new evolutionary algorithm primary works based on the concept of grey wolf society, is presented in 2014 [59]. Mirjalili et al. claimed that [66] GWO outperforms other evolutionary algorithms for solving large-scale engineering and science problems. Algorithm 5 is a sample process of solving

---

Algorithm 5 Implementation of GWO algorithm for feature selection [66]

Input: $S = \{x_0, x_1, x_2, \ldots, x_n\}$, $X_i = (i = 1, 2, \ldots n)$, A, t=0 , α, C, $max_{iteration} \geq 0$, $Best_{solution} \leq S_{length}$.

Output: $Best_{solution}$: *An optimal subset of features* (F) , $F = \{x_0, x_1, x_2, \ldots, x_m\}$, m$\leq$ n, ($\forall f_i \in F) \in S$, $F_{length} \leq S_{length}$.

1: Call fitness function for each search agent to evaluate the whole food source (S) (primary evaluation of each food)
2:    X_α = the best search agent
3:    X_β= the second best search agent
4:    X_δ= the third best search agent
5: **for** t=0 $\cdots$ $max_{iteration}$ **do**
6:      **for** each search agent **do**
7:          *Update the best position of current search agent using* $\overrightarrow{X_{t+1}} = \frac{\overrightarrow{X_1} + \overrightarrow{X_2} + \overrightarrow{X_3}}{3}$.
8:      **end for**
9:      Update α , A and C.
10:      Call fitness function to calculate the fitness of each search agent
11:      Update X_ω, X_β, X_δ
12:      **if** any solution obtained the best score **then**
13:          Update the $Best_{solution}$
14:      **end if**
15: **end for**

large-scale engineering problems and one of the evolutionary feature selection algorithm. The GWO algorithm inspired by the natural mechanism of animals. The most common behavior which almost wild animal inherited normally is their attitude to have a kingdom, rule others and having the same hunting mechanism. It solves the science problems through the following steps:

- First of all, it searches for some animal as prey. In other words, it tries to explore the area (food source);
- Then, it surrounds the possible prey(s) by exploitation, doing local search to find the border of sample space;
- Finally, it attacks the prey, doing local search to find the best value within a new area. "A" stands for the most important parameter in GWO and adjusts the step size towards the prey. Thus, "A" has positive impact on convergence of this algorithm to the global optimum by tuning step size which influences both exploitation and exploration. However, GWO still suffers from stalling in local minimum, so initializing the parameter "A" with a proper value helps it to prevent from stopping in local minimum.

3.3.7   Coyote Optimization Algorithm (COA)

Coyote optimization algorithm (COA) is another yet important population-based meta-heuristic algorithms which have been inspired from the Canis latrans species and natural coyotes' behavior. COA has a very certain procedure that works based on the way how these animals approaching other animals (preys) for catching them. Thus, COA seems to be one particular type of grey wolf optimizer (GWO) as COA just does the third step of GWO. COA is presented recently in [22] by Pierezan and Coelho in 2018 to solve large-scale optimization problems. Algorithm 6 presents a general overview of COA for feature selection.

3.3.8   Other Optimization Algorithms

Meng et al. [67] proposed chicken swarm optimization algorithm (CSO) in 2014. Algorithm 7 presents well-structured pseudocode of CSO for optimized feature selection. Based on performance of CSO, researchers have successfully solved and optimized engineering and science problems, directional reader antennas optimization [68], community detection in social networks [69], parameter optimization of a fuzzy logic system [70].

Algorithm 6 Implementation of COA algorithm for feature selection [22]

Input: $S = \{x_0, x_1, x_2, \ldots, x_n\}$ $which\ consists\ of\ N_p \in N^* and N_c \in N^* are\ initialized\ using\ soc_{c,j}^{p,t} = lower_{Boundj} + v_j \cdot (upper_{Boundj} - lower_{Boundj})$, t=0 , $max_{iteration} \geq 0$, $Best_{cayotes} = \emptyset$.

Output: $Best_{cayotes}$: $An\ optimal\ subset\ of\ features$ (F) , $F = \{x_0, x_1, x_2, \ldots, x_m\}$, m$\leq$ n , $(\forall f_i \in F) \in S$, $F_{length} \leq S_{length}$.

1: Call fitness function to calculate the coyote's fitness using:
2: $fit_c^{p,t} = f(soc_c^{p,t})$
3: **for** t=0 $\cdots$ $max_{iteration}$ **do**
4:     $alpha^{p,t} = \{soc_c^{p,t} | arg_{c=\{1,2,\ldots,N_c\}} minf(soc_c^{p,t})\}$
5:     Calculate the social tendency of the pack based on N_c as follows:
6:     **if** N_c is odd **then**
7:         $cult_j^{p,t} = \quad O_{\frac{(N_c+1)}{2},j}^{p,t}$
8:     **else :**
9:         $cult_j^{p,t} = \dfrac{O_{\frac{N_c}{2},j}^{p,t} + O_{(\frac{N_c}{2}+1),j}^{p,t}}{2}$
10:    **end if**
11:    **for** each c coyotes in the p pack **do**
12:        Update the social condition using:
13:        $\_soc_c^{p,t} = soc_c^{p,t} + r_1 \cdot \delta_1 + r_2 \cdot \delta_2$
14:        Examine the new social condition using:
15:        $new\_fit_c^{p,t} = f(new\_soc_c^{p,t})$
16:        update food source with respect to better fitness using:
17:        $soc_c^{p,t+1} = new\_soc_c^{p,t}$
18:    **end for**
19:    Birth and death using:
20:    $pup_j^{p,t} = \begin{cases} soc_{r_1,j}^{p,t}, & rnd_j < P_s\ or\ j = j_1 \\ soc_{r_2,j}^{p,t}, & rnd_j \geq P_s + P_a\ or\ j = j_2 \\ R_j, & otherwise \end{cases}$
21:    $Transition\ between N_c and N_p\ packs\ using\ P_e = 0.005 \cdot N_c^2$
22:    Update the coyotes' information with respect to the age
23:    **if** stop condition met **then**
24:        $Return\ the\ Best_{coyotes}$
25:    **end if**
26: **end for**

Li et al. introduced fish swarm algorithm (FSA) which is another population-based (or swarm-based) evolutionary algorithm [71]. FSA inspired from the behaviors of fish school. Algorithm 8 shows the process of feature selection using FSA. Research studies have applied FSA to optimize their solution such as neighborhood feature selection [72], multi-modal benchmark functions solver [73]. Learning to sense is referred to as meta-sensing [74].

Algorithm 7 Implementation of CSO algorithm for feature selection [67]

Input: $S = \{x_0, x_1, x_2, \ldots, x_n\}$, $N_p \in N^*$, $N_c \in N^*$ are done using $soc_{c,j}^{p,t} = lower_{Boundj} + v_j \cdot (upper_{Boundj} - lower_{Boundj})$, t=0, $rooster_{ratio}$, $chicks_{ratio}$, $hens_{ratio}$, $food_{position}$ C, $Random_{value}$, $min_{iteration}$, $max_{iteration}$, $chickenSwarm_{size}$.

Output: $Best_{solution}$: $Anoptimalsubsetoffeatures$(F), $F = \{x_0, x_1, x_2, \ldots, x_m\}$, m≤ n, ($\forall f_i \in F$) $\in S$, $F_{length} \leq S_{length}$.

1: **for** t=0 $\cdots$ $max_{iteration}$ **do**
2:     call fitness function to compute the fitness using chicken
3:     **if** fitness of chicken ==$best_{fitness}$ **then**
4:         *Update the Random_{value}*
5:         *Update the rooster position*
6:     **end if**
7:     **if** fitness of chicken ==$worst_{fitness}$ **then**
8:         Update the chicks position
9:     **end if**
10:    **if** fitness of chicken != $worst_{fitness}$ and fitness of chicken != $best_{fitness}$ **then**
11:        *Update the Random_{value}*
12:        *Update the hens position*
13:    **end if**
14:    *Update chicken position*
15:    **if** t==$chickenSwarm_{size}$ **then**
16:        Return the best position as the global optimum
17:    **end if**
18: **end for**

3.4 Conclusion

Both dimension reduction by generating new dimension of features and feature selection by eliminating irrelevant and redundant features take care of missing values and classify supervised/unsupervised datasets; all of these operations come together to solve emerging challenging Np-hard problems in engineering and sciences. A large number of datasets, particularly big data, are available to work on. The main problem, here, concerns their features and dimensionality, the curse of dimensionality (CoD), which causes yet another important problem, high time complexity. In this chapter, we addressed these problems and professional approaches using advanced machine learning algorithms. The studies prove that applying nature-inspired algorithms, together with machine learning techniques, enabled researchers' attempts to solve the CoD problem, which yields a proper running time with a lowest time complexity. It is noteworthy that evolutionary algorithms are non-dependent domain specific, which provides an optimized environment for researchers who want to solve their problems or optimize their approaches. In this chapter, we have explored evolutionary algorithms and their applications in solving large-scale optimization problems, especially the feature extraction process for data analytics. This chapter provides insightful information for researchers who are seeking for the application of evolutionary algorithms for engineering, optimization, and data science. Having said this, in [75], we address the emerging problem, CoD, and an evolutionary-based solution is presented to solve it. We discuss

---

Algorithm 8 Implementation of FSA algorithm for feature selection [22]

Input: $S = \{x_0, x_1, x_2, \ldots, x_n\}$, t=0 , $max_{iteration} \geq 0$, R_{min}, L_{min}, $\gamma_B(D) = \frac{|POS_B(D)_\gamma|}{|U|}$,

$Best_{cayotes} = \emptyset$.

Output: $best_{cayotes}$: An optimal subset of features (F) , $R_{min} = F = \{x_0, x_1, x_2, \ldots, x_m\}$, $m \leq$

n , $(\forall f_i \in F) \in S$, $F_{length} \leq S_{length}$.

1: R_{min}=C , L_{min}=C
2: **for** t=0 $\cdots$ $max_{iteration}$ **do**
3:     generate total fish (Fish)
4:     **for** each fish $K \in Fish$ **do**
5:         R_K=$\emptyset$, L_K=0
6:         $Choose\ a\ feature\ \alpha_k \in C(randomly)$
7:         $Update R_K, L_K by R_K \bigcup \alpha_K and |R_K|, respectively$
8:     **end for**
9:     **for** each fish$K \in Fish$ **do**
10:         $R_s = Search(R_k)$
11:         $R_\omega = Swarm(R_k)$
12:         $R_f = Follow(R_k)$
13:         UpdateR_K, L_K by seeking for the $max_{fitness}$ through (R_k, R_ω, R_f)
14:         **if** $\gamma_{R_k}(D)_\delta ==_{\gamma_C}(D)_\delta$ **then**
15:             $The\ fish_K\ obtained\ a\ local\ reduction\ and\ break$
16:         **end if**
17:         **if** $\gamma_{R_k}(D)_\delta ==_{\gamma_C}(D)_\delta and L_K \leq L_{min}$ **then**
18:             $update R_{min}, L_{min} by R_K and L_K, respectively$
19:         **end if**
20:     **end for**
21:     **if** stop condition met **then**
22:         $Return\ the\ R_{min}, L_{min}$
23:     **end if**
24: **end for**

---

the feature extraction optimization process in detail, leveraging feature extraction and evolutionary algorithms. Then, we provide detailed and practical examples of applying evolutionary algorithms with a wide variety of domains. We also classify all research studies based on the most common challenging issues such as stego image classification, network anomalies detection, network traffic classification, sentiment analysis, and supervised benchmark classification.

References

1. P. Marrow, Nature-inspired computing technology and applications. BT Technol. J. **18**(4), 13–23 (2000)
2. M.H. Amini, Distributed computational methods for control and optimization of power distribution networks, PhD Dissertation, Carnegie Mellon University, 2019
3. M.H. Amini, J. Mohammadi, S. Kar, Distributed holistic framework for smart city infrastructures: tale of interdependent electrified transportation network and power grid. IEEE Access **7**, 157535–157554 (2019)

4. M.H. Amini, J. Mohammadi, S. Kar, Distributed intelligent algorithm for interdependent electrified transportation and power networks, *Proceedings of the 9th ACM Symposium on Design and Analysis of Intelligent Vehicular Networks and Applications* (ACM, New York, 2019)

5. A. Imteaj, M.H. Amini, J. Mohammadi, Leveraging decentralized artificial intelligence to enhance resilience of energy networks. arXiv preprint arXiv:1911.07690 (2019)

6. M.H. Amini, B. Nabi, M.R. Haghifam, Load management using multi-agent systems in smart distribution network, in *2013 IEEE Power and Energy Society General Meeting* (IEEE, New York, 2013)

7. M.H. Amini, S. Bahrami, F. Kamyab, S. Mishra, R. Jaddivada, K. Boroojeni, P. Weng, Y. Xu, Decomposition methods for distributed optimal power flow: panorama and case studies of the DC model, in *Classical and recent aspects of power system optimization* (Academic Press, Cambridge, 2018), pp. 137–155

8. A. Imteaj, M.H. Amini, Distributed sensing using smart end-user devices: pathway to federated learning for autonomous IoT, in *2019 International Conference on Computational Science and Computational Intelligence* (Las Vegas, 2019)

9. F.G. Mohammadi, H.R. Arabnia, ISEA: image steganalysis using evolutionary algorithms. arXiv preprint, arXiv:1907.12914 (2019)

10. N. Maheswaranathan, L. Metz, G. Tucker, J. Sohl-Dickstein, Guided evolutionary strategies: escaping the curse of dimensionality in random search. arXiv preprint, arXiv:1806.10230 (2018)

11. M.J.L.F. Cruyff, U. Böckenholt, P.G.M. Van Der Heijden, L.E. Frank, A review of regression procedures for randomized response data, including univariate and multivariate logistic regression, the proportional odds model and item response model, and self-protective responses, in *Handbook of Statistics*, vol. 34 (Elsevier, Amsterdam, 2016), pp. 287–315

12. F. Zhuang, X. Cheng, P. Luo, S.J. Pan, Q. He, Supervised representation learning: transfer learning with deep autoencoders, in *Twenty-Fourth International Joint Conference on Artificial Intelligence* (2015)

13. J. Yang, S. Shebalov, D. Klabjan, Semi-supervised learning for discrete choice models, in *IEEE Transactions on Intelligent Transportation Systems* (2018)

14. N. Altman, M. Krzywinski, The curse (s) of dimensionality. Nat. Methods **15**, 399–400 (2018)

15. W. Guo, G. Lynch, J.P. Romano, A new approach for large scale multiple testing with application to FDR control for graphically structured hypotheses. arXiv preprint, arXiv:1812.00258 (2018)

16. S. Gupta, S. Bhardwaj, P.K. Bhatia, A reminiscent study of nature inspired computation. Int. J. Adv. Eng. Technol. **1**(2), 117 (2011)

17. R. Balamurugan, A.M. Natarajan, K. Premalatha, Stellar-mass black hole optimization for biclustering microarray gene expression data. Appl. Artif. Intell. **29**(4), 353–381 (2015)

18. C. Blum, A. Roli, Metaheuristics in combinatorial optimization: overview and conceptual comparison. ACM Comput. Surv. **35**(3), 268–308 (2003)

19. S.F. Razavi, H. Sajedi, SVSA: a semi-vortex search algorithm for solving optimization problems. Int. J. Data Sci. Anal. **8**, 1–18 (2018)

20. P. Moscato, C. Cotta, An accelerated introduction to memetic algorithms, in *Handbook of Metaheuristics* (Springer, Cham, 2019), pp. 275–309

21. T. Bäck, D.B. Fogel, Z. Michalewicz, *Evolutionary Computation 1: Basic Algorithms and Operators* (CRC Press, Boca Raton, 2018)

22. J. Pierezan, L.D.S. Coelho, Coyote optimization algorithm: a new metaheuristic for global optimization problems, in *2018 IEEE Congress on Evolutionary Computation (CEC)* (IEEE, Rio de Janeiro, 2018), pp. 1–8

23. F.G. Mohammadi, M.S. Abadeh, Image steganalysis using a bee colony based feature selection algorithm. Eng. Appl. Artif. Intell. **31**, 35–43 (2014)

24. F.G. Mohammadi, M.S. Abadeh, A new metaheuristic feature subset selection approach for image steganalysis. J. Intell. Fuzzy Syst. **27**(3), 1445–1455 (2014)

25. D. Wunsch, R. Nigro, G. Coussement, C. Hirsch, Uncertainty quantification in an engineering design software system, in *Uncertainty Management for Robust Industrial Design in Aeronautics* (Springer, Cham, 2019), pp. 747–754
26. D.L. Barbour, Precision medicine and the cursed dimensions. npj Digit. Med. **2**(1), 4 (2019)
27. S.N. Karpagam, S. Raghavan, Automated diagnosis system for Alzheimer disease using features selected by artificial bee colony. J. Comput. Theor. Nanosci. **16**(2), 682–686 (2019)
28. W.K. Vong, A.T. Hendrickson, D.J. Navarro, A. Perfors, Do additional features help or hurt category learning? The curse of dimensionality in human learners. Cogn. Sci. **43**(3), e12724 (2019)
29. N.P. Patel, E. Sarraf, M.H. Tsai, The curse of dimensionality. Anesthesiol. J. Am. Soc. Anesthesiol. **129**(3), 614–615 (2018)
30. M. Oudah, A. Henschel, Taxonomy-aware feature engineering for microbiome classification. BMC Bioinf. **19**(1), 227 (2018)
31. A. Serani, M. Diez, J. Wackers, M. Visonneau, F. Stern, Stochastic shape optimization via design-space augmented dimensionality reduction and RANS computations, in *AIAA SciTech 2019 Forum* (2019), pp. 2218
32. S.L. Gupta, A.S. Baghel, A. Iqbal, Big data classification using scale-free binary particle swarm optimization, in *Harmony Search and Nature Inspired Optimization Algorithms* (Springer, Singapore, 2019), pp. 1177–1187
33. H. Shi, H. Li, D. Zhang, C. Cheng, X. Cao, An efficient feature generation approach based on deep learning and feature selection techniques for traffic classification. Comput. Netw. **132**, 81–98 (2018)
34. U. Khurana, H. Samulowitz, D. Turaga, Feature engineering for predictive modeling using reinforcement learning, in *Thirty-Second AAAI Conference on Artificial Intelligence* (2018)
35. D. Zhang, J. Yin, X. Zhu, C. Zhang, Network representation learning: a survey, in *IEEE Transactions on Big Data* (2018)
36. R. Vanaja, S. Mukherjee, Novel wrapper-based feature selection for efficient clinical decision support system, in *International Conference on Intelligent Information Technologies* (Springer, Singapore, 2018), pp. 113–129
37. E. Hancer, B. Xue, M. Zhang, D. Karaboga, B. Akay, Pareto front feature selection based on artificial bee colony optimization. Inf. Sci. **422**, 462–479 (2018)
38. X.-Y. Liu, Y. Liang, S. Wang, Z.-Y. Yang, H.-S. Ye, A hybrid genetic algorithm with wrapper-embedded approaches for feature selection. IEEE Access **6**, 22863–22874 (2018)
39. V. Rostami, A.S. Khiavi, Particle swarm optimization based feature selection with novel fitness function for image steganalysis, in *2016 Artificial Intelligence and Robotics (IRANOPEN)* (IEEE, Qazvin, 2016), pp. 109–114
40. S. Jiang, S. Yang, A steady-state and generational evolutionary algorithm for dynamic multiobjective optimization. IEEE Trans. Evol. Comput. **21**(1), 65–82 (2016)
41. M.H. Amini, M.P. Moghaddam, O. Karabasoglu, Simultaneous allocation of electric vehicles parking lots and distributed renewable resources in smart power distribution networks. Sustain. Cities Soc. **28**, 332–342 (2017)
42. L.Y. Zhang, G. Luo, L.N. Lu, Genetic algorithms in resource optimization of construction project. J. Tianjin Univ. (Sci. Technol.) **34**(2), 188–192 (2001)
43. M.H. Amini, A. Islam, Allocation of electric vehicles' parking lots in distribution network, in *ISGT 2014* (IEEE, Washington, DC, 2014), pp. 1–5
44. J.R. Koza, *Genetic Programming: On the Programming of Computers by Means of Natural Selection*, vol. 1 (MIT Press, Cambridge, 1992)
45. I.L.S. Russo, H.S. Bernardino, H.J.C. Barbosa, Knowledge discovery in multiobjective optimization problems in engineering via genetic programming. Expert Syst. Appl. **99**, 93–102 (2018)
46. L.J. Fogel, Artificial intelligence through a simulation of evolution, in *Proceedings of the 2nd Cybernetics Science Symposium* (1965)
47. D. Karaboga, B. Gorkemli, C. Ozturk, N. Karaboga, A comprehensive survey: artificial bee colony (ABC) algorithm and applications. Artif. Intell. Rev. **42**(1), 21–57 (2014)

48. D. Karaboga, An idea based on honey bee swarm for numerical optimization. Technical report, Technical report-tr06, Erciyes University, Engineering Faculty, Computer (2005)
49. F. Zabihi, B. Nasiri, A novel history-driven artificial bee colony algorithm for data clustering. Appl. Soft Comput. **71**, 226–241 (2018)
50. Y. Cao, Y. Lu, X. Pan, N. Sun, An improved global best guided artificial bee colony algorithm for continuous optimization problems. Clust. Comput. **22**, 3011–3019 (2018)
51. Y. Xue, J. Jiang, B. Zhao, T. Ma. A self-adaptive artificial bee colony algorithm based on global best for global optimization. Soft Comput. **22**, 2935–2952 (2018)
52. F. Harfouchi, H. Habbi, C. Ozturk, D. Karaboga, Modified multiple search cooperative foraging strategy for improved artificial bee colony optimization with robustness analysis. Soft Comput. **22**(19), 6371–6394 (2018)
53. H. Wang, J.-H. Yi, An improved optimization method based on krill herd and artificial bee colony with information exchange. Memet. Comput. **10**(2), 177–198 (2018)
54. K. Chen, F.-Y. Zhou, X.-F. Yuan, Hybrid particle swarm optimization with spiral-shaped mechanism for feature selection. Expert Syst. Appl. **128**, 140–156 (2019)
55. J. Kennedy, R. Eberhart, Particle swarm optimization, in *Proceedings of IEEE International Conference on Neural Networks*, vol. 4 (IEEE Press, Perth, 1995), pp. 1942–1948
56. J. Kennedy, Particle swarm optimization, in *Encyclopedia of Machine Learning* (Springer, Boston, 2010), pp. 760–766
57. Z.-F. Hao, Z.-G. Wang, H. Huang, A particle swarm optimization algorithm with crossover operator, in *2007 International Conference on Machine Learning and Cybernetics*, vol. 2 (IEEE, Hong Kong, 2007), pp. 1036–1040
58. Y. Zhang, S. Wang, P. Phillips, G. Ji, Binary PSO with mutation operator for feature selection using decision tree applied to spam detection. Knowl. Based Syst. **64**, 22–31 (2014)
59. A. Agrawal, S. Tripathi, Particle swarm optimization with probabilistic inertia weight, in *Harmony Search and Nature Inspired Optimization Algorithms* (Springer, Singapore, 2019), pp. 239–248
60. M.A. Abido, Optimal design of power-system stabilizers using particle swarm optimization. IEEE Trans. Energy Convers. **17**(3), 406–413 (2002)
61. S. Naka, T. Genji, T. Yura, Y. Fukuyama, Practical distribution state estimation using hybrid particle swarm optimization, in *2001 IEEE Power Engineering Society Winter Meeting. Conference Proceedings (Cat. No. 01CH37194)*, vol. 2 (IEEE, Columbus, 2001), pp. 815–820
62. H. Yoshida, K. Kawata, Y. Fukuyama, S. Takayama, Y. Nakanishi, A particle swarm optimization for reactive power and voltage control considering voltage security assessment. IEEE Trans. Power Syst. **15**(4), 1232–1239 (2000)
63. M. Dorigo, G.D. Caro, Ant colony optimization: a new meta-heuristic, in *Proceedings of the 1999 Congress on Evolutionary Computation-CEC99 (Cat. No. 99TH8406)*, vol. 2 (IEEE, Washington, DC, 1999), pp. 1470–1477
64. K. Socha, M. Dorigo, Ant colony optimization for continuous domains. Eur. J. Oper. Res. **185**(3), 1155–1173 (2008)
65. M.M. Kabir, M. Shahjahan, K. Murase, A new hybrid ant colony optimization algorithm for feature selection. Expert Syst. Appl. **39**(3), 3747–3763 (2012)
66. S. Mirjalili, S.M. Mirjalili, A. Lewis, Grey wolf optimizer. Adv. Eng. Softw. **69**, 46–61 (2014)
67. X. Meng, Y. Liu, X. Gao, H. Zhang, A new bio-inspired algorithm: chicken swarm optimization, in *International Conference in Swarm Intelligence* (Springer, Cham, 2014), pp. 86–94
68. W. Shi, Y. Guo, S. Yan, Y. Yu, P. Luo, J. Li, Optimizing directional reader antennas deployment in UHF RFID localization system by using a MPCSO algorithm. IEEE Sensors J. **18**(12), 5035–5048 (2018)
69. K. Ahmed, A.E. Hassanien, E. Ezzat, P.-W. Tsai, An adaptive approach for community detection based on chicken swarm optimization algorithm, in *International Conference on Genetic and Evolutionary Computing* (Springer, Cham, 2016), pp. 281–288

70. X.-B. Meng, H.-X. Li, Dempster-Shafer based probabilistic fuzzy logic system for wind speed prediction, in *2017 International Conference on Fuzzy Theory and Its Applications (iFUZZY)* (IEEE, Pingtung, 2017), pp. 1–5
71. X.-L. Li, An optimizing method based on autonomous animats: fish-swarm algorithm. Syst. Eng. Theory Pract. **22**(11), 32–38 (2002)
72. Y. Chen, Z. Zeng, J. Lu, Neighborhood rough set reduction with fish swarm algorithm. Soft Comput. **21**(23), 6907–6918 (2017)
73. I. Rahman, J. Mohamad-Saleh, N. Sulaiman, Artificial fish swarm-inspired whale optimization algorithm for solving multimodal benchmark functions, in *10th International Conference on Robotics, Vision, Signal Processing and Power Applications* (Springer, Singapore, 2019), pp. 59–65
74. F.G. Mohammadi, M.H. Amini, Promises of meta-learning for device-free human sensing: learn to sense, in *Proceedings of the 1st ACM International Workshop on Device-Free Human Sensing (DFHS'19)* (ACM, New York, 2019), pp. 44–47. https://doi.org/10.1145/3360773.3360884
75. F.G. Mohammadi, M.H. Amini, H.R. Arabnia, Applications of nature-inspired algorithms for dimension reduction: enabling efficient data analytics. arXiv preprint, arXiv: 1908.08563 (2019)

Chapter 4
Applications of Nature-Inspired Algorithms for Dimension Reduction: Enabling Efficient Data Analytics

Farid Ghareh Mohammadi, M. Hadi Amini, and Hamid R. Arabnia

Abstract In Mohammadi et al. (Evolutionary computation, optimization and learning algorithms for data science. arXiv preprint, arXiv: 1908.08006, 2019), we have explored the theoretical aspects of feature selection and evolutionary algorithms. In this chapter, we focus on optimization algorithms for enhancing data analytic process, i.e., we propose to explore applications of nature-inspired algorithms in data science. Feature selection optimization is a hybrid approach leveraging feature selection techniques and evolutionary algorithms process to optimize the selected features. Prior works solve this problem iteratively to converge to an optimal feature subset. Feature selection optimization is a non-specific domain approach.

Data scientists mainly attempt to find an advanced way to analyse data n with high computational efficiency and low time complexity, leading to efficient data analytics. Thus, by increasing generated/measured/sensed data from various sources, analysis, manipulation and illustration of data grow exponentially. Due to the large scale datasets, curse of dimensionality (CoD) is one of the NP-hard problems in data science. Hence, several efforts have been focused on leveraging evolutionary algorithms (EAs) to address the complex issues in large scale data analytics problems. Dimension reduction, together with EAs, lends itself to solve CoD and solve complex problems, in terms of time complexity, efficiently. In this chapter, we first provide a brief overview of previous studies that focused on solving CoD using feature extraction optimization process. We then discuss

F. G. Mohammadi · H. R. Arabnia
Department of Computer Science, Franklin College of Arts and Sciences,
University of Georgia, Athens, GA, USA
e-mail: farid.ghm@uga.edu; hra@cs.uga.edu

M. H. Amini (✉)
School of Computing and Information Sciences, Florida International University, Miami, FL, USA

Sustainability, Optimization, and Learning for InterDependent Networks Laboratory (solid lab), Florida International University, Miami, FL, USA
e-mail: amini@cs.fiu.edu; hadi.amini@ieee.org; www.solidlab.network

© Springer Nature Switzerland AG 2020   67
M. H. Amini (ed.), *Optimization, Learning, and Control for Interdependent Complex Networks*, Advances in Intelligent Systems and Computing 1123,
https://doi.org/10.1007/978-3-030-34094-0_4

practical examples of research studies that successfully tackled some application domains, such as image processing, sentiment analysis, network traffics/anomalies analysis, credit score analysis and other benchmark functions/datasets analysis.

Keywords Dimension reduction · Data science · Hybrid optimization · Curse of dimensionality (CoD) · Supervised learning · Unsupervised learning · Wrapper feature selection · Classification · Evolutionary computation · Swarm intelligence · Filter feature selection

4.1 Introduction

4.1.1 Overview

Feature selection and evolutionary algorithms have been explored in [1]. In this chapter, we aim to focus on an optimization approach for enhancing data analytic process. Feature selection optimization is hybrid approach leveraging pure feature selection techniques and evolutionary algorithms process to optimize the selected features. Researchers try to iterate this process until to converge to an optimal feature subsets. Feature selection optimization is a non-specific domain approach which enables scientists to apply this to their data technically.

Data scientists always attempt to find an advanced way to work with data that are successfully conducted in a short time with high computational efficiency and low time complexity, leading to efficient data analytics. Thus, by increasing generated/measured/sensed data from various sources, analysis, manipulation and illustration of data grow exponentially. Due to the large scale datasets, curse of dimensionality (CoD) is one of NP-hard problems in data science. Hence, several efforts have been focused on leveraging evolutionary algorithms (EAs) to address the complexity issues in large scale data analytics problems. Dimension reduction, together with EAs, lends itself to solve CoD and solve complex problems, in terms of time complexity, efficiently. In this chapter, we first provide a brief overview of previous studies that focused on solving CoD using feature extraction optimization process. We then discuss practical examples of research studies that successfully tackled some application domains, such as image processing, sentiment analysis, network traffics/anomalies analysis, credit score analysis and other benchmark functions/datasets analysis.

In [1], we have comprehensively explored various evolutionary algorithms for data science, including artificial bee colony (ABC), ant colony optimization (ACO), Coyote Optimization Algorithm (COA), genetic algorithm (GA), grey wolf optimizer (GWO) and particle swarm optimization. These algorithms have real-world applications in the context of dimension reduction [2–10], interdependent smart city infrastructures [11, 12], optimization [13–15]. This chapter focuses on application of aforementioned evolutionary algorithms in dimension reductions [16–25]. As mentioned in [1], some studies deployed GA, PSO or their combination as effective tools for solving large scale optimization problems, including optimal allocation of electric vehicle charging station and distributed renewable resource in power distribution networks [26, 27], resource optimization in construction projects [15] and allocation of electric vehicle parking lots in smart grids [28]. Moreover, we discuss some decent optimization using evolutionary algorithms including butterfly optimization algorithm (BOA), chicken swarm optimization (CSO), coral reefs optimization (CRO) and whale optimization algorithm (WOA).

Table 4.1 presents a representative information about the feature selection techniques using evolutionary algorithms.

4.1.2 Organization

The rest of this study is organized as follows. In Sect. 4.2, we have discussed the applied evolutionary algorithms and their application. Then, we introduce hybrid feature selection methods using evolutionary algorithms and their application in solving engineering and science problems. In general, Fig. 4.1 represents the overall structure of this study.

4.2 Application of Evolutionary Algorithms

Engineering, industries, scientists consider EA at the very final plan. They try to solve their problems easily in low time complexity run time. They used plenty of algorithms to find an optimal solution for their problems. For instance, distributed optimization and learning algorithms have been deployed as promising solutions to deal with information privacy, scalability, as well as (near) real-time decision making capability; applications of such algorithms include optimal operation of smart city infrastructures, interdependent power and transportation networks [42–44], artificial intelligence for energy system resilience [45], energy management and optimal power flow problem [46, 47], and learning at the IoT device level [48]. Some problems they are struggling with are NP-hard problems which are required to ponder the problems deeply. Therefore, final plan for researchers which is left is adopting evolutionary algorithms. Therefore, research scientists do not use EAs for solving simple problems and only consider them for challenging issues and NP-hard problems. It means that EAs have wide variety of applications and are not limited to a specific problems. In this section we will review some applications of them in image processing [2, 31, 49], optimization problems [9]. Thus, evolutionary algorithms are impactful methods to address different NP-hard problems [50]. Table 4.1 presents a representative information about the hybrid (combined) feature selection techniques using evolutionary algorithms on supervised/unsupervised datasets. Based on [5, 22], hybrid (combined) methods are newly introduced, which mixes evolutionary algorithms together with filter-based or wrapper-based algorithms. So, all proposed methods you see in the table are hybrid

Table 4.1 An overview of previous studies on supervised/unsupervised datasets applying hybrid (combined) feature selection

Paper	Evolutionary algorithm	Proposed method	Feature extraction type	Problem	Domain	Year
[2]	Artificial bee colony	IFAB	FS (wrapper)	Supervised	Image processing	2014
[29]	Artificial bee colony	IFAB-KNN	FS (wrapper)	Supervised	Image processing	2014
[30]	Artificial bee colony	RISAB	FS (wrapper)	Supervised	Image processing	2017
[31]	Artificial bee colony	ISD–ABC	FS (wrapper)	Unsupervised	Image processing	2019
[32]	Artificial bee colony	ABCoDT	FS (wrapper)	Unsupervised	Benchmark analysis	2019
[33]	Artificial bee colony	TMABC-FS	FS (wrapper)	Unsupervised	Benchmark analysis	2019
[34]	Ant colony optimization	FACO	FS (wrapper)	Supervised	Network anomalies analysis	2018
[35]	Ant colony optimization	ACOAR	FS (filter)	Supervised	Benchmark analysis	2008
[36]	Ant colony optimization	UFSACO	FS (filter)	Unsupervised	Benchmark analysis	2014
[37]	Ant colony optimization	graph-ACO	FS (filter)	supervised	Benchmark analysis	2015
[38]	Ant colony optimization	ACOFS	FS (filter and wrapper)	Supervised	Benchmark analysis	2012
[39]	Butterfly optimization algorithm	bBOA	FS (wrapper)	Supervised	Benchmark analysis	2019
[40]	Coral reefs optimization	BCROSAT	FS (wrapper)	Supervised	Biomedical analysis	2019
[3]	Chicken swarm optimization	CCSO	FS (wrapper)	Supervised	Benchmark analysis	2017
[4]	Chicken swarm optimization	CSO-KNN	FS (wrapper)	Supervised	Benchmark analysis	2015
[5]	Genetic algorithms	HGAWE	FS (wrapper and embedded)	Supervised	Biological analysis	2019
[6]	Genetic algorithms	NSGA-II	FS (wrapper)	Supervised	Credit scoring analysis	2019
[16]	Grey wolf optimizer	LFGWO	FS (wrapper)	Supervised	Image processing	2019
[17]	Grey wolf optimizer and PSO	LFGWO	FS (wrapper)	Supervised	Benchmark analysis	2019
[17]	Fish swarm algorithm	FSANRSR	FS (wrapper)	Supervised	Rough set-benchmark analysis	2019
[18]	–	PCA	Dimension reduction	Supervised	Network traffics analysis	2017
[19]	–	PCA	Dimension reduction	Supervised	Network traffics analysis	2018

(continued)

Table 4.1 (continued)

Paper	Evolutionary algorithm	Proposed method	Feature extraction type	Problem	Domain	Year
[20]	Particle swarm optimization	HYBRID PSO	FS (filter and wrapper)	Supervised	Image processing	2016
[21]	Particle swarm optimization	APSO	FS (filter)	Supervised	Image analysis	2016
[41]	Particle swarm optimization	MI-APSO	FS (filter)	Supervised	Image analysis	2018
[23]	RelieF and PSO	RFPSO	FS (filter and wrapper)	Supervised	Network traffics analysis	2017
[24]	Whale optimization algorithm	WANFIS	FS (wrapper)	Supervised	Image processing	2019
[25]	Whale optimization algorithm	IWOA	FS (wrapper)	Supervised	Sentiment analysis	2019

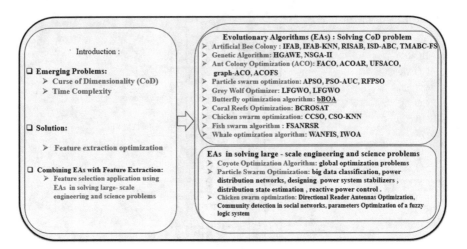

Fig. 4.1 Overall structure of this study

methods. Table 4.2 provides complete definitions for abbreviation which are used in this chapter.

4.2.1 Feature Extraction Optimization

Finding a proper subset of features or generating a new set of features while decreasing the dimension of datasets and improving the performance is still a NP-hard [51, 52] problem and scientists try to solve this problem. FE has been successfully done in several domains [19, 53, 54] and it is not domain specific. Here, we dig into some specific important feature selection algorithms like IFAB [2]. IFAB is a feature selection method applied on digital images. Figure 4.2 depicts a procedure of FE approach using EAs in an abstract way. It shows that EAs are adopted in pre-processing section to help scientists to reduce the number of feature properly. In general, we try to use a classification method to learn from train data and make a model. Then, leverage the generated model to predict unlabeled test data and calculate the performance of the classifier. As we discussed, classifiers fail to learn a large amount of data due to the CoD problem. So we discuss feature extraction optimization and how it helps classifier to learn the train data without struggling with over-fitting or under-fitting problems.

4.2.1.1 Feature Selection for Image Classification

Due to growth of generating images in different areas and networks, transferring, distributing images, image labeling and classification raised the majority of

Table 4.2 Abbreviation of words

Abb	Definition
ABC	Artificial bee colony
ACOAR	Ant colony optimization attribute reduction
BSO	Bat swarm algorithm
BCO	Bee colony optimization
BOA	Butterfly optimization algorithm
COA	Coyote optimization algorithm
CoD	Curse of dimensionality
CSO	Chicken swarm optimization
CCSO	Chaotic chicken swarm optimization
CRO	Coral reefs optimization
DA	Dragonfly algorithm
EAs	Evolutionary algorithms
FS	Feature selection
FSA	Fish swarm algorithm
GA	Genetic algorithm
GWO	Grey wolf optimizer
IFAB	Image steganalysis using FS based on ABC
IoT	Internet of things
ILS	Iterated local search
IWOA	Improved whale optimization algorithm
PCA	Principal component analysis
PEAs	Parallel evolutionary algorithms
RFPSO	RelieF and PSO algorithms
RL	Representation learning
RISAB	Region based image steganalysis using artificial bee colony
SLS	Stochastic local search
SSGA	Steady state genetic algorithm
SVD	Singular value dimension
SVM	Support vector machine
TMABC-FS	Two-archive multi-objective artificial bee colony algorithm for FS
WOA	Whale optimization algorithm
WANFIS	Whale adaptive neuro-fuzzy inference system

researchers' attention. To do that, scientists have implemented plenty of tools and packages and libraries, such as deep learning [55], generative adversarial networks (GANs) [56], convolutional neural network (CNN) [57], recurrent neural network (RNN) [58]. These provided algorithms have been successfully applied and determined very low loss in their accuracy. However, they still suffer from handling dig data, CoD problem and time complexity. Thus, researchers attempted to take advantage of leveraging evolutionary algorithms to address these problems properly. Image steganalysis is an image classification problem and has been quite an emerging challenge in large scale engineering and science. The importance

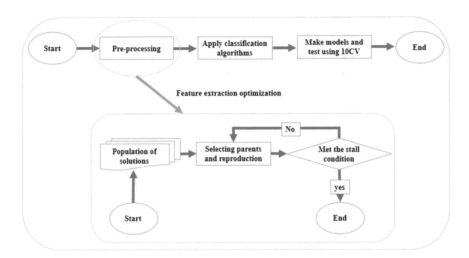

Fig. 4.2 Evolutionary algorithms process for feature extraction

of issue has proliferated significantly last 20 years after the tragedy happened in 2001. One of the promising methods to solve the problem using zero-shot learning, Mohammadi and Amini [59] presented a meta-sense algorithm to address the problem carefully.

- **Application of artificial bee colony in feature selection**:

Image steganalysis using feature selection based on artificial bee colony (IFAB) is presented by Mohammadi and Abadeh in [2]. IFAB has an optimized feature extraction process and is categorized into wrapper-based feature selection methods. This method works properly for examining input digital images and distinguishing cover images from stego images. A stego image is a cover image with an embedded message. The goal of IFAB is to decrease time complexity of training model while improving the classifier's performance.

Figure 4.3 presents IFAB's process in detail stating that how IFAB technically combines three types of bees to optimize the feature extraction process. First of all, employed bees are applied to generate food sources and a goodness of each food source is generated using support vector machine (SVM). Fitness function leverages SVM to compute the goodness of each food source. According to the (4.1) fitness function is calculated. (f_i) goes for each food source and P_i stands for the accuracy of SVM per each food source. Each food source stands for a solution to the problem. A solution would present the number of features selected to reduce the dimension of given dataset. This step is done for each employed bee population. The employed bees and onlooker bees have the same population and are equal to the dimension of input data. For instance, SPAM dataset has 686 features which is equal to the population of them.

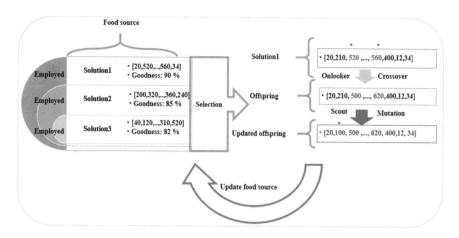

Fig. 4.3 An overview of image steganalysis using Feature selection based on Artificial Bee colony (IFAB) [29]

$$\text{Fitness}(f_i) = \begin{cases} \frac{1}{1+P_i} & \text{if } P_i > 0 \\ 1 + \text{abs}(P_i) & \text{if } P_i < 0 \end{cases} \quad (4.1)$$

Second step is choosing a food source based on (using Eq. (4.2)) the goodness of the food source to exploit it by onlooker bees. Once offspring (new solution) is generated, it is time for checking the condition, limit.

$$V_i = f_i + v * (f_i - f_j)., v = [-1, 1] \quad (4.2)$$

Where j is a random number between 1 and N. i, j stands for features index in an input dataset. N is the upper bound of number of features. If the solution's performance stopped improving within the pre-defined iteration equal to the limit, then, in third step, scout bee is selected to explore using Eq. (4.3) the area of the food source, and tries to update the solution by a pre-adjusted rate. X_{max}, X_{min} shows the upper bound and lover bound of population, respectively.

$$X_i = X_{max} + v' * (X_{max} - X_{min}), v' = [0, 1] \quad (4.3)$$

In each iteration the best food source with respect to the goodness value is reserved and ABC updates food sources by replacing the minimum food source(minimum value of goodness) with new food source if its goodness outperforms minimum food source. ABC process is complete the condition is met or when iteration reaches max iteration. Finally, ABC returns the best food source during whole iterations. This food source involves a number of relevant and important features and yields a better result.

IFAB-KNN is an advanced ABC for image steganalysis to enhance IFAB performance proposed by Mohammadi and Abadeh in [29]. IFAB-KNN stands for one of the wrapper-based feature selection algorithm. Fitness function takes advantage of a lazy algorithm, k-nearest neighbour (KNN), within ABC and enables ABC to examine each subset of features deeply. Keeping the same number of selected features, IFAB-KNN outperforms IFAB with the tuned hyper-parameters. Xie et al. [31] produced another unsupervised feature selection algorithm using ABC to classify hyperspectral images, ISD–ABC.

Mohammadi and Sajedi [30] presented another hybrid approach to feature selection, combination of data and images. They proposed RISAB, region based image steganalysis using artificial bee colony by leveraging the IFAB characteristics. The goal of RISAB is to find the location of image that does not follow the harmony of the whole images. By finding special pixel or sets of pixels, RISAB could distinguish stego images from cover images. In RISAB, the researchers first applied ABC to find the most probable sub-image that carries the embedded data, which would be messages, images, etc. ABC was tailored to focus on image spatial domain which was one of the important challenging issues. After that, there are one given input data and one sub-image of input data. Then, researchers tried to apply IFAB on both of them. They were able to improve the performance of feature extractors like SPAM and CC-PEV, even IFAB. Figure 4.4 presents overall steps of RISAB, for more information you may read this paper [30].

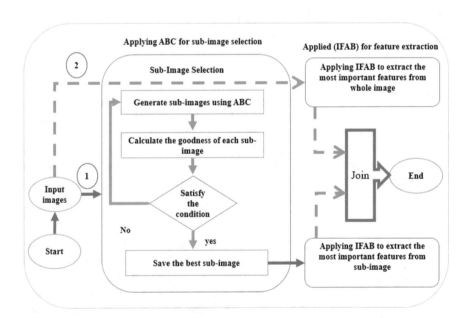

Fig. 4.4 The process of feature selection using RISAB

- **Application of particle swarm optimization in feature selection**:

Chhikara et al. [20] proposed HYBRID, a new approach using PSO for solving CoD problem in image steganalysis. They proposed a combined filter and wrapper based feature selection approach to deal with high dimensionality problem in image steganalysis. Authors tested HYBRID on data which had been extracted using feature extraction methods, using images which were attacked by steganography algorithms. The HYBRID enhanced the classification accuracy image steganalysis. Chikara and Kumari in [60] presented a new wrapper-based feature selection for image steganalysis, named global local PSO (GLBPSO) which leverages backpropagation neural networks to evaluate the selected feature subsets.

Adeli and Broumandnia in [21] introduced another filter-based feature selection algorithm for steganalysis named, an adaptive inertia weight-based PSO (APSO) where the inertia weight of PSO is adaptively adjusted leverage of three components, such as average distance of particles, the swarm diameter, and average velocity of particles. Rostami and Khiavi [22] take advantage of a novel fitness function which uses area under curve (AUC) to evaluate feature subset. The final accuracy of APSO yields a better result in comparison with IFAB; however, their time complexity problem still remains for CoD problems.

- **Application of grey wolf optimizer in feature selection**:

Pathak et al. [16] proposed a new feature selection algorithm, LFGWO which has been used to classify stego images from cover images. GWO has been widely used to solve large scale optimization, engineering, science problems, such as global optimization tasks [13, 17] optimized feature selection algorithm using combining GWO and PSO called PSOGWO.

4.2.2 Feature Selection for Network Traffic Classification

Internet traffics growth has increased unexpectedly due to expanding new technologies and data comes from everywhere using internet of things (IoT). However, as lack of traditional traffic classification approaches, researchers have been applying traffic classification using ML techniques. Current and simple ML techniques may not turn into an optimum solution because of very high dimensionality.

Having multi-class imbalance datasets becomes another emerging challenging issues for scientists. ML algorithms struggles with the data and do not yield high recall for the minority classes. Researchers in their studies have proposed different hybrid approaches to address these problems. Scientists worked on data in pre-processing phase by using re-sampling approaches, cost-sensitive approaches and feature extraction approaches which has very high impact on the classification process. Dong et al. [23] proposed a new hybrid method using PSO to address the aforementioned problems. Researchers in this paper presented a new hybrid feature selection algorithm combining RelieF and PSO algorithms called RFPSO.

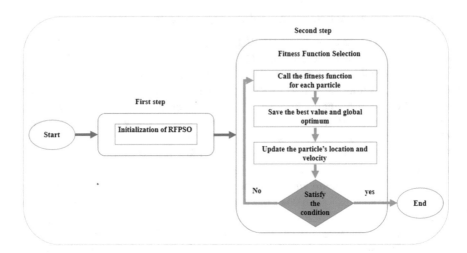

Fig. 4.5 The process of feature selection using RelieF and PSO algorithms (RFPSO)

According to the figure, RFPSO has two main steps to follow. First step goes for initialization of RFPSO, and second step is fitness function selection (Fig. 4.5).

Relief is an feature selection algorithm proposed by Kira and Rendell in 1992 [61]. It is one of the filter-based feature selection that works deeply on relationship among features. It was firstly presented to address binary classification problems including discrete or numerical features. RFPSO calls the fitness function which stands for the inconsistency rate defined in Eq. (4.4) [61]. T stands for the total number of instances and the number of inconsistencies goes for N. They considered three popular criteria to evaluate their work: one recall, one precision and one F_{measure}.

$$\text{fitness} = \frac{N}{T}. \tag{4.4}$$

Shi et al. [18, 19] aimed to classify traffic data by proposing a robust feature extraction and feature selection algorithms which leverage customized PCA. Hamamoto et al. [62] used GA to detect network anomaly detection. GA is applied to create a digital signature of network segment by leveraging flow analysis. The information are provided and extracted from network flows which have been used to predict networks traffic action within short time period. They combined GA with fuzzy logic to improve their result. Actually, fuzzy logic enables the ML algorithm to decide whether a sample represents an anomaly or not.

4.2.3 Feature Selection Benchmarks

Ahmed et al. [3] presented an advanced feature selection algorithm using chaotic chicken swarm optimization (CCSO). They used CCSO to cover the problem of stalling in local minimum which is one of the important and common problems of traditional evolutionary and swarm algorithm. Unlike other evolutionary algorithms, it remembers both minimum and maximum value of each solution to optimize the search step. They used five different datasets, such as spambase, WBDC, ionosphere, lung and sonar. They compared their result with other evolutionary algorithms like PSO, binary CSO, bat swarm algorithm (BSO) and dragonfly algorithm (DA).

Ant colony optimization attribute reduction (ACOAR) is another feature selection algorithm which reduces the number features using filtering based algorithm. ACOAR is introduced in [35] and leverages the ACO process to improve the performance of algorithm in rough set theory. Moreover, Tabakhi et al. [36] proposed another yet novel approach for feature selection algorithm for unsupervised data, called UFSACO. UFSACO looks for the optimal and the most relevant feature subset; however, it does not take advantage of learning algorithms to that end. Furthermore, its goodness of features relation is computed with respect to the similarity among features. UFSACO aims to minimize the redundancy of features. Moreover, ACOFS, a hybrid ant colony optimization algorithm presented in [38], graph-based feature selection using ACO [37].

- **Recent advances in feature selection optimization**:

We have included and classified the most popular and important research studies with respect to the evolutionary algorithms. In this section, we would like to add last, but not the least, yet cutting-edge studies attempted to solve the NP-hard problems. Rao et al. [32] presented another hybrid feature selection method using bee colony and leveraging gradient boosting decision tree to address the NP-hard problems like curse of dimensionality. Bui et al. [24] proposed a hybrid feature selection for land pattern classification, using whale optimization algorithm (WOA) and adaptive neuro-fuzzy inference concepts. Another improved WOA is presented in [25] to Arabic sentiment analysis by feature selection. Kozodoi et al. [6] leveraged profit measures to propose a wrapper feature selection algorithm using NSGA-II genetic algorithm. Yan et al. [40] improved yet another evolutionary algorithm, coral reefs optimization (CRO), to select the best matched feature subsets, called BCROSAT to apply on biomedical data. Arora and Anand [39] introduced a binary variants of a new evolutionary algorithm, the butterfly optimization algorithm (BOA), to select the most important features. Zhang et al. [33] proposed new cost-sensitive multi-objective ABC-based feature selection, TMABC-FS. Pierezan and Coelho [63] proposed new optimization algorithm based on COA, for solving global optimization problems.

4.3 Discussion

NP-hard problems always become the most challenging issues, most frequently seen in engineering and science. In this chapter, we address one of the emerging NP-hard problems in data science, the curse of dimensionality (CoD). A large number of research studies have attempted to solve this problematic issue. Researchers continue to publish new papers on this problem since they have not found an algorithm that performs in an accurate and robust way. Researchers did not obtain a better result by only using basic machine learning algorithms, such as support vector machine (SVM) or K-nearest neighbour (KNN). During the last two decades, scientists started to apply evolutionary algorithms to this issue and will continue publishing unlimited papers on this topic. Furthermore, data scientists have significantly increased the number of domains to which they could apply evolutionary algorithms. So, lately, the number of data has proliferated and lead scientist to find a new type of data called big data. Thus, working with data including manipulating data and finding the pattern governing whole data becomes harder. Not only do we have different kinds of supervised dataset with ground truth, but also this new type of data introduced a new concept, the unsupervised dataset. Scientists proposed a technical way to deal with different groups of data, meanwhile their approaches should be able to learn unsupervised datasets, too. The future work would be trying to propose a solution to apply evolutionary algorithms to help representation learning which enables scientists to find patterns and important features independent of the ground truth and enables classifiers to learn supervised and unsupervised data properly to avoid over-fitting.

4.4 Conclusion

Data science has some defects because of some data problems due to unexpected growing amount of it. The main problem of data is the curse of dimensionality (CoD), which causes yet another important problem, high time complexity. To solve CoD problem, researchers have been proposed different feature selection optimization techniques by selecting the most relevant and optimal features, tries to evaluate supervised/unsupervised datasets. In this chapter, we provided the practical examples of applied evolutionary algorithm for feature selection optimization and reviewed emerging optimization application. Furthermore, scientists have adopted evolutionary algorithms like IFAB and tailored them towards their goals. Moreover, having applied evolutionary algorithms to select the most important features, researchers improved the performance of classification algorithms. For instance, IFAB decreased the dimension of a given dataset intensely, while also enhancing the performance of the support vector machine (SVM) significantly. In addition, IFAB decreased the time to learn a model, and improved time complexity. In this chapter, we categorized optimization method according to the evolutionary algorithms and their applications. It is noteworthy that evolutionary algorithms have been applied successfully on most of engineering, science and even biology and medical domains which make them powerful and robust enough. This study provides the required information for the researchers who plan to pursue research on dimension reduction and large scale optimization problems.

References

1. F.G. Mohammadi, M.H. Amini, H.R. Arabnia, Evolutionary computation, optimization and learning algorithms for data science. arXiv preprint, arXiv: 1908.08006 (2019)
2. F.G. Mohammadi, M.S. Abadeh, Image steganalysis using a bee colony based feature selection algorithm. Eng. Appl. Artif. Intell. **31**, 35–43 (2014)
3. K. Ahmed, A.E. Hassanien, S. Bhattacharyya, A novel chaotic chicken swarm optimization algorithm for feature selection, in *2017 Third International Conference on Research in Computational Intelligence and Communication Networks (ICRCICN)* (IEEE, Piscataway, 2017), pp. 259–264
4. A.I. Hafez, H.M. Zawbaa, E. Emary, H.A. Mahmoud, A.E. Hassanien, An innovative approach for feature selection based on chicken swarm optimization, in *2015 7th International Conference of Soft Computing and Pattern Recognition (SoCPaR)* (IEEE, Fukuoka, 2015), pp. 19–24
5. X.-Y. Liu, Y. Liang, S. Wang, Z.-Y. Yang, H.-S. Ye, A hybrid genetic algorithm with wrapper-embedded approaches for feature selection. IEEE Access **6**, 22863–22874 (2018)
6. N. Kozodoi, S. Lessmann, K. Papakonstantinou, Y. Gatsoulis, B. Baesens, A multi-objective approach for profit-driven feature selection in credit scoring. Decis. Support Syst. **120**, 106–117 (2019)
7. Y. Xue, J. Jiang, B. Zhao, T. Ma, A self-adaptive artificial bee colony algorithm based on global best for global optimization. Soft Comput. **22**, 2935–2952, 2018.
8. F. Harfouchi, H. Habbi, C. Ozturk, D. Karaboga, Modified multiple search cooperative foraging strategy for improved artificial bee colony optimization with robustness analysis. Soft Comput. **22**(19), 6371–6394 (2018)
9. H. Wang, J.-H. Yi, An improved optimization method based on krill herd and artificial bee colony with information exchange. Memet. Comput. **10**(2), 177–198 (2018)
10. Y. Cao, Y. Lu, X. Pan, N. Sun, An improved global best guided artificial bee colony algorithm for continuous optimization problems. Clust. Comput. **22**, 3011–3019 (2018). https://doi.org/10.1007/s10586-018-1817-8
11. M.H. Amini, K.G. Boroojeni, S.S. Iyengar, F. Blaabjerg, P.M. Pardalos, A.M. Madni, A panorama of future interdependent networks: from intelligent infrastructures to smart cities, in *Sustainable Interdependent Networks* (Springer, Cham, 2018), pp. 1–10
12. M.H. Amini, H. Arasteh, P. Siano, Sustainable smart cities through the lens of complex interdependent infrastructures: panorama and state-of-the-art, in *Sustainable Interdependent Networks II* (Springer, Cham, 2019), pp. 45–68
13. S. Gupta, K. Deep, An opposition-based chaotic grey wolf optimizer for global optimisation tasks. J. Exp. Theor. Artif. Intell. **31**, 751–779 (2018)
14. X.-L. Li, An optimizing method based on autonomous animats: fish-swarm algorithm. Syst. Eng. Theory Pract. **22**(11), 32–38 (2002)
15. L.Y. Zhang, G. Luo, L.N. Lu, Genetic algorithms in resource optimization of construction project. J. Tianjin Univ. (Sci. Technol.) **34**(2), 188–192 (2001)
16. Y. Pathak, K.V. Arya, S. Tiwari, Feature selection for image steganalysis using levy flight-based grey wolf optimization. Multimed. Tools Appl. **78**(2), 1473–1494 (2019)

17. Q. Al-Tashi, S.J. Abdul Kadir, H. Md Rais, S. Mirjalili, H. Alhussian, Binary optimization using hybrid grey wolf optimization for feature selection. IEEE Access **7**, 39496–39508 (2019)
18. H. Shi, H. Li, D. Zhang, C. Cheng, W. Wu, Efficient and robust feature extraction and selection for traffic classification. Comput. Netw. **119**, 1–16 (2017)
19. H. Shi, H. Li, D. Zhang, C. Cheng, X. Cao, An efficient feature generation approach based on deep learning and feature selection techniques for traffic classification. Comput. Netw. **132**, 81–98 (2018)
20. R.R. Chhikara, P. Sharma, L. Singh, A hybrid feature selection approach based on improved PSO and filter approaches for image steganalysis. Int. J. Mach. Learn. Cybern. **7**(6), 1195–1206 (2016)
21. A. Adeli, A. Broumandnia, Image steganalysis using improved particle swarm optimization based feature selection. Appl. Intell. **48**, 1609–1622 (2018)
22. V. Rostami, A.S. Khiavi, Particle swarm optimization based feature selection with novel fitness function for image steganalysis, in *2016 Artificial Intelligence and Robotics (IRANOPEN)* (IEEE, Qazvin, 2016), pp. 109–114
23. Y. Dong, Q. Yue, M. Feng, An efficient feature selection method for network video traffic classification, in *2017 IEEE 17th International Conference on Communication Technology (ICCT)* (IEEE, Chengdu, 2017), pp. 1608–1612
24. Q.-T. Bui, M.V. Pham, Q.-H. Nguyen, L.X. Nguyen, H.M. Pham, Whale optimization algorithm and adaptive neuro-fuzzy inference system: a hybrid method for feature selection and land pattern classification. Int. J. Remote Sens. **40**(13), 5078–5093 (2019)
25. M. Tubishat, M.A.M. Abushariah, N. Idris, I. Aljarah, Improved whale optimization algorithm for feature selection in Arabic sentiment analysis. Appl. Intell. **49**(5), 1688–1707 (2019)
26. M.H. Amini, M.P. Moghaddam, O. Karabasoglu, Simultaneous allocation of electric vehicles parking lots and distributed renewable resources in smart power distribution networks. Sustain. Cities Soc. **28**, 332–342 (2017)
27. M.R. Mozafar, M.H. Moradi, M.H. Amini, A simultaneous approach for optimal allocation of renewable energy sources and electric vehicle charging stations in smart grids based on improved GA-PSO algorithm. Sustain. Cities Soc. **32**, 627–637 (2017)
28. M.H. Amini, A. Islam, Allocation of electric vehicles' parking lots in distribution network, in *ISGT 2014* (IEEE, 2014), pp. 1–5
29. F.G. Mohammadi, M.S. Abadeh, A new metaheuristic feature subset selection approach for image steganalysis. J. Intell. Fuzzy Syst. **27**(3), 1445–1455 (2014)
30. F.G. Mohammadi, H. Sajedi, Region based image steganalysis using artificial bee colony. J. Vis. Commun. Image Represent. **44**, 214–226 (2017)
31. F. Xie, F. Li, C. Lei, J. Yang, Y. Zhang, Unsupervised band selection based on artificial bee colony algorithm for hyperspectral image classification. Appl. Soft Comput. **75**, 428–440 (2019)
32. H. Rao, X. Shi, A.K. Rodrigue, J. Feng, Y. Xia, M. Elhoseny, X. Yuan, L. Gu, Feature selection based on artificial bee colony and gradient boosting decision tree. Appl. Soft Comput. **74**, 634–642 (2019)
33. Y. Zhang, S. Cheng, Y. Shi, D.-W. Gong, X. Zhao, Cost-sensitive feature selection using two-archive multi-objective artificial bee colony algorithm. Expert Syst. Appl. **137**, 46–58 (2019)
34. H. Peng, C. Ying, S. Tan, B. Hu, Z. Sun, An improved feature selection algorithm based on ant colony optimization. IEEE Access **6**, 69203–69209 (2018)
35. L. Ke, Z. Feng, Z. Ren, An efficient ant colony optimization approach to attribute reduction in rough set theory. Pattern Recogn. Lett. **29**(9), 1351–1357 (2008)
36. S. Tabakhi, P. Moradi, F. Akhlaghian, An unsupervised feature selection algorithm based on ant colony optimization. Eng. Appl. Artif. Intell. **32**, 112–123 (2014)
37. P. Moradi, M. Rostami, Integration of graph clustering with ant colony optimization for feature selection. Knowl. Based Syst. **84**, 144–161 (2015)
38. M.M. Kabir, M. Shahjahan, K. Murase, A new hybrid ant colony optimization algorithm for feature selection. Expert Syst. Appl. **39**(3), 3747–3763 (2012)

39. S. Arora, P. Anand, Binary butterfly optimization approaches for feature selection. Expert Syst. Appl. **116**, 147–160 (2019)
40. C. Yan, J. Ma, H. Luo, A. Patel, Hybrid binary coral reefs optimization algorithm with simulated annealing for feature selection in high-dimensional biomedical datasets. Chemom. Intell. Lab. Syst. **184**, 102–111 (2019)
41. J. Kaur, S. Singh, Feature selection using mutual information and adaptive particle swarm optimization for image steganalysis, in *2018 7th International Conference on Reliability, Infocom Technologies and Optimization (Trends and Future Directions) (ICRITO)* (IEEE, Noida, 2018), pp. 538–544
42. M.H. Amini, Distributed computational methods for control and optimization of power distribution networks, PhD Dissertation, Carnegie Mellon University, 2019
43. M.H. Amini, J. Mohammadi, S. Kar, Distributed holistic framework for smart city infrastructures: tale of interdependent electrified transportation network and power grid. IEEE Access **7**, 157535–157554 (2019)
44. M.H. Amini, J. Mohammadi, S. Kar, Distributed intelligent algorithm for interdependent electrified transportation and power networks, *Proceedings of the 9th ACM Symposium on Design and Analysis of Intelligent Vehicular Networks and Applications* (ACM, New York, 2019)
45. A. Imteaj, M.H. Amini, J. Mohammadi, Leveraging decentralized artificial intelligence to enhance resilience of energy networks. arXiv preprint arXiv:1911.07690 (2019)
46. M.H. Amini, B. Nabi, M.R. Haghifam, Load management using multi-agent systems in smart distribution network, in *2013 IEEE Power and Energy Society General Meeting* (IEEE, New York, 2013)
47. M.H. Amini, S. Bahrami, F. Kamyab, S. Mishra, R. Jaddivada, K. Boroojeni, P. Weng, Y. Xu, Decomposition methods for distributed optimal power flow: panorama and case studies of the DC model, in *Classical and recent aspects of power system optimization* (Academic Press, Cambridge, 2018), pp. 137–155
48. A. Imteaj, M.H. Amini, Distributed sensing using smart end-user devices: pathway to federated learning for autonomous IoT, in *2019 International Conference on Computational Science and Computational Intelligence* (Las Vegas, 2019)
49. M.A. Elaziz, K.M Hosny, I.M. Selim, Galaxies image classification using artificial bee colony based on orthogonal Gegenbauer moments. Soft Comput. **23**, 9573–9583 (2018)
50. M. Gong, Y. Wu, Q. Cai, W. Ma, A.K. Qin, Z. Wang, L. Jiao, Discrete particle swarm optimization for high-order graph matching. Inf. Sci. **328**, 158–171 (2016)
51. S. Ermon, C. Gomes, A. Sabharwal, B. Selman, Taming the curse of dimensionality: discrete integration by hashing and optimization, in *International Conference on Machine Learning* (2013), pp. 334–342
52. N. Kouiroukidis, G. Evangelidis, The effects of dimensionality curse in high dimensional kNN search, in *2011 15th Panhellenic Conference on Informatics* (IEEE, 2011), pp. 41–45
53. R. Vanaja, S. Mukherjee, Novel wrapper-based feature selection for efficient clinical decision support system, in *International Conference on Intelligent Information Technologies* (Springer, Singapore, 2018), pp. 113–129
54. E. Hancer, B. Xue, M. Zhang, D. Karaboga, B. Akay, Pareto front feature selection based on artificial bee colony optimization. Inf. Sci. **422**, 462–479 (2018)
55. S. Khan, N. Islam, Z. Jan, I.U. Din, J.J.P.C. Rodrigues, A novel deep learning based framework for the detection and classification of breast cancer using transfer learning. Pattern Recogn. Lett. **125**, 1–6 (2019)
56. S. Roy, E. Sangineto, N. Sebe, B. Demir, Semantic-fusion Gans for semi-supervised satellite image classification, in *2018 25th IEEE International Conference on Image Processing (ICIP)* (IEEE, Athens, 2018), pp. 684–688
57. A. Zhang, X. Yang, L. Jia, J. Ai, Z. Dong, SAR image classification using adaptive neighborhood-based convolutional neural network. Eur. J. Remote Sens. **52**(1), 178–193 (2019)

58. R. Hang, Q. Liu, D. Hong, P. Ghamisi, Cascaded recurrent neural networks for hyperspectral image classification. IEEE Trans. Geosci. Remote Sens. **57**, 5384–5394 (2019)
59. F.G. Mohammadi, M.H. Amini, Promises of meta-learning for device-free human sensing: learn to sense, in *Proceedings of the 1st ACM International Workshop on Device-Free Human Sensing* (DFHS'19) (ACM, New York, 2019), pp. 44–47. https://doi.org/10.1145/3360773.3360884
60. M. Kumari, R. Chhikara, Blind image steganalysis using neural networks and wrapper feature selection, in *2017 International Conference on Computing, Communication and Automation (ICCCA)* (IEEE, Noida, 2017), pp. 1065–1069
61. K. Kira, L.A. Rendell, A practical approach to feature selection, in *Machine Learning Proceedings 1992* (Elsevier, Oxford, 1992), pp. 249–256
62. A.H. Hamamoto, L.F. Carvalho, L.D.H. Sampaio, T. Abrão, M.L. Proença Jr., Network anomaly detection system using genetic algorithm and fuzzy logic. Expert Syst. Appl. **92**, 390–402 (2018)
63. J. Pierezan, L.D.S. Coelho, Coyote optimization algorithm: a new metaheuristic for global optimization problems, in *2018 IEEE Congress on Evolutionary Computation (CEC)* (IEEE, Rio de Janeiro, 2018), pp. 1–8

Chapter 5
Feature Selection in High-Dimensional Data

Amirreza Rouhi and Hossein Nezamabadi-Pour

Abstract Today, with the increase of data dimensions, many challenges are faced in many contexts including machine learning, informatics, and medicine. However, reducing data dimension can be considered as a basic method in handling high-dimensional data, because by reducing dimensions, applying many of the existing operations on data is facilitated.

Microarray data are derived from tissues and cells considering differences in the gene, which can be useful for diagnosing disease and tumors. Due to the large number of features (genes) and small number of samples in microarray datasets, selecting the most salient genes is a difficult task. Among the many techniques of machine learning, feature selection and data classification play a very important and widespread role in enhancing human life, from detecting voice emotion to detecting illness in the body. In medicine, an effective gene selection can greatly enhance the process of prediction and diagnosis of cancer. After selecting effective genes, the duty of a specific classifier is usually to discriminate healthy people from patients that are suffering from cancer based on their expression of the selected genes.

A vast body of feature selection methods has been proposed for high-dimensional microarray data. Traditionally, these methods fall into three categories including filter, wrapper, and hybrid approaches. Furthermore, new techniques such as ensemble methods have recently been developed to improve the process of feature selection and classification.

A. Rouhi (✉)
Data Science and Bioinformatics Laboratory, Politecnico di Milano, Department of Electronics, Information and Bioengineering, Milan, Italy

Intelligent Data Processing Laboratory (IDPL), Department of Electrical Engineering, Shahid Bahonar University of Kerman, Kerman, Iran
e-mail: amirreza.rouhi@mail.polimi.it

H. Nezamabadi-Pour
Intelligent Data Processing Laboratory (IDPL), Department of Electrical Engineering, Shahid Bahonar University of Kerman, Kerman, Iran
e-mail: nezam@mail.uk.ac.ir

© Springer Nature Switzerland AG 2020
M. H. Amini (ed.), *Optimization, Learning, and Control for Interdependent Complex Networks*, Advances in Intelligent Systems and Computing 1123,
https://doi.org/10.1007/978-3-030-34094-0_5

This chapter presents an overview of the most popular feature selection methods to deal with high-dimensional data and analyze their performance under different conditions. The chapter starts with a global overview of the high-dimensional data and feature selection (Sects. 5.2 and 5.3). Then, in Sect. 5.4 we review the state-of-the-art methods on filter algorithms. In the next three Sects. (5.5, 5.6 and 5.7) we describe the wrapper, hybrid, and embedded methods and in each section, an overview of several works performed on these methods is discussed. Sect. 5.8 describes the ensemble techniques recently considered by the researchers and summarizes the works done based on these techniques. In Sect. 5.9, we present the experimental results of the most significant methods on high-dimensional data. Finally, Sect. 5.10 summarizes this chapter.

Keywords Data science · Dimension reduction · Classification · Feature selection · High-dimensional data

5.1 Overview

Today, with the increase of data dimensions, many challenges are faced in many contexts including machine learning, informatics, and medicine. However, reducing data dimension can be considered a basic method in handling high-dimensional data, because by reducing dimensions, applying many of the existing operations on data is facilitated.

Microarray data are derived from tissues and cells considering differences in the gene, which can be useful for diagnosing disease and tumors. Due to the large number of features (genes) and small number of samples in microarray datasets, selecting the most salient genes is a difficult task. Among the many techniques of machine learning, feature selection and data classification play a very important and widespread role in enhancing human life, from detecting voice emotion [1, 2] to detecting illness in the body [3]. In medicine, an effective gene selection can greatly enhance the process of prediction and diagnosis of cancer. After selecting effective genes, the duty of a specific classifier is usually to discriminate healthy people from patients that are suffering from cancer based on their expression of the selected genes.

A vast body of feature selection methods has been proposed for high-dimensional microarray data. Traditionally, these methods fall into three categories including filter, wrapper, and hybrid approaches. Furthermore, new techniques such as ensemble methods have recently been developed to improve the process of feature selection and classification.

This chapter presents an overview of the most popular feature selection methods to deal with high-dimensional data and analyze their performance under different conditions. The chapter starts with a global overview of the high-dimensional data and feature selection (Sects. 5.2 and 5.3). Then, in Sect. 5.4 we review the state-of-the-art methods on filter algorithms. In the next three Sects. (5.5, 5.6 and 5.7) we describe the wrapper, hybrid, and embedded methods and in each section, an overview of several works performed on these methods is discussed. Sect. 5.8 describes the ensemble techniques recently considered by the researchers and summarizes the works done based on these techniques. In Sect. 5.9, we present the experimental results of the most significant methods on high-dimensional data. Finally, Sect. 5.10 summarizes this chapter.

5.2 Intrinsic Characteristics of High-Dimensional Data

Today, data is crucial resources in scientific research and industrial production. Data is obtained through different ways at various costs. Indeed, finding the proper and optimal data extraction method is a key step to work. Each method has its benefits and disadvantages. High costs and the addition of noise are two important examples of the negative points that these techniques may have.

Moreover, since the quality and characteristics of data may affect the results of the classification, the most important prerequisite is to fully understand the investigated data. The purpose of this introductory section is to briefly present some characteristics that high-dimensional data may present.

5.2.1 Large Number of Features

In pattern recognition, a set of quantitative attributes is measured and extracted from real-world patterns as a set of features describing the pattern in the digital world. Various attributes can be generally measured from the real-world patterns, among which only some quantities are measured and recorded depending on their necessity, and accessibility to their extraction resources and storage devices.

As mentioned, various attributes and features can be measured and recorded from each pattern. Evidently, recording all the extracted features is not a cost-effective approach. The objective of feature selection is to select a small subset from these features, in such a way that the considered patterns can be ideally described and optimally classified the patterns. From a theoretical perspective, higher discriminating power can be achieved by increasing the number of features [4]. In practice, however, given the limited number of training data, a large number of features can considerably lengthen the learning process, and also increases the computational complexity of the problem. That is due to the fact that the irrelevant or redundant features confuse the learning algorithm [4].

Today, the number of features in data has considerably grown. As an example, the prostate dataset, as a two-class microarray data structure, includes 10,509 features. Additionally, the 11_Tumor dataset contains 11 classes including 12,533 features (or genes). Studies have shown that despite the large number of features, a large portion of these features contain no useful information since they are either redundant or irrelevant to the class. In other words, the effectiveness of the learning process can only be increased by considering relevant and non-redundant features.

On the other hand, given an excessively complex model, for example, when the number of features is considerably greater than that of the observations, the overfitting phenomenon may occur. In cases where increasing the number of training samples is not practical, reducing the number of samples is needed for the training dataset and, consequently this improves the overall performance of the classification algorithm.

Hence, feature selection can be defined as the problem of reducing the data dimensions through identifying the subset of features most required for the classification process from among numerous features [5]. Feature selection should be conducted such that the reduced data contains as much information from the main dataset as possible [6]. In other words, only features containing redundant, noisy, and unnecessary information should be eliminated.

In general, the number of possible feature subsets increases exponentially (2^D, where D is the number of features) as the dimension increases [7]. Hence, finding an optimal subset is mostly a difficult task, which is why the feature selection problems are categorized as NP-hard type.

5.2.2 Small Number of Samples

According to [8], for a classification problem with D dimensions (the number of features) and C classes, at least $10 \times D \times C$ training samples are required. For instance, at least 40,000 training samples are required for the colon microarray as a two-class dataset containing a total of 2000 features, while only 62 samples are available. Therefore, the scarcity of samples is the most important challenge regarding high-dimensional data.

As explained in [9], the problem of small sample size occurs when only m samples are available in an D-dimensional vector space, such that $m < D$. When facing scarcity of samples in a high-dimensional data, the data is referred to as high dimension, low sample size (HDLSS) data, which is common in many different fields such as microarray data, medical imaging, text recognition, and face recognition. As mentioned in the previous section, the scarcity of samples can increase the risk of overfitting.

5.2.3 Class Imbalance

Class imbalance occurs when in a dataset, the number of samples in a class is significantly greater or smaller than the other classes. In such data, the class with the lowest number of samples is referred to as minority classes. It is noteworthy that, for the two-class cases, the positive class and the negative class are assumed to be the minority class and the majority class, respectively.

Table 5.1 Imbalance ratio of the 11 most useful high-dimensional datasets

Dataset	Data type	Imbalance ratio
Brain_Tumors1	Microarray	15.00
Breast_Cancer	Microarray	1.10
CNS	Microarray	1.857
Colon	Microarray	1.818
Leukemia	Microarray	1.780
Prostate_Cancer	Microarray	1.04
SRBCT	Microarray	2.64
Gli	Microarray	2.27
Lung_Cancer	Microarray	23.17
Ovarian	Microarray	1.78
RELATHE	Text data	1.2

In most classification algorithms, the classifiers presume an equal number of training samples in each class. Hence, when these algorithms are applied to imbalanced data, the classifier is mostly trained based on the samples in the majority class, which consequently leads to a very poor prediction of the samples in the minority class given its improper training.

The ratio of the cardinality of the majority class, $Nmaj$, to the cardinality of the minority class, $Nmin$, is called the ratio of imbalance and is expressed as [8]:

$$IR = \frac{Nmaj}{Nmin} \tag{5.1}$$

The main challenge in imbalanced data is to correctly identify the samples in the minority classes and the minority class is of greater significance and misclassification of minority samples leads to elevated risk [8]. For instance, the number of positive samples in the diagnosis of diseases such as cancer is considerably smaller than the negative ones, while identifying the samples in the positive class is greatly important.

Table 5.1 reported the imbalance ratio of the 11 most useful high-dimensional datasets.

Hence, numerous methods have been proposed to improve the performance of classifiers facing this problem [9–13]. These methods are generally classified into three categories [8]:

1. Classifier-independent preprocessing methods:

Two different approaches are employed in these methods to rebalance the distribution of classes in the training data. In the first approach, referred to as over-sampling, samples are added to the minority class. In the second approach, referred to as under-sampling, some of the samples are removed from the majority class. Evidently, both approaches attempt to somehow balance the classes. Authors of [8] proposed an under-sampling method called HTSS to select the best training set in imbalanced datasets.

2. Modifications of algorithms

Employing specific solutions, these methods attempt to modify and improve the considered classifier, matching it with the imbalanced data. The proposed method in [14] is one of these methods which uses the ensemble technique to improve the performance of the classifier when dealing with imbalance data.

3. Ensemble learning methods

These methods are based on the use and incorporation of multiple classifier results. For instance, in [15] an ensemble-based method is proposed which, in the first step imbalanced data is converted into some balanced ones and then by applying multiple classifiers, the results are combined.

5.2.4 Label Noise

Today, the presence of noise in the data obtained from practical real-world applications using common methods is a well-known fact. Noises can be caused due to different reasons, ranging from faults in the measurement devices to transmission irregularities. Noisy data can significantly reduce the performance of the classifier. In other words, the quality of the training data can highly influence the classifier's accuracy.

Labeling noises or mislabeling can happen for different reasons. As the first reason, the available data may not be sufficient for reliable labeling [14, 16, 17], an example of which includes a significantly low data quality [18, 19]. As another reason, since the data is collected and labeled by an expert, the labeling phase is prone to human errors. Moreover, the labeling process is dependent on the expert's opinion, and two different experts may label the same data differently.

As previously mentioned, the performance of the classifier is negatively affected by noise. Besides, the label noise may cause the number of observations in the class to change, which is a frequently occurring problem in medical fields. The importance of this lies in the fact that in medical researches, measuring the number of occurrences of a disease in a population is an important objective, while the population size is generally not that large, consequently leading to a bias in the measurements due to the label noise.

Feature selection, specifically when a ranking method is used, is among the most important cases negatively affected by label noise, in which case the feature selection algorithm may overlook an important feature or select an irrelevant feature as an appropriate feature.

Various methods have been proposed to cope with label noise. These methods are generally classified into three categories [20]:

1. Label-noise robust methods: By reducing overfitting, these methods attempt to deal with the label noise. Examples include the bagging and boosting methods [21].

2. Data cleansing methods: This method eliminates those samples which seem to be mislabeled. Various methods are used to identify and filter the mislabeled samples. For example, the outlier detection method and the removal of misclassified data points can be pointed out.
3. Label noise-tolerant learning algorithms: In these methods, the noise is coped within the modeling step. For instance, in Reference [22], the loss function was modified to deal with noise.

5.2.5 Intrinsic Characteristics of Microarray Data

In genetics, gene expression is one of the most fundamental issues. Genes are the atomic units of genetic inheritance within the genome, holding info concerning all the corporal traits of an individual. The expression of a gene in DNA can transfer or reject a property to a person. In bioinformatics, the study of gene expression is very important because it helps to diagnose and predict a variety of diseases, such as cancer.

Microarray data is one of the most important data in bioinformatics, which is used in many areas, including cancer detection. Such data are extracted from tissues and cells considering variations within the gene which may be helpful for disease and tumors diagnosis. The challenges of these data include a large number of genes, as well as a very small number of samples, which makes it difficult to select effective genes in this data. Moreover, they have the risk of overfitting due to the small sample size of microarray data.

For example, breast data consists of 24,481 genes and only 60 samples of people with this cancer and healthy people and shows the amount of expression of various genes in them.

"Dimensionality curse" is one of the other difficulties in this data. This challenge arises when a feature vector's length is so large that the classifier is confused, and its efficiency and extensibility are reduced.

Like most other high-dimensional data, microarray data is also affected by the class imbalance problem. As mentioned, the classes are assumed to be balanced in the standard data classification algorithms, which is why their application to imbalanced data will not yield acceptable results, since the classifier tends to the training samples in the majority class, consequently increasing the number of errors in identifying positive samples.

Among the many techniques of machine learning, feature selection plays an important role, especially in data classification issues. In medicine, an effective gene selection can greatly enhance the process of prediction and diagnosis of cancer. After the selection of effective genes, the duty of a specific classifier is usually to discriminate healthy people from patients suffering from cancer based on their expression of the selected genes.

The advantages of gene selection come with the extra effort of trying to get an optimal subset of genes that will be a true representation of the original dataset.

One of the researchers' goals in bioinformatics is to achieve an optimal gene selection method that has minimal computational complexity. Today, many studies have been conducted to select important genes in medical data.

An optimal gene selection method is a method that selects an effective set of genes that does not reduce the performance of the classifier, but also improves the results of its classification.

5.3   Feature Selection

Today, feature selection has become one of the most important and fundamental contexts in high-dimensional data. In the past, there were several feature selection methods for classic data, but as the data dimension increases, many challenges emerge in the context of feature selection. Traditional data whose features are limited to a few tens of features have been replaced with high-dimensional data which have thousands of features. High-dimensional data are usually found in text processing, combinational chemistry, or bioinformatics.

Among thousands of features in these data, many of them are irrelevant to the class labels or involve redundant information. Thus, data preprocessing is one of the most essential steps for achieving an accurate and reliable classifier in high-dimensional data [23]. Amidst, selecting suitable features by eliminating irrelevant and redundant ones can be one of the most difficult and important steps for obtaining a suitable classifier.

In bioinformatics, different researches have shown that most measured features in a microarray experiment are not associated with classification validity of output groups of the problem [20], thus, in order to prevent the curse of dimensionality, eliminating irrelevant and excess features is one of the most important preprocessing steps [18]. Therefore, feature selection is a vital preprocessing step in bioinformatics and medicine. For example, diagnosing risk factor of deaths caused by cancer and selecting effective features in cancer diagnosis is one of the most important applications of feature selection in medicine [21]. Thus, finding a desirable feature selection method for choosing suitable genes among thousands of existing ones is very valuable in such a way that the best classification accuracy is obtained. An appropriate attribute set is highly correlated with class labels and has very little correlation with other features.

Recent studies indicate the importance of feature selection methods for selecting informative genes before classifying microarray data for predicting and diagnosing cancer [22].

Several feature selection methods have been proposed for high-dimensional data. In general, the relation between evaluation function and the classifier can be categorized into four main groups: filter methods, wrapper methods, embedded methods, and hybrid methods. In the following, we will outline these four methods and the proposed algorithms based on them.

5.4   Filter Methods

Filter methods select a subset of features from the main dataset through employing specific evaluation metrics which are mainly based on statistical and independent methods. These methods select suitable features based on inherent characteristics of data without involving data-mining algorithms. In other words, no feedback from the learning algorithm is used. Figure 5.1 shows the flowchart of feature selection using filter methods.

Filter approaches are fast; thus, they are suitable for high-dimensional data [24], but the classification accuracy of the feature set provided by these methods is low [25].

Filter methods can be divided into univariate and multivariate classes. In univariate methods, the relation of a feature is measured using one evaluation metric, then a subset of features with highest ranking values are selected as the final subset, while in the multivariate method, the relation among features affects the selection of genes [23]. In general, multivariate filter methods have lower speed compared to univariate filter methods. Information gain (IG) [26] and F-score [27] are of famous univariate filter methods [28] and ReliefF [29], FCBF [30], mRMR [31], and CFS [32] are of prominent multivariate filter methods.

Filter methods can also be divided into three main groups: similarity based, statistical based, information theoretical based.

5.4.1   Similarity-Based Methods

Each feature selection method uses a distinct benchmark to specify the relevance of features. Similarity-based methods evaluate the importance of features by their ability to preserve data similarity [33]. Most of the similarity-based methods cannot handle feature redundancy due to the fact that these methods often evaluate the importance of features individually.

For supervised feature selection, data similarity can be extracted from label information; whereas for unsupervised feature selection techniques, most techniques use different distance metrics to achieve data similarity [33].

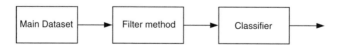

Fig. 5.1 The flowchart of the filter methods

5.4.1.1 Relief and ReliefF

Relief [29] is one of the most well-known filter methods which can be applied to nominal and numerical features. Relief searches for the features which are statistically associated to a group; according to this algorithm, a feature is more desirable which has more difference among samples of different groups and the same values for samples of similar groups [29].

This method first selects a random sample and finds the "near Hit" and "near Miss" based on Euclidean distance. It should be mentioned that near Hit is referred to samples which have minimum Euclidean distance among samples of the same class and near Miss is referred to samples which have minimum Euclidean distance among samples of different classes [34]. Weight of all weights are zero at first which are updated at each execution of the algorithm using Eq. (5.1) [35]:

$$Wi = W_{i-1} - (x_i - \text{nearHit}_i)^2 + (x_i - \text{nearMiss}_i)^2 \tag{5.2}$$

Weight of each feature increases if its difference from near samples of the same class is less than near samples in another class and vice versa. m relevance vectors are constructed by m times execution of the method on m different random samples and dividing each component of weight W on m, which is a metric for selecting superior features. In other words, features whose relevance vector is greater than the predetermined threshold are selected as the final feature [35]. One of the shortcomings of Relief is that it cannot handle incomplete and noisy data; moreover, this algorithm is only defined for two-group problems and cannot handle multiple-group problems. Thus, in the ReliefF algorithm [29] which is an extension of Relief, these problems are resolved and ReliefF is able to handle multiple-group problems, noisy and incomplete data.

5.4.1.2 Fisher Score

The idea of Fisher score selection algorithm is to find subsets of features in which distance between data points in different groups is as high as possible and distance between data points in a group is as low as possible [28].

Consider feature Xi of an m-group dataset. If the sample set of features i is in kth group X_i^k and $\left| X_i^k \right| = n_k$ where $k = 1, 2, \ldots, m$ and $\overline{X}_i^k$ and $\overline{X}_i$ are mean of features in X_i^k and X_i, then F-score of a feature is defined as follows [28]:

$$F(X_i) = \frac{\sum_{k=1}^m n_k \left(\overline{X}_i^k - \overline{X}_i \right)^2}{\sum_{k=1}^m \sum_{x \in X_i^k} \left(x - \overline{X}_i^k \right)^2} \tag{5.3}$$

where the numerator indicates discrimination between two groups and the denominator shows scattering in each group. The higher is F-score of a feature, discrimination of that feature would be higher. After calculating F-score of each feature, a number of features with a higher score are selected as the final feature subset based on a predetermined threshold.

5.4.1.3 Laplacian Score

Laplacian score [36] is one of the most common similarity-based methods of evaluation features which selects features that can best preserve the data manifold structure. Applying the algorithm requires three steps. On the first step, the affinity matrix is constructed as follows:

$$S(i, j) = \begin{cases} e^{-\frac{\|x_i - x_j\|^2}{t}} & if \ x_i \text{ is among of } p - \text{ nearest nighbor of } x_j \\ 0 & \text{Otherwise} \end{cases}$$

(5.4)

where t is a suitable constant. Then the diagonal matrix D is defined as follows:

$$D(i, i) = \sum_{j=1}^{n} S(i, j)$$

(5.5)

Then the matrix L is expressed as follows:

$$L = D - S$$

(5.6)

Finally, the Laplacian score for each feature f_i is calculated as follows:

$$\text{Laplacian\_Score} (f_i) = \frac{\tilde{f}_i' L \tilde{f}_i}{\tilde{f}_i' D \tilde{f}_i}$$

(5.7)

where

$$\tilde{f}_i = f_i - \frac{f_i' D1}{1' D1} 1$$

(5.8)

Since this algorithm is a ranking algorithm, the top k features with the smallest Laplacian scores will be selected as the best features.

5.4.2 Statistical-Based Methods

Statistical-based methods select the appropriate features based on different statistical criteria. Most of these methods are based on predefined straightforward statistical criteria to remove undesirable features. These methods are very useful in high-dimensional data even as a preprocessing step because their computational cost is very low.

Also, as similarity-based methods, these methods are often not able to consider the feature redundancy.

5.4.2.1 Correlation-Based Feature Selection (CFS)

This method is a multivariate filter feature selection method proposed in [37]. This method evaluates features according to correlation measure based on a heuristic evaluation metric which is biased towards subsets which include uncorrelated features with high correlation with the group.

The heuristic "merit" of the feature subset S with k features is expressed as follows:

$$CFS\_Score(S) = \frac{k\overline{r_{cf}}}{\sqrt{k + k\,(k-1)r_{ff}}} \tag{5.9}$$

where $\overline{r_{cf}}$ is the mean feature class correlation and $\overline{r_{ff}}$ is the average feature-feature correlation. To calculate $\overline{r_{ff}}$ and $\overline{r_{cf}}$, CFS uses symmetrical uncertainty [38].

5.4.2.2 Low Variance

Low variance filter is a useful dimensionality reduction algorithm. Low variance eliminates features whose variance is below a predefined threshold.

As noted earlier, if a feature is not able to differentiate between samples of different classes, it is irrelevant and cannot be a proper feature alone. Consequently, in this method, if a feature has a constant value for all samples, then the attribute is irrelevant, and the variance associated with that attribute will be zero.

5.4.2.3 T-Score

T-score [39] is one of the most widely used methods of feature selection. T-score is calculated by using the sample, mean and standard deviation values of the features for each class.

$$t\_score(fi) = \frac{|\mu_i^+ - \mu_i^-|}{\sqrt{\frac{n_i^+(\sigma_i^+)^2 + n_i^-(\sigma_i^-)^2}{n_i^+ + n_i^-}}} \tag{5.10}$$

which + and – are class labels, μ_i^+ and μ_i^- are means of class labels, μ_i^+ and μ_i^- are sample size of classes, σ_i^+ and σ_i^- are sample size of classes. It should be noted that this method is only used for classification in data with two classes (binary classification).

This method is also a ranking method. As a result, the higher the T-score value, the more important the corresponding feature is.

5.4.2.4 Information Theoretical-Based Methods

This method evaluates the importance of features by using different heuristic filter criteria. The basic characteristics of information theoretical-based methods are to maximize feature relevance and minimize feature redundancy [40]. It should be noted that most of the proposed algorithms in this method is only applied to discrete data. Consequently, if the properties values are continuous, the discretization process is a necessary preprocessing step.

Unlike similarity-based methods and statistical-based methods, which were mostly unable to address redundancy between features, these methods have the ability to handle the features redundancy.

5.4.2.5 FCBF

Fast correlation-based filter method (FCBF) [33] is one of the multivariate filter methods which is designed based on mutual information for handling high-dimensional data. This method is employed symmetrical uncertainty (SU) measure (Eq. (5.11)) to detect relevant and redundant features and evaluate the correlation between feature-class and feature-feature [30].

$$SU\,(X, Y) = 2 \left[\frac{IG\,(X, Y)}{H(X) + H(Y)} \right]$$ (5.11)

where in Eq. (5.11), $H(X)$ and $H(Y)$ are entropies of two features and $IG(X, Y)$ is information gain. FCBF first selects a set of features which have a high correlation with the group based on SU measure and after eliminating redundant features, it keeps the features relevant to the group.

5.4.2.6 Minimum-Redundancy-Maximum-Relevance (mRMR)

mRMR [31] is also a multivariate feature selection method which is based on mutual information and measures correlation among features and correlation between features and group using this measure. This method selects features which have maximum relevance with class and minimum redundancy (maximum dissimilarity among features). The feature score for a new unselected feature X_k is

$$J_{mRMR}(X_k) = I(X_k; Y) - \frac{1}{|S|} \sum_{X_j \in S} I(X_k; X_j) \tag{5.12}$$

where in Eq. (5.12) the feature relevance is evaluated by $I(X_k; Y)$ and $I(X_k; X_j)$ is information gain or mutual information between feature X_k and feature X_j.

5.4.2.7 Information Gain

IG method is one of the most applicable filter feature selection methods among existing methods. This method which is a univariate method evaluates features based on information gain. Information gain is based on entropy concept in information theory. In other words, IG of feature xi in Sx is written as Eq. (5.13):

$$IG(Sx, xi) = H - \sum_{v=\text{values}(xi)}^{\frac{|Sxi=0|}{|Sx|}} H(Sxi = v) \tag{5.13}$$

In Eq. (5.13), values(xi) is the set of values which xi can take. Entropy $H(Sxi = v)$ is also defined as follows:

$$H(S) = -(p_+) \log_2(p_+) - (p_-) \log 2(p_-) \tag{5.14}$$

which $p+$ is the ratio of positive samples to total samples and p_- is the ratio of negative samples to total samples. After calculating information gain of each feature, features are sorted based on their rank; desired features are selected using a threshold.

5.5 Wrapper Methods

Wrapper methods select desired features through employing results and performance of the classifier for evaluating the importance of the feature subsets. In these methods, a search mechanism is used to find the best subset of features among all possible subsets. Two main search mechanisms which are used are greedy search and random (stochastic) search. Each subset, which is proposed by an employed search algorithm, is evaluated by a classifier and the correct classification rate of the classifier is used as the fitness value for the corresponding feature subset [41]. Greedy search methods are single-track search methods which are easily trapped into local optima. Sequential forward selection and sequential backward elimination are among greedy search methods. Stochastic search methods select the subset of features randomly. Figure 5.2 shows the flowchart of feature selection using wrapper methods.

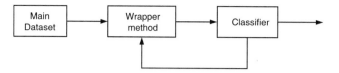

Fig. 5.2 The flowchart of the wrapper methods

The main random search algorithms are metaheuristics including particle swarm optimization (PSO), ant colony optimization (ACO), genetic algorithm (GA), and gravitational search algorithm (GSA). Since wrapper methods use the accuracy of the classifier as an evaluation metric, their accuracy is higher and their speed is lower than filter methods [42]. Moreover, wrapper methods cannot be employed alone due to the high computational complexity and low speed in high-dimensional data [43, 44].

We subsequently discuss some representative metaheuristics algorithms that have recently been used in high-dimensional data.

5.5.1 ABACO_H and ACO

Ants colony optimization algorithm was first introduced in 1991 inspired by the behavior of ants searching for food [45]. Despite being blind and dumb, ants are able to find the shortest path from nest to food by tracking remained pheromone through communicating with each other and transferring path information. In other words, pheromone intensity and its evaporation in paths which are rarely used allow the ants to choose the shortest path.

Using this algorithm for feature selection, the problem should be defined as a graph in which features are used as nodes of the graph. Location of ants is selected on the graph randomly, then each ant selects its subsequent node using the following equation:

$$
P_{ij}^k(t) = \begin{cases} \dfrac{\tau_{ij}^\alpha \eta_{ij}^\beta}{\sum_l \tau_{il}^\alpha \eta_{il}^\beta} & \text{if } i \text{ and } j \text{ are addmissible node} \\ 0 & \text{otherwise} \end{cases} \tag{5.15}
$$

If k^{th} ant is at position i at time t, it would be at location j at time $t + 1$ with probability P_{ij}^k. n_{ij} is cost of moving from node i to node j. α and β are parameters of the problem which control importance of trace against vision. τ_{ij}^α is pheromone intensity of the edge between node i and node j. The trace which k^{th} ant adds to edge (ij) is obtained using the following equation [46]:

$$\Delta\tau^k_{ij} = \begin{cases} \frac{Q}{F_k} & \text{if the } k^{\text{th}} \text{ ant traverse arc } (i,\, j) \text{ in } T_k \\ 0 & \text{otherwise} \end{cases} \tag{5.16}$$

F_k is the cost of passing the path which kth ant has passed and branch (i, j) is in that path. T^k is tour of the kth ant or movement path of this ant. Therefore, the trace which all ants add to edge (ij) is equal to

$$\Delta\tau^k_{ij} = \sum_1^m \Delta\tau^k_{ij} \tag{5.17}$$

In Eq. (5.17), m is the number of ants. Considering the above relations, new pheromone intensity created on the edge between nodes i and j is calculated using Eq. (5.18):

$$\tau_{ij}(new) = (1 - \rho)\,\tau_{ij}(old) + \Delta\tau^k_{ij} \tag{5.18}$$

In Eq. (5.18), ρ is the evaporation coefficient of pheromone intensity which prevents excess accumulation of trace. If a lot of ants pass a path, trace of that path is increased and trace of paths which few ants have passed through evaporates gradually [46].

In recent years, several methods have been proposed for feature selection based on ants colony algorithm. Binary ants colony optimization algorithm which has been proposed based on ants algorithm has advantages including more desirable classification rate and higher speed compared to ACO. This method was first introduced to solve binary space as TACO; later it was used for optimization.

Main problem of BACO in feature selection is that each ant at node i is only able to decide about the next feature and if it ignores this feature, it cannot investigate presence of this feature in subsequent nodes. Moreover, this method is able to offer a comprehensive solution for machine vision [46].

In 2013, a version of BACO called ABACO$_H$ was proposed in [46], in which BACO and discrete ACO are used in combination. In this algorithm, P^k_{ij} is defined as follows:

$$P^k_{i\_x,j\_y} \begin{cases} \dfrac{\tau_{ix,jy}\,\eta_{ix,jy}}{\sum_j \tau_{ix,j0}\,\eta_{ix,j0} + \tau_{ix,j1}\,\eta_{ix,j1}} & j \in \text{addmissible node} \\ 0 & \text{otherwise} \end{cases} \tag{5.19}$$

This equation specifies the probability of selecting bit of value $y \in \{0, 1\}$ in the next node for kth ant at time t at location $x \in \{0,1\}$ of node i. In addition, $\tau_{i0,j0}$, $\tau_{i1,j0}$, τ_{i0}, j_1, $\tau_{i0,j0}$ indicate pheromone intensity between paths which connect nodes i and j on (0 to 0), (1 to 0), (0 to 1), and (1 to 1) edges [38]. This algorithm has resolved the shortcomings of ACO. ABACO$_H$ allows the ant to search among all existing features, which is an important characteristic of ACO. In this method, unlike previous ACO methods, where the observed feature was the selected feature, the ant can select or reject the observed features. Moreover, in this method, ants can see all the features which have not been observed previously which gives better results in the process of selecting features.

5.5.2   PSO

Particle swarm optimization (PSO) [47] algorithm is based on population technique inspired by the social behavior of birds. Suitable computational complexity, few parameters, and global search ability are among advantages of this algorithm which make PSO one of the most successful existing algorithms.

In this algorithm, each solution is a particle in the swarm which has a location in the search space which is represented with vector x_i:

$$xi = \left(x_i^1, x_i^2, \ldots, x_i^d \right) \tag{5.20}$$

where d is the dimension of the search space. Particles move in the search space to find the optimal solution and their velocity is shown with vector v_i:

$$vi = \left(v_i^1, v_i^2, \ldots, v_i^d \right) \tag{5.21}$$

Then the location and velocity of each particle are updated considering the experience of the particle itself and its neighbors according to the following equations:

$$x_i^d (t + 1) = x_i^d (t) + v_i^d (t + 1) \tag{5.22}$$

$$v_i^d (t + 1) = w * v_i^d (t) + c1 * r1 \left(pbest_i^d - x_i^d(t) \right) + c2 * r2 \left(gbest^d - x_i^d(t) \right) \tag{5.23}$$

In Eq. (5.22) and Eq. (5.23), t indicates t^{th} iteration and d indicates d^{th} dimension of the search space. w is the weight of inertia which controls the effect of previous velocity on new velocity. C1 and C2 are speedup constants, $r1$ and $r2$ are random numbers with uniform distribution in $[0, 1]$. $pbest_i^d$ shows best location achieved by particle i in the dth dimension (best solution achieved by the particle i) and $gbest^d$ is the best location achieved by the whole swarm at d^{th} dimension (best solution obtained by the whole swarm). The algorithm is stopped when the desired solution is obtained, or the number of iterations reaches the predetermined value (maximum iteration).

5.5.3 IBGSA

The gravitational search algorithm is an example of metaheuristic methods proposed in 2009 which are inspired by mass and gravity [48]. According to Newton's law, each particle in the universe applies force to other particles which is directly proportional to multiplication of their masses and inversely proportional to the square root of their distance [49].

In recent years, this algorithm has attracted attentions due to its high efficiency in solving different optimization problems. In 2010, the binary version of this algorithm called BGSA was proposed in [50]. Then, in 2014, an improved version of this algorithm called IBGSA was proposed in [51] to prevent being trapped in local optimums in solving feature selection problems. In the following, metaheuristic IBGSA algorithm is described.

In a system which has s agents, the position of ith agent in the mentioned algorithm is described as in Eq. (5.24):

$$X_i = \left(x_i^1, \ldots, x_i^d, \ldots, x_i^n \right) \qquad i = 1, 2, \ldots, s \tag{5.24}$$

where x_i^d shows the position of the dimension d of mass i. n indicated dimension of search space. Mass of each agent is also calculated after calculating the fitness of current population using Eq. (5.25):

$$M_i(t) = \frac{\mathrm{fit}_i(t) - \mathrm{worst}(t)}{\sum_{j=1}^{S} \mathrm{fit}_i(t) - \mathrm{worst}(t)} \tag{5.25}$$

where $M_i(t)$ and $\mathrm{fit}_i(t)$ are mass and fitness of ith agent at time t and $\mathrm{worst}(t)$ is described as Eq. (5.26):

$$\mathrm{worst}(t) = \max \mathrm{fit}_j(t) \qquad j \in \{1, 2, \ldots, s\} \tag{5.26}$$

Resultant forces applied to i^{th} agent from heavier agents are calculated according to gravity law as in Eq. (5.27):

$$F_i^d(t) = \sum_{j \in kbest, j \neq i} \mathrm{rand}\, j G(t) \frac{M_i(t) M_j(t)}{R_{ij}(t) + \varepsilon} \left(x_j^d(t) - x_i^d(t) \right) \tag{5.27}$$

kbest includes k superior agents with more fitness which is a function of time which starts with $k0$ and reduces with time.

Resultant forces applied to an agent are calculated according to accelerating movement law of the agent using Eq. (5.28):

$$a_i^d(t) = \frac{F_i^d(t)}{M_i(t)} = \sum_{j \in kbest, j \neq i} \text{rand}_j G(t) \frac{M_j(t)}{R_{ij}(t) + \varepsilon} \left(x_j^d(t) - x_i^d(t) \right) \tag{5.28}$$

Finally, speed of each agent is updated using Eq. (5.29):

$$v_i^d(t+1) = \text{rand}_i \times v_i^d(t) + a_i^d(t) \tag{5.29}$$

In the above equations, $rand_i$ and $rand_j$ are two random numbers with uniform distribution on $[0, 1]$ and ε is a small value. $R_{ij}(t)$ is hamming distance of agents i and j which is calculated using Eq. (5.30):

$$R_{ij}(t) = \frac{1}{n} \sum_{d=1}^{n} \left| x_j^d(t) - x_i^d(t) \right| \tag{5.30}$$

Gravitational constant G is also a function of time which is initialized with $G0$ and reduces with time. Location of agents changes with a probability according to Eq. (5.31) which is known as transfer function:

$$Tf\left(v_i^d(t)\right) = A + (1 - A) \times \left| \tanh v_i^d(t) \right| \tag{5.31}$$

In the above equation, A is calculated using Eq. (5.32):

$$A = k_1 \left(1 - \exp \frac{Fc}{k_2} \right) \tag{5.32}$$

where k_1 is a constant parameter and k_2 is time constant defined based on the algorithm. Fc is the failure counter. Failure occurs when the best-observed solution does not change after an iteration. Finally, agents move according to Eq. (5.33):

$$x_i^d(t+1) = \begin{cases} \text{complement } x_i^d(t), & if \text{ rand} < Tf\left(v_i^d(t+1)\right) \\ x_i^d(t), & else \end{cases} \tag{5.33}$$

One of the most important differences of IBGSA compared to BGSA is its elitism property. In this version, the location of the agent changes only when the new location has better, less than or equal fitness to previous fitness. Mathematically, this characteristic is defined as follows:

$$X_i(t+1) = \begin{cases} X_i(t+1), & \text{if fit}(X_i(t+1)) < \text{fit}(X_i(t)) \\ X_i(t), & \text{otherwise} \end{cases} \tag{5.34}$$

To stop the algorithm, different criteria are considered.

5.6 Hybrid Method

Hybrid methods combine filter and wrapper methods; at the first step, the dimension of the main feature set is reduced using filter methods and then wrapper methods are implemented on the reduced feature set. Accuracy of these methods is usually higher than filter methods [52]. Moreover, these methods have lower computational complexity and higher speed compared to wrapper methods, thus they can be a proper option for feature selection in high dimensions [53]. Figure 5.3 displays the flowchart of feature selection using hybrid methods.

Authors of [54] proposed a hybrid method combining SVM-RFE and mRMR. Results of this method have given better results compared to SVM-RFE, mRMR, and some other methods. In 2011, authors of [55] have proposed a hybrid method for text categorization in which information gain filter method is applied to data and then genetic algorithm is applied to selected data of the first step to select the final features. Chuang et al. [56] proposed a hybrid method called CFS-TGA which combines correlation-based feature selection (CFS) and Taguchi-genetic algorithm and then results are applied to 11 microarray datasets.

Authors of [57] have proposed a type of hybrid method for gene selection in microarray data in which a genetic algorithm with dynamic parameter setting (GADP) is applied to data to generate a number of subsets of genes and then $\chi 2$ is applied to selected features of the first step to select the final features. Shreem et al. [58] proposed a hybrid method called R-m-GA which combines ReliefF, mRMR, and GA.

In 2015, authors of [59] proposed a hybrid method which in the first step, the number of features is reduced by a filter algorithm based on information gain values and then by applying binary DE-based wrapper method, the final features are selected.

In 2019, authors of [60] proposed a hybrid method by combination of mutual information (MI) and recursive feature elimination (RFE) for reducing the dimensions of the three benchmark datasets from the UCI repository.

FSCBAS, which is a hybrid method, was proposed by [61]. This method is obtained by combination of clustering and modified binary ant system (BAS).

In [62] the authors developed another hybrid-based feature selection algorithm. It first uses a V-WSP based filter method to select the top features. Then, a particle swarm optimization (PSO) based method is used to select the final features.

In [63] the authors introduced a new hybrid method using the heuristic search. Five filter algorithms including CFS, mRMR, oneRFeatureEval, corrFeatureEval, and IG are used along with the genetic algorithm for dimensionality reduction in three biomedical data.

Fig. 5.3 The flowchart of the hybrid methods

Table 5.2 Hybrid methods used on high-dimensional data

Method	Data category	Data type	Year	Ref
SVM-RFE + mRMR	Microarray data	Binary	2009	[54]
IG-PCA/GA	Text data	Multiclass	2011	[55]
CFS-TGA	Microarray data	Multiclass	2011	[56]
GADP	Microarray data	Multiclass	2011	[57]
R-m-GA	Microarray data	Binary	2012	[58]
BDE-X$_{Rank}$	Microarray data	Binary	2015	[59]
MI-RFE	Physical data	Binary	2019	[60]
FSCBAS	Physical/life/computer/microarray data	Multiclass	2019	[61]
V-WSP-PSO	Spectra data	Multiclass	2019	[62]
5Filter + ga	Biomedical data	Multiclass	2019	[63]
RFACO-GS	Microarray data	Multiclass	2019	[64]

Authors of [64] have proposed a hybrid method using Relief algorithm and ant colony optimization algorithm to reduce dimensionality in microarray data and tumor data classification.

Table 5.2 presents a summary of the hybrid methods described, along with the original reference, the data category, and the type of data.

5.7 Embedded Methods

Embedded methods perform feature selection as an inseparable part of the machine learning algorithm [65]. In these methods, learning and feature selection are two impartible components. Embedded methods have shorter execution time compared to wrapper methods. The computational complexity of these methods is more than filter methods and lower than wrapper methods [66].

Among methods which are applied on high-dimensional data based on embedded methods, the method proposed in [67] which is proposed for cancer data classification can be mentioned. This method performs feature selection through repeated training of support vector machine with existing features set and eliminating the least significant features.

In 2010, kernel-penalized SVM (KP-SVM) was proposed in [68]. This method selects relevant features through penalizing use of the feature in the dual formula of SVM. In 2012, authors of [69] proposed iterative feature perturbation (IFP) method. This embedded method employs backward elimination and a metric for determining the least significant features and investigates the effect of each feature on the performance of classifier while being disturbed by noise.

In 2018, KP-CSSV was proposed in [70]. This method is inspired by the KP-SVM method to solve the class imbalance problem in microarray data.

In 2019, authors of [71] proposed an embedded method based on the weighted Gini index method to deal with the imbalanced classification problems.

Table 5.3 Embedded methods used on high-dimensional data

Method	Data	Data type	Year	Ref
KP-SVM	Microarray data	Multiclass	2010	[68]
IFP	Microarray data	Binary	2012	[69]
KP-CSSV	Microarray data	Multiclass	2018	[70]
GI − FS$^\rho$	Life/computer data	Binary	2019	[71]
MGRFE	Microarray data	Multiclass	2019	[72]

Authors of [72] proposed an embedded method called MGRFE, based on an embedded integer-coded genetic algorithm to select the top genes in microarray data. Table 5.3 presents a summary of embedded methods described.

5.8 Ensemble Techniques

Today, many methods are proposed for feature selection in high-dimensional data. High-dimensional data not only may be very large in terms of number of features and data, but also they may face problems in terms of redundancy, noise, and nonlinearity, thus it cannot be claimed that a method gives good results in all data because many of the methods are not able to handle these problems alone. Thus, researchers are attracted towards the ensemble feature selection/classification techniques. Thus, the probability of selecting a wrong solution is decreased and learning algorithms which are trapped in local optimums create better estimations [73]. In ensemble techniques, instead of considering the results of a single method as the final results, results of applying several methods on data are combined.

Among several frameworks which are based on ensemble on high-dimensional data, ensemble methods shown in Figs. 5.4 and 5.5 can be mentioned. In Fig. 5.4, as can be seen, the results of several filter methods on high-dimensional data are combined in different ways, thus the final selected subset is obtained. This method does require an integration method to combine the results obtained from each filter method.

In Fig. 5.5, several filter methods are applied to high-dimensional data, independently. Each filter method selects its selected subset, then results of each one are given to the classifier; finally, after the classification stage, an integration method combines the results of each classifier [74].

Yang et al. [75] have proposed a method called multi-criteria fusion-based recursive feature elimination (MCF-RFE) based on ensemble concept which has provided desirable results for microarray data; their motivation is announced to be combining multiple criteria and recursive feature elimination (RFE) search strategy. Bolon et al. [76] have proposed a filter ensemble framework and an ensemble of classifiers in which several filter methods like CFS, INTERACT, and information gain are utilized. Authors of [35] have proposed a hybrid-ensemble method for

Fig. 5.4 Instance 1 of
ensemble techniques
(ensemble feature selection)

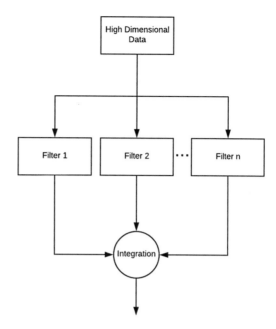

feature selection in microarray data in which results obtained from ReliefF, IG, and F-score are combined; then, the improved binary gravitational search algorithm (IBGSA) is applied to the results obtained from the mentioned methods.

In 2014, four ensemble methods called E1-cp, E1-nk, E1-ns, and E2 were proposed in [74]. Advantages of this method include a variety of responses in filter methods which makes the final response desirable. In [73], a method based on ensemble is proposed which combines IG, CFS, and Relief. Comparing the results obtained from applying the above methods to high-dimensional data with filter methods and two other ensemble methods show that this algorithm outperforms other methods in terms of classification. Authors of [34] have proposed a hybrid-ensemble method called HM-ABACO$_H$. They have combined the results of three different filter methods: ReliefF, FCBF, IG and then the ABACO$_H$ metaheuristic method is applied to results of the combination of filter methods. Finally, this method is applied to seven microarray datasets. In [74], CFS, Cons, IG, and Relief filter methods are combined with different classifiers using ensemble methods and compared the results on several microarray datasets.

It can be said while working with high-dimensional data, the first step is to reduce data dimension, because high dimension increases computational cost and causes the curse of dimensionality. As mentioned, in order to reduce dimension in such data, filter methods are the simplest and fastest methods, but these methods usually do not have high classification accuracy. On the other hand, wrapper methods typically have low speed but high classification accuracy. As mentioned above,

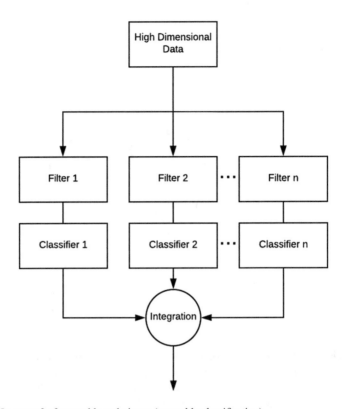

Fig. 5.5 Instance 2 of ensemble techniques (ensemble classification)

hybrid methods using both filter and wrapper methods have reasonable speed and relatively satisfactory classification accuracy. Hence, one of the new approaches proposed for dimensionality reduction in high-dimensional data is the use of hybrid methods in the form of ensemble techniques.

In 2017, [22] introduced a new type of hybrid-ensemble methods. In this method, after applying the filter method to the data, the ensemble technique is used to combine the results of the metaheuristic methods. In this method, the FCBF algorithm was used as a filtering method and the ABACO and IBGSA methods were used as wrapper methods.

Authors of [42] proposed a framework based on hybrid-ensemble methods. In this framework, each filter technique generates its output. Then the outcomes of each technique are provided in several wrapper techniques. Finally, the outcomes of different wrapper methods are combined, and the final output of this phase is provided. Figure 5.6 shows the flowchart of this method.

In 2018, authors of [77] proposed an ensemble feature selection technique based on t-test and nested genetic algorithm for feature selection in high-dimensional data.

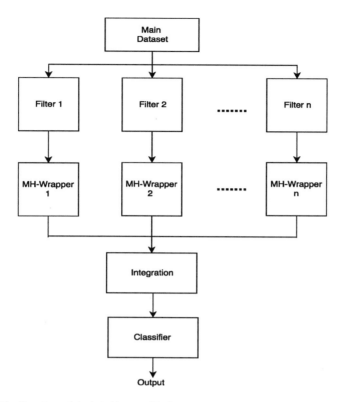

Fig. 5.6 The flowchart of the hybrid-ensemble framework proposed in [42]

In 2019, authors of [78] by examining several different methods, including filter methods, embedded methods, and univariate/multivariate techniques have shown that ensemble techniques are more robust than other single methods.

Authors of [79] proposed an ensemble-based method using bits from k-mean techniques called feature co-association ensemble (CFE) to select the best features on several types of data in UCI repository. In [80], the authors present an ensemble technique using the three feature selection methods of maximum information coefficient, XGBoost, and chi-square for feature selection in high-dimensional two-class data. Authors of [81] developed an ensemble feature selection method by combining four state-of-the-art filter methods to identify robust diabetic kidney disease (DKD) risk factors by balancing predictability and stability. Table 5.4 summarizes methods based on ensemble techniques.

In [82], the authors present an ensemble technique for feature selection in microarray data. In this method, they have implemented a multistage neural network ensemble to combine the outputs.

Table 5.4 Ensemble-based methods used on high-dimensional data

Method	Framework	Data	Data type	Year	Ref
MCF-RFE	Ensemble	Text data	Multiclass	2010	[75]
Ensemble of filters/classifiers	Ensemble	Microarray data	Multiclass	2012	[76]
ReliefF-IG-Fscore +IBGSA	Hybrid-ensemble	Microarray data	Multiclass	2014	[35]
E1-cp	Ensemble	Microarray data	Multiclass	2014	[74]
E1-nk	Ensemble	Microarray data	Multiclass	2014	[74]
E1-ns	Ensemble	Microarray data	Multiclass	2014	[74]
E2	Ensemble	Microarray data	Multiclass	2014	[74]
IG-CFS-ReliefF	Ensemble			2013	[73]
HM-ABACO$_H$	Hybrid-ensemble	Microarray data	Multiclass	2016	[34]
FCBF +ABACO-IBGSA	Hybrid-ensemble	Microarray data	Multiclass	2017	[22]
PFW-ensemble	Hybrid-ensemble	Microarray data	Multiclass	2017	[42]
Nested-GA	Ensemble	Microarray data	Binary	2018	[77]
Ensemble empirical study	Ensemble	Microarray/biomedical data	Multiclass	2019	[78]
CFE	Hybrid-ensemble	Microarray/biomedical/life/physical data	Multiclass	2019	[79]
SA-EFS	Ensemble	Physical data	Binary	2019	[80]
Ensemble-IDKD	Ensemble	Biomedical data	Binary	2019	[81]
Ensemble neural network	Ensemble	Microarray data	Binary	2019	[82]

5.9   Practical Evaluation

5.9.1   Dataset

In order to perform the experiments, ten microarray data whose general characteristics are given in Table 5.5 are employed. This table indicates the variety of samples, number of features, and number of classes. It should be noted that all datasets used in this chapter are available for download in [83, 84].

As mentioned, microarray data have a large number of features and a few numbers of samples which can be seen in Table 5.5. Among datasets of Table 5.5, the minimum number of features is 2000 which corresponds to colon dataset which has 62 samples and the maximum number of features is 24,481 which corresponds to Breast_Cancer dataset with 97 samples. It should be mentioned that in these experiments, multi-class datasets like Brain_Tumor1 and SRBCT are investigated.

5.9.2   Performance Evaluation Criteria

In this experiment, seven measurement criteria have been used to compare the methods, including the correct classification rate, the number of selected features, sensitivity (Se) and specificity (Sp), MCC (Matthews correlation coefficient), GM (geometric mean), and GMEAN.

The correct classification rate is the number of test specimens properly classified divided by the total number of test specimens, which can be obtained using the following equation:

$$ACC = \frac{\text{The number of correctly classifed test samples}}{\text{The total number of test samples}} \tag{5.35}$$

Table 5.5 Microarray datasets used for benchmarking

No	Dataset	# of features	# of samples	# of classes
1	Brain_Tumors1	5920	90	5
2	Breast_Cancer	24,481	97	2
3	CNS	7129	60	2
4	Colon	2000	62	2
5	Leukemia	7129	72	2
6	Prostate	10,509	102	2
7	Prostate_Cancer	12,600	21	2
8	Lung_Cancer	12,533	181	2
9	SRBCT	2308	83	4
10	Ovarian	15,154	253	2

The higher the correct classification rate, the more significant the role of the subset of selected features in the correct classification rate; therefore, they will be considered more appropriate features.

Sensitivity and specificity are also designed to evaluate the performance of binary (binomial) classifications. In general, if we consider classes of two-class data as positive and negative classes, assuming that:

TP: Number of test samples which are classified correctly.

FP: Number of test samples which are classified as positive incorrectly.

TN: Number of test samples which are classified correctly.

FN: Number of test samples which are classified as negative incorrectly.

In this case, the sensitivity, the specificity, the geometric mean of sensitivity and specificity, and the Matthews correlation coefficient are expressed as follows:

$$\text{Sensitivity}(SN) = \frac{TP}{TP + FN} \tag{5.36}$$

$$\text{Specificity}(SP) = \frac{TN}{TN + FP} \tag{5.37}$$

The feature reduction (Fr) parameter is the ratio of the number of selected features by the algorithm to the total number of features and is given by the following equation:

$$Fr = \frac{p - q}{p} \tag{5.38}$$

where p is the total number of features, and q is the number of selected features. According to this equation, the closer the Fr value to 1, the greater the reduction in the number of features and more preferred. Note that the lower the number of selected features, the lower the computational complexity.

Gmean and MCC are expressed as:

$$\text{Gmean} = \sqrt{\text{Sensitivity} \times \text{Specificity}} \tag{5.39}$$

$$MCC = \frac{TP \times TN - FP \times FN}{\sqrt{(TP + FP)(TP + FN)(TN + FP)(TN + FN)}} \tag{5.40}$$

The authors state that since the number of selected features and the Fr parameter alone cannot measure the strength or weakness of a method, we need to use criteria such as the classifier accuracy rate as well. Therefore, the geometric mean between two criteria has been used to consider the effect of the classification accuracy and the feature reduction parameter criteria, as follows:

$$GM = \sqrt{ACC \times Fr} \tag{5.41}$$

5.9.3 Data Normalization

Data normalization can be considered as one of the preprocessing stages in each experiment. With normalization, the values of continuous features of each dataset are placed in the interval [0, 1]. In the normalization process, all features are given the same weight when calculating the distance between datasets.

5.9.4 Analysis of Filter Algorithms

As mentioned, filter methods are the most popular and applicable feature selection methods. These methods operate independently of learning algorithm; thus, their speed is higher than other feature selection methods. In the following, we compare the results of applying the filter methods mentioned in Sect. 5.4 on several high-dimensional datasets.

In this subsection, the results were evaluated by both SVM and KNN classifiers, as well as by 10 CV evaluation method.

It is noteworthy that the results related to multi-class datasets are not addressed in the results of the T-score filter method because the simple version of this method is only applicable to two-class data.

Tables 5.6 and 5.7 show the results of applying nine filter-based methods on 8 well-known microarray datasets, in terms of average classification rates over ten different runs, by applying the SVM and KNN classifiers, respectively.

All of the methods mentioned in in Tables 5.6 and 5.7 are ranker methods, except the two methods, i.e., FCBF and CFS which return a set of selected features, meaning that they assign a rank to each feature which is the importance of that feature. Therefore, among the ranker methods mentioned, 100 top features have been studied as selected features.

According to the results of Tables 5.6 and 5.7, it can be found that the Prostate_Cancer data is the most challenging available data compared to other data provided. One of the most important reasons why this data is the most challenging is the enormous number of features of this data (i.e., 12,600) compared to the negligible number of its samples (i.e., 21). Moreover, the other major reason why applying classification on this dataset is difficult is that its test data have been extracted from several different experiments and has the dataset shift problem. However, the T-score and FCBF methods have been able to obtain acceptable results on this data.

As can be seen, the SVM classifier yields better classification accuracy than the KNN classifier. In addition, according to the results corresponding to this classifier, we can say that the FCBF method, which is one of the information theoretical-based methods, has been able to obtain better results than other methods. However, it should be noted that in this experiment, only 100 features are considered as the number of selected features for the ranker methods. These methods may be able

Table 5.6 Experimental results for the SVM classifier after performing regular tenfold cross-validation

Method		Brain_Tumor1	Breast_Cancer	CNS	Colon	Leukemia	Prostate_Cancer	SRBCT	Ovarian
Similarity-based methods	ReliefF	0.862	0.702	0.646	0.798	0.978	0.5611	0.952	1
	Fisher score	0.884	0.642	0.554	0.782	0.967	0.542	0.981	1
	Laplacian score	0.860	0.770	0.662	0.761	0.85	0.560	0.958	0.898
Statistical-based methods	CFS	0.902	0.813	0.873	0.841	0.94	0.682	0.962	0.991
	Low variance	0.866	0.663	0.681	0.81	0.9421	0.543	0.993	1
	T-score	–	0.562	0.816	0.762	0.9428	0.811	–	1
Information theoretical-based methods	FCBF	0.961	0.825	0.912	0.824	0.982	0.887	0.992	1
	mRMR	0.887	0.802	0.836	0.786	0.972	0.710	0.978	0.998
	Information gain	0.911	0.800	0.795	0.801	0.9819	0.799	1	1

Table 5.7 Experimental results for the KNN classifier after performing regular tenfold cross-validation

Method		Brain_Tumor1	Breast_Cancer	CNS	Colon	Leukemia	Prostate_Cancer	SRBCT	Ovarian
Similarity-based methods	ReliefF	0.830	0.654	0.544	0.778	0.957	0.5322	0.898	0.992
	Fisher score	0.861	0.6513	0.64	0.792	0.960	0.542	1	0.992
	Laplacian score	0.831	0.602	0.688	0.650	0.835	0.511	0.973	0.878
Statistical-based methods	CFS	0.883	0.781	0.84	0.812	0.941	0.590	0.934	0.970
	Low variance	0.844	0.581	0.61	0.760	0.849	0.553	0.858	0.932
	T-score	–	0.525	0.809	0.781	0.981	0.855	–	0.990
Information theoretical-based methods	FCBF	0.962	0.798	0.89	0.811	0.980	0.841	0.993	1
	mRMR	0.853	0.802	0.812	0.715	0.972	0.688	0.942	0.992
	Information gain	0.900	0.800	0.58	0.80	0.972	0.730	0.980	1

to obtain better results by changing this value. In general, one of the challenges of ranker feature selection methods is to find the optimal threshold value for the appropriate feature selection.

Hence, one can say that achieving an optimal filter-based feature selection and then optimal classification on a particular dataset depends on factors such as the feature selection algorithm, the threshold value considered for the ranker methods, and the classifier used.

5.9.5 Analysis of Hybrid-Ensemble Methods

As mentioned earlier, hybrid feature selection methods can be highly efficient and produce desirable results while dealing with high-dimensional data due to enjoying the advantages of both filter and wrapper methods. In terms of time complexity, these methods significantly outperform the wrapper methods because the data dimension has declined dramatically before entering the wrapper phase. On the other hand, the accuracy of these methods can be much better than the filter methods because the high-accuracy wrapper methods have been utilized after applying the filter method.

As mentioned previously, in ensemble techniques, the results of several different approaches are mixed, and the response of each method affects the final results. Consequently, these approaches can be highly effective in the selection of desirable features as well as eliminating redundant and unrelated features.

Meanwhile, using ensemble techniques in the form of hybrid methods can also be a useful approach to handle high-dimensional data. As a result, researchers have been attracted to these two techniques.

In this section, we will investigate and analyze the results of applying hybrid-ensemble methods on high-dimensional data.

5.9.5.1 Hybrid-Ensemble 1

The proposed method in [52] is based on hybrid-ensemble, which has produced satisfactory results on high-dimensional data. In this method, as mentioned earlier, data dimensionality reduction is carried out by a filter method; then, the data of two metaheuristic algorithms are applied independently on the dimensionality reduced data to choose effective features with high precision. Finally, the results of the two metaheuristic methods are combined with an integration method. The authors' proposed framework is shown in Fig. 5.7.

In this study, the ABACO$_H$ and IBGSA algorithms, which showed an acceptable performance on the high-dimensional data in [34, 35], have been used for the two metaheuristic algorithms. The results of several different filter methods have been compared to select a suitable filter method. Furthermore, the results of the OR and AND operators have been compared to choose the optimum integration approach.

Fig. 5.7 Flowchart of the
hybrid-ensemble method
proposed in [39]

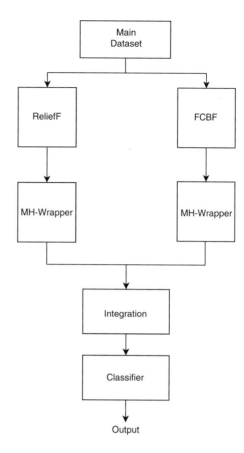

5.9.5.1.1   Find the Optimal Filter Method and Integration Approach

In order to obtain the optimal filter method as well as the proper integration
method, the results of applying several different filter-based methods and two
integration methods (AND and OR) are investigated and compared. The results of
this comparison are shown in Table 5.8. In this experiment, the validation method
has been the data dividing with the ratios of 2/3 and 1/3 for the training and test
data, respectively.

All of the filter methods listed in Table 5.8, except the FCBF method, act as
ranker methods. As already mentioned, these methods assign a rank to each feature
and indeed rank all the available features in the data after they are applied on
the data. As a result, these methods require a threshold limit to select the desired
features. In the research mentioned, the threshold limit of 0.004 has been chosen for
all these methods. The results were evaluated by KNN ($K = 1$) classifier.

According to the results of Table 5.8, for colon data with 2000 features and 62
samples, the HEMO-FCBF method could achieve the highest values for the four
classification criteria, e.g., the classifier accuracy rate, the Matthews correlation

Table 5.8 Experimental results for the KNN classifier after performing hybrid-ensemble methods to find the optimal filter method

Integration method	Method name	Measure	Colon	Leukemia	Prostate	Lung_Cancer	Ovarian	Average
AND	HMEBA-ReliefF	ACC	0.7714	0.9167	0.8824	0.8512	0.9959	0.88352
		FS	2.8	7.2	15.4	15	13.6	10.8
		SN	0.8645	0.9398	0.9017	0.862	0.9962	0.91284
		SP	0.6248	0.9448	0.892	0.8217	0.9938	0.85542
		GMEAN	0.734942	0.942297	0.896837	0.841609	0.994999	0.883664
		MCC	0.6821	0.9028	0.8012	0.638	0.9899	0.8028
		GM	0.878	0.957	0.939	0.922	0.997	0.938
	HMEBA-IG)	ACC	0.809	0.950	0.894	0.894	0.998	0.909
		FS	3.600	7.600	11.400	16.200	4.600	8.680
		SN	0.902	0.936	0.940	0.940	0.996	0.943
		SP	0.781	0.910	0.915	0.902	0.999	0.901
		GMEAN	0.840	0.923	0.928	0.921	0.997	0.922
		MCC	0.761	0.909	0.821	0.820	0.990	0.860
		GM	0.899	0.974	0.945	0.945	0.998	0.952
	HMEBA-F-score	ACC	0.752	0.983	0.859	0.866	0.986	0.889
		FS	2.600	6.200	9.600	15.000	14.000	9.480
		SN	0.802	0.973	0.912	0.919	1.000	0.921
		SP	0.780	0.960	0.849	0.850	0.960	0.880
		GMEAN	0.791	0.966	0.880	0.884	0.980	0.900
		MCC	0.651	0.924	0.807	0.791	0.941	0.823
		GM	0.867	0.991	0.926	0.930	0.992	0.941
	HMEBA-FCBF	ACC	0.712	0.959	0.853	0.861	0.993	0.876
		FS	4.000	4.000	12.800	14.100	8.400	8.660
		SN	0.708	0.971	0.898	0.851	1.000	0.886
		SP	0.691	0.942	0.897	0.891	1.000	0.884
		GMEAN	0.699	0.956	0.897	0.871	1.000	0.885
		MCC	0.611	0.924	0.781	0.752	0.985	0.810
		GM	0.843	0.979	0.923	0.927	0.996	0.933
OR	HMEBO-ReliefF	ACC	0.795	0.946	0.921	0.908	0.993	0.913
		FS	4.400	20.200	32.800	34.100	43.800	27.060
		SN	0.836	0.967	0.946	0.908	0.989	0.929
		SP	0.792	0.920	0.912	0.821	0.998	0.889
		GMEAN	0.813	0.943	0.929	0.864	0.994	0.909
		MCC	0.601	0.915	0.893	0.798	0.984	0.838
		GM	0.891	0.971	0.958	0.952	0.995	0.953
	HMEBO-IG	ACC	0.805	0.938	0.874	0.926	0.998	0.908
		FS	4.000	20.600	30.000	32.800	13.200	20.120
		SN	0.842	0.941	0.863	0.922	0.996	0.913
		SP	0.749	0.928	0.872	0.842	1.000	0.878
		GMEAN	0.794	0.934	0.867	0.881	0.998	0.895
		MCC	0.669	0.921	0.832	0.782	0.995	0.840
		GM	0.896	0.967	0.933	0.961	0.998	0.951

(continued)

Table 5.8 (continued)

Integration method	Method name	Measure	Colon	Leukemia	Prostate	Lung_Cancer	Ovarian	Average
	HMEBO-F-score	ACC	0.800	0.983	0.844	0.935	0.998	0.912
		FS	3.400	23.100	29.600	32.200	45.000	26.660
		SN	0.838	0.980	0.909	0.946	0.996	0.934
		SP	0.783	0.935	0.812	0.901	1.000	0.874
		GMEAN	0.810	0.957	0.859	0.923	0.998	0.903
		MCC	0.651	0.931	0.758	0.892	0.995	0.845
		GM	0.894	0.990	0.917	0.966	0.997	0.953
	HMEBO-FCBF	ACC	0.814	0.941	0.929	0.943	1.000	0.925
		FS	7.000	5.000	26.400	18.750	17.200	14.870
		SN	0.855	0.975	0.946	0.952	1.000	0.946
		SP	0.785	0.965	0.931	0.904	1.000	0.915
		GMEAN	0.820	0.970	0.938	0.928	1.000	0.930
		MCC	0.772	0.912	0.902	0.891	1.000	0.895
		GM	0.901	0.970	0.963	0.970	0.999	0.961

coefficient (MCC), specificity, and geometric mean (GM). The HMEBA-IG method has achieved the highest value for the sensitivity (SN) and Gmean of this data.

For leukemia data with 7129 features and 72 samples, the HEMO-F-score method could achieve the highest values for the classifier accuracy rate, geometric mean (GM), and the Matthews correlation coefficient (MCC). The HEMO-FCBF method has achieved the highest value for the specificity and Gmean.

The prostate data with 10,509 features and 102 samples is one of the hardest microarray datasets because its test dataset has been extracted from several different tests and has the dataset shift problem. In this data, the HEMO-FCBF method could achieve the highest values for the classifier accuracy rate, the Matthews correlation coefficient (MCC), specificity, geometric mean (Gmean), and geometric mean (GM). The HEMO-Relief method has achieved the highest value for sensitivity.

In the Lung_Cancer data with 12,533 features and 181 samples, the HEMO-FCBF method has managed to obtain the best results for the criteria of classifier accuracy, sensitivity, specificity, geometric mean, and GM. The HEMO-F-score method has achieved the highest value for the Matthews correlation coefficient.

For ovarian data with 15,154 features and 253 samples, which has the largest number of features among the data investigated, the HEMO-FCBF method has managed to obtain the score of 1.00 for the criteria of classifier accuracy, sensitivity, specificity, geometric mean, and the Matthews correlation coefficient (MCC). For the geometric mean, GM, the value of the method applied to this data is 0.999, which is the best result compared to other methods applied to this data. The average values obtained from the five criteria mentioned in Table 5.8 also indicate that the HEMO-FCBF method has managed to gain optimum dominance in the five datasets.

According to the results, the FCBF filter method with OR integration method has produced the best results.

Fig. 5.8 Execution time of
different hybrid-ensemble
methods

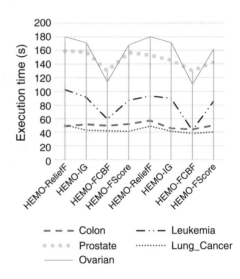

5.9.5.1.2   Analyze the Execution Time

Another important criterion which is very important when it comes to comparing different methods is the running time of an algorithm. The lower the computational complexity of an algorithm, the lower its running time.

Unlike normal data, the attention to the problem of the running time is more critical in high-dimensional data. That is because if a method is highly complex, it may take even months and years to run the algorithm on significantly high-dimensional data. As a result, one of the criteria that should be considered when selecting an appropriate method is the running time of the algorithm.

In Fig. 5.8, these methods have also been compared in terms of running time. As it can be observed, the HEMO-FCBF and HEMA-FCBF have lower running times compared to other methods.

Finally, the hybrid-ensemble methods have been employed to evaluate this sample.

In the reference mentioned, this method has been compared with several hybrid-ensemble, filter, and metaheuristic methods and obtained acceptable results. Furthermore, the method has achieved good results regarding the data like Lung_Cancer, which suffers from high time complexity due to high dimensionality.

5.9.5.2   Hybrid-Ensemble 2

Reference [42] has discussed the issue of dimensionality reduction in high-dimensional microarray data by offering a framework derived from hybrid-ensemble methods. As mentioned previously, in this method, each line initially reduces the dimensionality of the high-dimensional data. Each filter method gives its selected

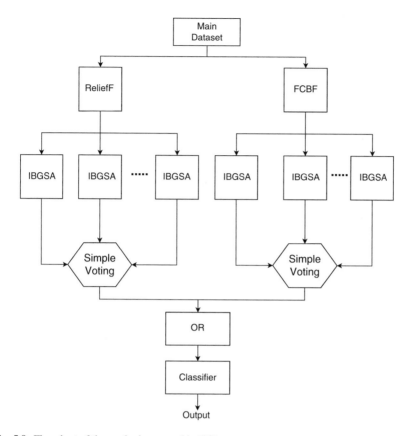

Fig. 5.9 Flowchart of the method proposed in [38]

features as the output, in which the data dimensionality is drastically reduced. Then, the desirable features with optimal classification accuracy will be selected using metaheuristic methods. In the next stage, the results of feature selection in each line will be combined using different methods such as the AND and OR logical operators. Finally, the accuracy of the proposed method is calculated by the classifier. The authors' proposed framework is shown in Fig. 5.6.

In [42] it is noted that after the investigation of several filter methods, the FCBF and ReliefF methods have been selected as optimal filter methods and the IBGSA algorithm has been used as the metaheuristic method. The final flowchart is shown in Fig. 5.9.

The simple voting algorithm performance in feature selection is as follows:

Consider that the algorithm repeats for *n* times, if the number of times in which the feature is selected as the desired feature is more than the times in which it is not selected, the result of this feature will be selected as the selected feature; otherwise, it will not be selected as the selected feature.

5.9.5.2.1   Find the Optimal Threshold

As already mentioned, the ranking filter methods like the ReliefF method require a threshold to select the number of features. In this work, different thresholds to achieve the optimal threshold for the Relief algorithm have been investigated, whose results are mentioned using decision-tree, SVM, and KNN classifiers in Table 5.9. In this table, the Th_1, Th_2, and Th_3 values are considered 0.0066, 0.009, and 0.02, respectively. The Thf is also equal to the number of features selected by the FCBF algorithm.

Table 5.9 shows the error rate of the classifiers as well as the number of features selected by each of the hybrid-ensemble methods. As can be seen, the investigated method using the KNN classifier yields better results than two other classifiers. By reviewing the four scenarios listed in the table, it can be found that scenario 1 and scenario 2 have the least selected features as well as the highest classification accuracy compared to scenario 3 and scenario 4.

Based on the results, with the increase of the threshold, the accuracy of the classifier is reduced, which indicates the existence of a large number of irrelevant and redundant features in the high-dimension data.

In the mentioned reference, after selecting scenario 1 and scenario 2 as optimal cases, these methods are compared with other ensemble methods such as E1-sv, E1-cp, E1-ni, E1-ns, and E2. The hybrid-ensemble framework proposed in [42] using the mentioned algorithms as well as the two thresholds mentioned, achieved better classification accuracy than the other five methods. Besides, regarding the comparison of the number of selected features, scenario1 and scenario 2 have slightly much fewer features than the other five methods.

In short, we can say that the hybrid-ensemble method presented in reference [42], with a lesser number of selected features, has managed to obtain a more desirable classifier accuracy.

5.10   Summary

In an era in which the size of data is growing significantly, feature selection in high-dimensional data is one of the most challenging subjects in the fields of medicine, biology, and bioinformatics.

With the emergence of medicine data including microarray data, reducing data dimension and selecting effective features have become necessary. Microarray data are high-dimensional data with a few numbers of samples which has become one of the challenges in informatics and machine vision. Thus, feature selection in such data is one of the most difficult and most important processes which has attracted many researchers.

In this chapter, we first looked at the challenges of high-dimensional data including their large number of features, small number of samples, class imbalance, and label noise.

Table 5.9 Experimental results after performing hybrid-ensemble method proposed in [42] to find the optimal threshold for ReliefF algorithm

	Threshold	Classifier	Colon	Leukemia	SRBCT	Lung cancer	Ovarian
Scenario 1	Th_f	DT	0.2333 (11.1)	0.292 (9)	0.1714 (54)	0.1456 (313)	0.286 (27.6)
		KNN	0.1095 (11.1)	0.022 (9)	0.0107 (54)	0.0221 (313)	0.00 (27.6)
		SVM	0.1667 (11.1)	0.0345 (9)	0.00 (54)	0.0309 (313)	0.00 (27.6)
Scenario 2	Th_1	DT	0.2333 (10.8)	0.0458 (24.4)	0.1607 (37.1)	0.1588 (219.5)	0.0262 (56.8)
		KNN	0.1286 (10.8)	00042 (24.4)	0.00 (37.1)	0.0279 (219.5)	0.00 (56.8)
		SVM	0.1429 (10.8)	0.0272 (24.4)	0.00 (37.1)	0.0456 (219.5)	0.0012 (56.8)
Scenario 3	Th_2	DT	0.2476 (12.7)	0.0375 (31.8)	0.1464 (37.9)	0.1553 (227.9)	0.0286 (77.5)
		KNN	0.1429 (12.7)	0.0046 (31.8)	0.0143 (37.9)	0.0176 (227.9)	0.00 (77.5)
		SVM	0.1524 (12.7)	0.025 (31.8)	0.00 (37.9)	0.0368 (227.9)	0.00 (77.5)
Scenario 4	Th_3	DT	0.2095 (21.6)	0.0167 (68.4)	0.1459 (68.1)	0.1618 (296.1)	0.024 (156.8)
		KNN	0.1381 (21.6)	0.017 (68.4)	0.00 (68.1)	0.0294 (296.1)	0.00 (156.8)
		SVM	0.1571 (21.6)	0.00 (68.4)	0.00 (68.1)	0.0529 (296.1)	0.00 (156.8)

The classification error rates are reported in this table. The number in parenthesis is the number of features selected by the method

Then, filter methods were described and evaluated. These methods are of high speed and do the process of selecting the feature, regardless of the learning algorithm, and only with respect to the inherent characteristics of the data.

In the next three sections, wrapper, hybrid, and embedded methods are discussed. Wrapper methods often have higher classification accuracy compared to filtering methods, but due to their high computational complexity, they are usually not directly applied to high-dimensional data. In the meantime, hybrid methods can be a good solution for working with high-dimensional data. In these methods, if suitable filter and wrapper techniques are used, optimal solutions can be found for choosing features in high-dimensional data.

In the next section, after a review of ensemble and hybrid-ensemble techniques, several hybrid-ensemble methods have been investigated.

In the next section, the results of applying several filter and hybrid-ensemble methods on several high-dimensional data are discussed. First, by applying nine effective filter methods to eight high-dimension data, the results were presented. Also, two classifiers were selected to measure the effectiveness of the selected features. Finally, the evaluation of the results of the two recently proposed hybrid-ensemble methods has been presented on several microarray data.

References

1. C.E. Crangle, R. Wang, M. Perreau-Guimaraes, M.U. Nguyen, D.T. Nguyen, P. Suppes, Machine learning for the recognition of emotion in the speech of couples in psychotherapy using the Stanford Suppes Brain Lab Psychotherapy Dataset. arXiv preprint arXiv:1901.04110 (2019)
2. A. Rouhi, M. Spitale, F. Catania, G. Cosentino, M. Gelsomini, F. Garzotto, Emotify: emotional game for children with autism spectrum disorder based-on machine learning, in *Proceedings of the 24th International Conference on Intelligent User Interfaces: Companion* (ACM, New York, 2019), pp. 31–32
3. U. Shruthi, V. Nagaveni, B. Raghavendra, A review on machine learning classification techniques for plant disease detection, in *2019 5th International Conference on Advanced Computing & Communication Systems (ICACCS)*, (IEEE, Piscataway, 2019), pp. 281–284
4. R.O. Duda, P.E. Hart, D.G. Stork, *Pattern Classification* (Wiley, Hoboken, 2012)
5. M. Fernandes, A. Canito, V. Bolón-Canedo, L. Conceição, I. Praça, G. Marreiros, Data analysis and feature selection for predictive maintenance: A case-study in the metallurgic industry. Int. J. Inf. Manag. **46**, 252–262 (2019)
6. H. Liu, H. Motoda, *Feature Selection for Knowledge Discovery and Data Mining* (Springer, Berlin, 2012)
7. H. Handels, T. Roß, J. Kreusch, H.H. Wolff, S.J. Poeppl, Feature selection for optimized skin tumor recognition using genetic algorithms. Artif. Intell. Med. **16**(3), 283–297 (1999)
8. B. Nikpour, H. Nezamabadi-pour, HTSS: a hyper-heuristic training set selection method for imbalanced data sets. Iran J. Comput. Sci. **1**(2), 109–128 (2018)
9. K. Borowska, J. Stepaniuk, A rough–granular approach to the imbalanced data classification problem. Appl. Soft Comput. **83**, 105607 (2019)

10. A. Reyes-Nava, H. Cruz-Reyes, R. Alejo, E. Rendón-Lara, A. Flores-Fuentes, and E. Granda-Gutiérrez, Using deep learning to classify class imbalanced gene-expression microarrays datasets, in *Iberoamerican Congress on Pattern Recognition* (Springer, Berlin, 2018), pp. 46–54
11. P.B. andLuis Torgo, R. Ribeiro, A survey of predictive modeling under imbalanced distributions. ACM Comput. Surv. **49**(2), 1–31 (2016)
12. H. He, E.A. Garcia, Learning from imbalanced data. IEEE Trans. Knowl. Data Eng. **9**, 1263–1284 (2008)
13. J. Błaszczyński, J. Stefanowski, Improving bagging ensembles for class imbalanced data by active learning, in *Advances in Feature Selection for Data and Pattern Recognition*, (Springer, Berlin, 2018), pp. 25–52
14. R.J. Hickey, Noise modelling and evaluating learning from examples. Artif. Intell. **82**(1–2), 157–179 (1996)
15. Z. Sun, Q. Song, X. Zhu, H. Sun, B. Xu, Y. Zhou, A novel ensemble method for classifying imbalanced data. Pattern Recogn. **48**(5), 1623–1637 (2015)
16. C.E. Brodley, M.A. Friedl, Identifying mislabeled training data. J. Artif. Intell. Res. **11**, 131–167 (1999)
17. B. Frénay, A. Kabán, A comprehensive introduction to label noise, in *ESANN* (2014)
18. F. Barani, M. Mirhosseini, H. Nezamabadi-Pour, Application of binary quantum-inspired gravitational search algorithm in feature subset selection. Appl. Intell. **47**(2), 304–318 (2017)
19. A.P. Dawid, A.M. Skene, Maximum likelihood estimation of observer error-rates using the EM algorithm. J. R. Stat. Soc. Ser. C Appl. Stat. **28**(1), 20–28 (1979)
20. T.R. Golub et al., Molecular classification of cancer: class discovery and class prediction by gene expression monitoring. Science **286**(5439), 531–537 (1999)
21. I. Kamkar, S.K. Gupta, D. Phung, S. Venkatesh, Stable feature selection for clinical prediction: exploiting ICD tree structure using tree-lasso. J. Biomed. Inform. **53**, 277–290 (2015)
22. A. Rouhi and H. Nezamabadi-Pour, A hybrid feature selection approach based on ensemble method for high-dimensional data, in *2017 2nd Conference on Swarm Intelligence and Evolutionary Computation (CSIEC)* (IEEE, Piscataway, 2017), pp. 16–20
23. S. Tabakhi, A. Najafi, R. Ranjbar, P. Moradi, Gene selection for microarray data classification using a novel ant colony optimization. Neurocomputing **168**, 1024–1036 (2015)
24. M.K. Ebrahimpour, H. Nezamabadi-Pour, M. Eftekhari, CCFS: a cooperating coevolution technique for large scale feature selection on microarray datasets. Comput. Biol. Chem. **73**, 171–178 (2018)
25. A. Rouhi and H. Nezamabadi-Pour, Filter-based feature selection for microarray data using improved binary gravitational search algorithm, in *2018 3rd Conference on Swarm Intelligence and Evolutionary Computation (CSIEC)* (IEEE, Piscataway, 2018), pp. 1–6
26. J.R. Quinlan, Induction of decision trees. Mach. Learn. **1**(1), 81–106 (1986)
27. Y.-W. Chen, C.-J. Lin, Combining SVMs with various feature selection strategies, in *Feature Extraction*, (Springer, Berlin, 2006), pp. 315–324
28. Q. Gu, Z. Li, J. Han, Generalized fisher score for feature selection. arXiv preprint arXiv:1202.3725, 2012
29. I. Kononenko, Estimating attributes: analysis and extensions of RELIEF, in *European Conference on Machine Learning* (Springer, Berlin, 1994), pp. 171–182
30. L. Yu, H. Liu, Feature selection for high-dimensional data: a fast correlation-based filter solution, in *Proceedings of the 20th International Conference on Machine Learning (ICML-03)* (2003), pp. 856–863
31. H. Peng, F. Long, C. Ding, Feature selection based on mutual information: criteria of max-dependency, max-relevance, and min-redundancy. IEEE Trans. Pattern Anal. Mach. Intell. **8**, 1226–1238 (2005)
32. M. A. Hall, Correlation-based feature selection for machine learning (1999)
33. J. Li et al., Feature selection: a data perspective. ACM Comput. Sur. (CSUR) **50**(6), 94 (2018)

34. A. Rouhi and H. Nezamabadi-Pour, A hybrid method for dimensionality reduction in microarray data based on advanced binary ant colony algorithm, in *2016 1st Conference on Swarm Intelligence and Evolutionary Computation (CSIEC)* (IEEE, Piscataway, 2016), pp. 70–75
35. N. Taheri, H. Nezamabadi-Pour, A hybrid feature selection method for high-dimensional data, in *2014 4th International Conference on Computer and Knowledge Engineering (ICCKE)* (IEEE, Piscataway, 2014), pp. 141–145
36. X. He, D. Cai, P. Niyogi, Laplacian score for feature selection, in *Advances in Neural Information Processing Systems*, (ACM, New York, 2006), pp. 507–514
37. M.A. Hall, L.A. Smith, Feature selection for machine learning: comparing a correlation-based filter approach to the wrapper, in *FLAIRS Conference,* vol. 1999 (1999), pp. 235–239
38. W.H. Press, S.A. Teukolsky, W.T. Vetterling, B.P. Flannery, Numerical recipes in C++. Art Sci. Comput. **2**, 1002 (1992)
39. J.C. Davis, R.J. Sampson, *Statistics and Data Analysis in Geology* (Wiley, New York, 1986)
40. H. Lee et al., Feature selection practice for unsupervised learning of credit card fraud detection. J. Theor. Appl. Inf. Technol. **96**(2), 408–417 (2018)
41. Y. Saeys, I. Inza, P. Larrañaga, A review of feature selection techniques in bioinformatics. Bioinformatics **23**(19), 2507–2517 (2007)
42. A. Rouhi, H. Nezamabadi-pour, A hybrid-ensemble based framework for microarray data gene selection. Int. J. Data Min. Bioinform. **19**(3), 221–242 (2017)
43. S. Kashef, H. Nezamabadi-pour, B. Nikpour, Multilabel feature selection: a comprehensive review and guiding experiments. Wiley Interdiscip. Rev. Data Min. Knowl. Disc. **8**(2), e1240 (2018)
44. M. Dowlatshahi, V. Derhami, H. Nezamabadi-Pour, Ensemble of filter-based rankers to guide an epsilon-greedy swarm optimizer for high-dimensional feature subset selection. Information **8**(4), 152 (2017)
45. M. Dorigo, G. di Caro, Ant colony optimization: a new meta-heuristic, in *Proceedings of the 1999 Congress on Evolutionary Computation-CEC99 (Cat. No. 99TH8406)*, vol. 2 (IEEE, Piscataway, 1999), pp. 1470–1477
46. S. Kashef, H. Nezamabadi-pour, An advanced ACO algorithm for feature subset selection. Neurocomputing **147**, 271–279 (2015)
47. J. Kennedy, Particle swarm optimization. Enc. Mach. Learn., 760–766 (2010)
48. E. Rashedi, H. Nezamabadi-Pour, S. Saryazdi, GSA: a gravitational search algorithm. Inf. Sci. **179**(13), 2232–2248 (2009)
49. A. Mahanipour, H. Nezamabadi-Pour, A multiple feature construction method based on gravitational search algorithm. Expert Syst. Appl. **127**, 199–209 (2019)
50. E. Rashedi, H. Nezamabadi-Pour, S. Saryazdi, BGSA: binary gravitational search algorithm. Nat. Comput. **9**(3), 727–745 (2010)
51. E. Rashedi, H. Nezamabadi-pour, Feature subset selection using improved binary gravitational search algorithm. J. Intell. Fuzzy Syst. **26**(3), 1211–1221 (2014)
52. A. Rouhi, P.H. Nezamabadi, A Hybrid-Based Feature Selection Method for High-Dimensional Data Using Ensemble Methods (2018)
53. V. Bolón-Canedo, N. Sánchez-Marono, A. Alonso-Betanzos, J.M. Benítez, F. Herrera, A review of microarray datasets and applied feature selection methods. Inf. Sci. **282**, 111–135 (2014)
54. P.A. Mundra, J.C. Rajapakse, SVM-RFE with MRMR filter for gene selection. IEEE Trans. Nanobioscience **9**(1), 31–37 (2009)
55. H. Uğuz, A two-stage feature selection method for text categorization by using information gain, principal component analysis and genetic algorithm. Knowl.-Based Syst. **24**(7), 1024–1032 (2011)
56. L.-Y. Chuang, C.-H. Yang, K.-C. Wu, C.-H. Yang, A hybrid feature selection method for DNA microarray data. Comput. Biol. Med. **41**(4), 228–237 (2011)
57. C.-P. Lee, Y. Leu, A novel hybrid feature selection method for microarray data analysis. Appl. Soft Comput. **11**(1), 208–213 (2011)

58. S.S. Shreem, S. Abdullah, M.Z.A. Nazri, M. Alzaqebah, Hybridizing ReliefF, MRMR filters and GA wrapper approaches for gene selection. J. Theor. Appl. Inf. Technol. **46**(2), 1034–1039 (2012)
59. J. Apolloni, G. Leguizamón, E. Alba, Two hybrid wrapper-filter feature selection algorithms applied to high-dimensional microarray experiments. Appl. Soft Comput. **38**, 922–932 (2016)
60. B. Venkatesh, J. Anuradha, A hybrid feature selection approach for handling a high-dimensional data, in *Innovations in Computer Science and Engineering*, (Springer, Berlin, 2019), pp. 365–373
61. Z. Manbari, F. AkhlaghianTab, C. Salavati, Hybrid fast unsupervised feature selection for high-dimensional data. Expert Syst. Appl. **124**, 97–118 (2019)
62. C. Yan, J. Liang, M. Zhao, X. Zhang, T. Zhang, H. Li, A novel hybrid feature selection strategy in quantitative analysis of laser-induced breakdown spectroscopy. Anal. Chim. Acta **1080**, 35–42 (2019)
63. T. Gangavarapu, N. Patil, A novel filter-wrapper hybrid greedy ensemble approach optimized using the genetic algorithm to reduce the dimensionality of high-dimensional biomedical datasets. Appl. Soft Comput. **81**, 105538 (2019)
64. L. Sun, X. Kong, J. Xu, R. Zhai, S. Zhang, A hybrid gene selection method based on ReliefF and ant colony optimization algorithm for tumor classification. Sci. Rep. **9**(1), 8978 (2019)
65. W. You, Z. Yang, G. Ji, PLS-based recursive feature elimination for high-dimensional small sample. Knowl.-Based Syst. **55**, 15–28 (2014)
66. T. Prasartvit, A. Banharnsakun, B. Kaewkamnerdpong, T. Achalakul, Reducing bioinformatics data dimension with ABC-kNN. Neurocomputing **116**, 367–381 (2013)
67. I. Guyon, J. Weston, S. Barnhill, V. Vapnik, Gene selection for cancer classification using support vector machines. Mach. Learn. **46**(1–3), 389–422 (2002)
68. S. Maldonado, R. Weber, J. Basak, Simultaneous feature selection and classification using kernel-penalized support vector machines. Inf. Sci. **181**(1), 115–128 (2011)
69. J. Canul-Reich, L.O. Hall, D.B. Goldgof, J.N. Korecki, S. Eschrich, Iterative feature perturbation as a gene selector for microarray data. Int. J. Pattern Recognit. Artif. Intell. **26**(05), 1260003 (2012)
70. S. Maldonado, J. López, Dealing with high-dimensional class-imbalanced datasets: embedded feature selection for SVM classification. Appl. Soft Comput. **67**, 94–105 (2018)
71. H. Liu, M. Zhou, Q. Liu, An embedded feature selection method for imbalanced data classification. IEEE/CAA J. Autom. Sin. **6**(3), 703–715 (2019)
72. C. Peng, X. Wu, W. Yuan, X. Zhang, Y. Li, MGRFE: multilayer recursive feature elimination based on an embedded genetic algorithm for cancer classification. IEEE/ACM Trans. Comput. Biol. Bioinform. (2019). https://doi.org/10.1109/TCBB.2019.2921961
73. A.B. Brahim, M. Limam, Robust ensemble feature selection for high dimensional data sets, in *2013 International Conference on High Performance Computing & Simulation (HPCS)* (IEEE, Piscataway, 2013), pp. 151–157
74. V. Bolón-Canedo, N. Sánchez-Marono, A. Alonso-Betanzos, Data classification using an ensemble of filters. Neurocomputing **135**, 13–20 (2014)
75. F. Yang, K. Mao, Robust feature selection for microarray data based on multicriterion fusion. IEEE/ACM Trans. Comput. Biol. Bioinform. **8**(4), 1080–1092 (2010)
76. V. Bolón-Canedo, N. Sánchez-Maroño, A. Alonso-Betanzos, An ensemble of filters and classifiers for microarray data classification. Pattern Recogn. **45**(1), 531–539 (2012)
77. S. Sayed, M. Nassef, A. Badr, I. Farag, A nested genetic algorithm for feature selection in high-dimensional cancer microarray datasets. Expert Syst. Appl. **121**, 233–243 (2019)
78. B. Pes, Ensemble feature selection for high-dimensional data: a stability analysis across multiple domains. Neural Comput. Applic., 1–23 (2019)
79. B. Singh, K. Kumar, S. Mohan, R. Ahmad, Ensemble of clustering approaches for feature selection of high dimensional data. Available at SSRN 3349018 (2019)
80. J. Wang, J. Xu, C. Zhao, Y. Peng, H. Wang, An ensemble feature selection method for high-dimensional data based on sort aggregation. Syst. Sci. Control Eng. **7**(2), 32–39 (2019)

81. X. Song, L.R. Waitman, Y. Hu, A.S. Yu, D. Robins, M. Liu, Robust clinical marker identification for diabetic kidney disease with ensemble feature selection. J. Am. Med. Inform. Assoc. **26**(3), 242–253 (2019)
82. V.P. Singh, D.J. Kalita, S. Tripathi, Classifying gene expression data of cancer using multistage ensemble of neural networks. Available at SSRN 3349578 (2019)
83. Feature Selection at Arizona State University. http://featureselection.asu.edu/datasets.php
84. B. Institute. *Cancer Program Data Sets*. http://www.broadinstitute.org/cgi-bin/cancer/datasets.cgi

Chapter 6
An Introduction to Advanced Machine Learning: Meta-Learning Algorithms, Applications, and Promises

Farid Ghareh Mohammadi, M. Hadi Amini, and Hamid R. Arabnia

Abstract In Chaps. 3 and 4, we have explored the theoretical aspects of feature extraction optimization processes for solving large-scale problems and overcoming machine learning limitations. Majority of optimization algorithms that have been introduced in Mohammadi et al. (Evolutionary computation, optimization and learning algorithms for data science, 2019. arXiv preprint arXiv: 1908.08006; Applications of nature-i nspired algorithms for dimension reduction: enabling efficient data analytics, 2019. arXiv preprint arXiv: 1908.08563) guarantee the optimal performance of supervised learning, given offline and discrete data, to deal with curse of dimensionality (CoD) problem. These algorithms, however, are not tailored for solving emerging learning problems. One of the important issues caused by online data is lack of sufficient samples per class. Further, traditional machine learning algorithms cannot achieve accurate training based on limited distributed data, as data has proliferated and dispersed significantly. Machine learning employs a strict model or embedded engine to train and predict which still fails to learn unseen classes and sufficiently use online data. In this chapter, we introduce these challenges elaborately. We further investigate meta-learning (MTL) algorithm, and their application and promises to solve the emerging problems by answering *how autonomous agents can learn to learn?*

F. G. Mohammadi · H. R. Arabnia
Department of Computer Science, Franklin College of Arts and Sciences, University of Georgia, Athens, GA, USA
e-mail: farid.ghm@uga.edu; hra@cs.uga.edu

M. H. Amini (✉)
School of Computing and Information Sciences, Florida International University, Miami, FL, USA

Sustainability, Optimization, and Learning for InterDependent Networks Laboratory (solid lab), Florida International University, Miami, FL, USA
e-mail: amini@cs.fiu.edu; hadi.amini@ieee.org; www.solidlab.network

© Springer Nature Switzerland AG 2020
M. H. Amini (ed.), *Optimization, Learning, and Control for Interdependent Complex Networks*, Advances in Intelligent Systems and Computing 1123, https://doi.org/10.1007/978-3-030-34094-0_6

Keywords Meta-learning · Machine learning · Online learning · Online optimization · Model-based learning · Metric-based learning · Gradient descent · Low-shot learning · Few-shot learning · One-shot learning

6.1 Introduction

Machine learning algorithms enable researchers to learn from supervised/unsupervised data. Collected data is mainly offline and it is not evolving over time. Hence, behavior of the ever-increasing data sets introduces uncertainty to data analytic solutions and methods. Conventionally, it is not possible to have the entire behavior learned [1] using the traditional machine learning, evolutionary algorithms, and optimization algorithms discussed earlier [2, 3].

Last decade, researchers have studied advanced research paradigms to solve optimization and learning problems efficiently. In this context, distributed optimization and learning algorithms lend themselves as promising solutions to deal with large scale nature of data, information privacy, scalability, as well as (near) real-time decision making capability; applications of such algorithms include optimal operation of smart city infrastructures, interdependent power and transportation networks [4–6], artificial intelligence for energy system resilience [7], energy management and optimal power flow problem [8, 9], and learning at the IoT device level [10]. They aim to learn using prior tasks or experiences and leverage them for future learning. One of the promising paradigms is meta-learning (MTL). Prior studies investigated MTL methods that learn to update a function or learning rule [1, 11]. MTL differs from classic machine learning with respect to the level of adaptation [12]. MTL is the process of learning to learn. It leverages past experiences to ascertain a prior model's parameters and learning process, i.e., algorithm. MTL investigates how to choose the right bias non-fixed, unlike base-learning where the bias is fixed a priori [12]. Concretely, MTL studies a setting where a set of tasks ($\mathcal{T}_i$) are made available together upfront. However, it cannot handle sequential and dynamic aspects of problems properly. To make it easy for readers to understand this study, we provide Table 6.1, abbreviation of words have been used during this study.

In contrast, online learning is the process of learning sequentially, however, it does not leverage past experiences like MTL, i.e., it may not consider how past experience can help to enhance the adaptation to a new task. The earliest research studies introduced sequential learning [13, 14] where tasks are revealed one after another repeatedly. The aim of learning is to learn as independent as possible to attain zero-shot learning with non-task-specific adaptation. We argue that neither setting is ideal for studying continual lifelong learning. MTL deals with learning to learn, but neglects the sequential and non-stationary aspects of the problem. Online learning offers an appealing theoretical framework, but does not generally consider how past experience can accelerate adaptation to a new task. In this work, we motivate and present the online MTL problem setting, where the agent

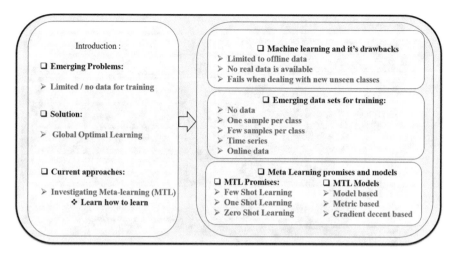

Fig. 6.1 Overall structure of this study

simultaneously uses past experiences in a sequential setting to learn good priors, and also adapt quickly to the current task at hand.

The rest of this chapter is organized as follows. In Sect. 6.2 emerging challenges in machine learning are discussed . After that, applications of MTL using transfer learning are covered. Finally we have promises of MTL. Figure 6.1 represents the overall structure of this study.

6.2   Machine Learning: Challenges and Drawbacks

Prior works on learning process, regression, and optimization problems have attempted to learn the behavior of input data, analyze and categorize it to attain high performance algorithms. Machine learning (ML) has been strongly applied to solve supervised and unsupervised problems. ML deploys different algorithms, such as online learning, multi-task learning, and supervised algorithms, including rule based [15, 16], function based [17, 18], lazy [19], and bootstrap [20]. Some of them are used to transform data, special example would be dimension reduction for optimization, some to build classifiers like supervised algorithms, others for prediction like regression, etc. Machine learning still yields subtle drawbacks for time-varying input data which restrict it to consider properly future and unseen classes to provide general idea and knowledge from data.

Traditionally, machine learning is a structured process that learns from training datasets and examine the learned model on test datasets. The learned model follows the rule of the equation $\mathcal{P}_i \times \mathcal{D} \longrightarrow \mathcal{M}$, where $\mathcal{P}_i$ stands for the specific supervised algorithm parameters, $\mathcal{D}$ represents the space of training data distribution, and $\mathcal{M}$ defines the space of generated models which will be applied on test data to evaluate the supervised algorithm performance.

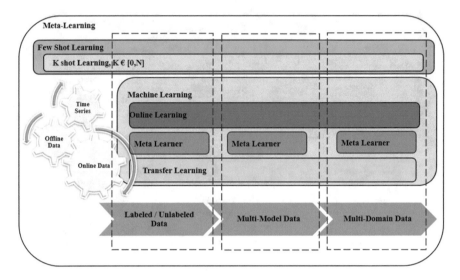

Fig. 6.2 The general overview of learning against emerging data

Figure 6.2 presents few machine learning approaches and algorithms which provide different applications with respect to the wide variety of data such as offline data vs. online data, labeled data vs. unlabeled data, multi-model data vs. single model data, and multi-domain data vs. single domain data. As it shows, machine learning has critical drawbacks which cannot handle whole data once. Moreover, it just considers each data as a new model and each model is separate from previous ones.

Furthermore, Fig. 6.2 depicts the relationship between traditional machine learning and advance machine learning. In traditional machine learning we have to deal with offline and limited amount of data and the number ground-truth. However, in the world of technology, where data growth has proliferated significantly and is coming from wherever technology exists, it is very critical to get to know the pattern and rules that govern whole data and learn the trend of the generated data for a specific domain. For that end, we need to classify data into three big categories, time series data, offline data, and online data. These all categories are shown in three different aspects: supervised and unsupervised; multi-model data; and multi-domain data.

Machine learning involves transfer learning and online learning, which is compatible to learn tasks and classes, which are consequential. Transfer learning is the theory of transferring knowledge from one task to another and learning from non-randomness. Meta-learner also is one of the bootstrap algorithms which learn data by sampling given dataset and generating different datasets. Meta-learner leverages boosting techniques by using several algorithm to make votes, then Meta-learning select the majority of the vote as final decision.

6.3   Meta-Learning Algorithms

Meta-learning (MTL) is firstly presented in [1] and [21]. After a decade gap, lately research studies have tried to deploy MTL again. MTL is a machine which learns the variety of input data. ML methods need to learn new tasks faster by leveraging previous experiences. MTL does not consider past experiences separately.MTL is the process of learning how to learn. MTL is an emerging learning algorithm with new challenges and research questions. It is an extension of transfer learning, which is one of the multi-task learning algorithms. MTL covers three different aspects as illustrated in Fig. 6.5: few-shot learning (FSL), one-shot learning (OSL), and zero-shot learning (ZSL). FSL and OSL yield highly accurate results as compared with traditional machine learning algorithms. However, they still have a critical challenge which limits them from converging to optimal results. Limits of ZSL have been addressed using domain semantic space, where includes all information system as presented in Fig. 6.5.

6.3.1   Model-Based MTL

Model-based MTL depends on a model and no conditional probabilistic method which enable it to be the best match for fast learning model where it updates its hyper-parameters so fast by training just few examples. The process of updating their hyper-parameters is done either internal architecture or external meta-learner. The concept of model-based MTL is having one neural network interact with sequential neural networks to accelerate the learning process. In other words, it tries to learn a model per each label using pixel by pixel value, according to Fig. 6.3. In other words, this model's algorithms try to train a recurrent model like the

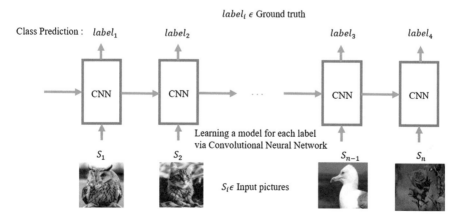

Fig. 6.3 Model-based MTL

work presented [11], which proposed long short-term memory (LSTM). Hochreiter and Schmidhuber [22] proposed for the first time in 1997 the theory of LSTM. Model-based algorithms take the dataset sequentially and analyze instances one by one. Since model-based algorithms leverage RNN for learning, they become the most least efficient model in comparison with other models. Nagabandi et al. [23] proposed online deep learning using MTL towards continual adaptation for model-based reinforcement learning. Santoro et al. [24] proposed memory-augmented neural network using MTL.

6.3.2 Metric-Based Learning

Metric-based learning leverages metric space learning, which leads to efficient data processing and is suitable for few-shot learning. Let us consider that our goal is image classification. As model-based learning tries to learn each image pixel by pixel which takes long time and time consuming, metric-based learning overcomes this limitation by leveraging comparing given two images to the network. The output per each input yields a vector, comparing these two vector states that whether they are similar or not (Fig. 6.4). One of the most common applications of metric-based learning is Siamese network presented in [25]. Koch et al. presented Siamese neural network (SNN) for one-shot learning which achieved strong and better results. The idea behind SNN is that it tries to use twin or half-twin network to compare the input images. Note that one of the input is already computed and we only need to take the second image and try to go through the layers and compute the vector. Then, SNN tries to compute the distance between them, if the result is small they are similar otherwise they are different. Another application of metric-based learning is [26], where Vinyals et al. proposed a matching network(MN) for one-shot learning.

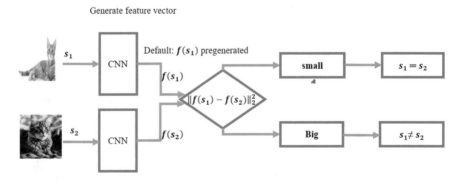

Fig. 6.4 Metric-based MTL

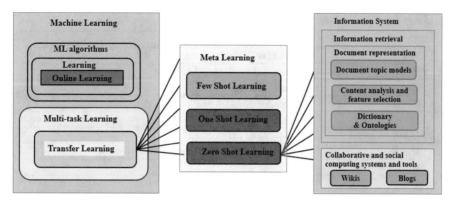

Fig. 6.5 The relation among machine learning, meta-learning, and information system

6.3.3 Gradient Decent-Based Learning

This model of MTL is also known as optimization-based model for tuning the parameter (θ). The idea here is to leverage stochastic gradient-decent (SGD) and for new given sample, it updates the parameters to be a universal learner. It may not converge to a local optimal since it does not rely on small number of samples.

Although gradient-based learning model works good, it still has some drawbacks. Ravi and Larochelle [11] addressed these problems carefully and provided LSTM-based MTL to overcome those problems. Finn [27] presented MAML to improve the accuracy of LSTM-based MTL (Fig. 6.5, Tables 6.1 and 6.2).

6.4 Promises of Meta-Learning

Learning to learn is an advance process which provides three promises: one few-shot learning (FSL), one one-shot learning (OSL), one zero-shot learning (ZSL). Figure 6.6 presents a general view of each promises. we have three layers: input data, meta-training, and meta-testing. Input data for FZL and OSL are the same type, particularly images for particular image classification aims. Further, ZSL becomes an independent learning MTL algorithm which evaluates input data based on domain semantic space and visual information of that domain.

In second layer, FSL tries to learn k-shot tasks, which means MTL is training by leveraging k different training dataset. K-shots had generated in advance before learning process have started. Thus, MTL is known as a certain type of bootstrap algorithms, however, in k-shot dataset we only have specific number of K instances per class. However, the bootstrap algorithm tries to split given dataset with different rate and would be keeping the ratio of number of classes. MTL attempts to calculate

Table 6.1 Abbreviation of words

Abb	Definition
CNN	Convolutional neural network
EGNN	Edge-labeling graph neural network
fSL	Few-shot learning
GSP	Goal-conditioned skill policy
LSL	Low-shot learning
MAML	Model-agnostic MTL
MANN	Memory-augmented neural network
MIL	Meta imitation learning
MTL	Meta-learning
ML	Machine learning
MN	Matching network
OSL	One-shot learning
PN	Prototypical network
RN	Relation network
RNN	Recurrent neural network
SAE	Semantic autoencoder
SNN	Siamese neural network
TL	Transfer learning
ZSL	Zero-shot learning
ZSL-FGVD	ZSL fine-grained visual descriptions
ZSL-H	Zero-shot learning by mitigating the hubness problem
ZSL-KT	Zero-shot learning and knowledge transfer

the loss of each shot using loss function. Furthermore, OSL also ties to learn the task based on $k = 1$ shot learning which means that OSL only has one shot at a moment. In other words, when it starts bootstrapping, it only selects one sample per class as a training set. Note that OSL represents a special kind of K-shot or few-shot learning. Both FSL and OSL use the equation below from [27]. $\mathcal{T}_i = \sum_{i=1}^{k} (\mathcal{L}(x_i, y_i), q(x_i), q(x_{t+1}|x_t, q_t))$

Further, in second lay, ZSL unlike FSL and OSL algorithms ties to work with domain semantic space rather than domain files like images. The goal here is to find an optimal mapping from semantic space to vector space. ZSL tries to map given extracted features to a new space called vector space.

Finally, last layer stands for the meta-testing which is responsible to predict the given test data and analyze them. First two algorithms try to predict unseen data using $f(\theta)$, however, ZSL attempts to solve the problem by mapping the unseen data to the new vector space.

Table 6.2 An overview of previous studies on meta-learning

Paper	Meta-learning	Proposed method	Meta-learning models	Conference/journal	Domain	Year
[27]	Few-shot learning	MAML	Gradient decent based	ICML	Image classification	2017
[28]	Few-shot learning	MAML++	Gradient decent based	ICRL	Image classification	2019
[29]	Few-shot learning	Probabilistic MAML	Gradient decent based	NIPS	Image classification	2018
[30]	Few-shot learning	PN	Metric based	NIPS	Image classification	2017
[31]	Few-shot learning	HML	Model based	arXiv	Image classification	2019
[32]	Few-shot learning	RN	Metric based	CVPR	Image classification	2018
[11]	Few-shot learning	LSTM-Metalearner	Gradient decent based	ICLR	Image classification	2017
[33]	Few-shot learning	EGNN	Model based	CVPR	Image classification	2019
[34]	Few-shot learning	LSL	Metric based	CVPR	Image classification	2019
[31]	One-shot learning	HML	Model based	arXiv	Image classification	2019
[24]	One-shot learning	MANN	Model based	ICML	Image classification	2016
[26]	One-shot learning	MN	Metric based	NIPS	Image classification	2016
[27]	One-shot learning	MAML	Gradient decent based	ICML	Image classification	2017
[35]	One-shot learning	MIL	Gradient decent based	CoRL	Visual imitation	2017
[25]	One-shot learning	MIL	Metric based	ICML	Image recognition	2015
[36]	Zero-shot learning	ZSL-FGVD	Metric based	CVPR	Image classification and retrieval	2016
[37]	Zero-shot learning	ZSL-H	Metric based	ICLR-workshop	Hubness problem	2015
[32]	Zero-shot learning	RN	Metric based	CVPR	Image classification	2018
[38]	Zero-shot learning	SAE	Metric based	CVPR	Image classification	2017
[39]	Zero-shot learning	Meta sensing	Metric based	ACM Buildsys	Device free human sensing	2019
[40]	Zero-shot learning	ZSL-KT	Metric based	arXiv	Music classification	2019
[41]	Zero-shot learning	GSP	Metric based	CVPR-workshop	Visual imitation	2018
[42]	Zero-shot learning	GSP	Metric based	ICLR	Visual imitation	2018

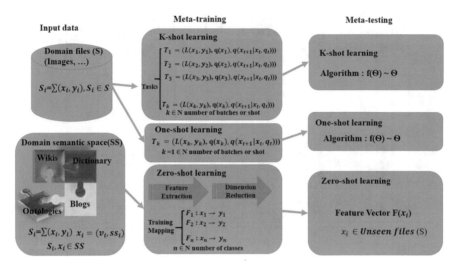

Fig. 6.6 Structure of meta-learning models

6.4.1 Few-Shot Learning

The first and one of the most common promises of MTL is few-shot learning (FSL). Few-shot classification is a specific extension of MTL in supervised learning. Lake et al. [43] challenged traditional machine learning algorithms by enabling them to learn every concept from one or few shot of that dataset. The idea behind that, MTL tries to re-sample the given input dataset for training using only K samples per each class. In other words, meta-training process is accomplished by learning k shot meta sets which are selected by replacement. Although few-shot learning outperforms traditional machine learning algorithms, it has an explanatory challenge, called task ambiguity. This problem happens when a small task is generated from large input dataset, to learn via few-shot learning. After taking a new task, the learned model based on that looks over fit which does not yield a promising result on test datasets.

The majority of MTL algorithms leverage few-shot learning. FSL has decent important extension: Finn et al. proposed model-agnostic MTL (MAML) [27], which adapts to new tasks via gradient descent-based MTL. In [29], Finn et al. re-sampled models for a new task using a model distribution. This paper extends MAML to conduct a parameter distribution that is trained through different lower bound. In [29] Finn et al. addressed the ambiguity problem by proposing probabilistic MAML.

Second important extension is online learning which is learning process of training data sequentially and continuously. The next one is online MTL. Finn et al. [44] proposed online MTL based on the regret-based meta-learner. Kim et al. [33] proposed EGNN, which applied a deep neural network on a certain model, edge-labeling graph. Furthermore, Sun et al. [45] proposed an advanced meta-transfer

$S_i \epsilon$ Data sets of pictures Categories (C) (birds, cows, flowers, cats), $Set_i \subset P(S_i)$

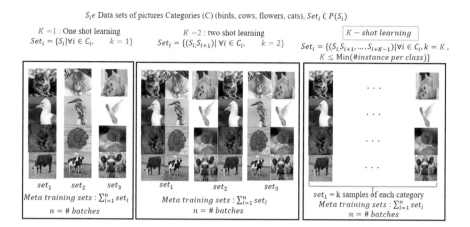

Fig. 6.7 Few-shot learning structure

learning for few-shot learning. Zou and Feng [31] introduced new type of MTL which works based on hierarchy structure, called hierarchical MTL(HML). MTL are limited to the tasks where training datasets where the tasks may different from each other. HML improve MTL result by overcoming the limitation. HML enables MTL to optimize adaptability of meta-model to tasks that are similar. Figure 6.7 provides a general view of few-shot learning, one-shot learning, 2-shot learning, and generalized k-shot learning.

6.4.2   One-Shot Learning

One-shot learning (OSL) is a critical challenge in the applications of deep neural networks. OSL is a special type of few-shot learning or k-shot learning in which it choose $k = 1$ shot for training section. In other words, when the algorithm starts training, they only leverage from one instance per class at a time with different batches. The research studies done for one-shot learning are listed as following: Matching networks [26] which is a metric-based MTL.

6.4.3   Zero-Shot Learning

Zero-shot learning (ZSL) is an emerging paradigm of machine learning which is recently proposed [32, 36, 40] to yield a better result than supervised learning algorithms by covering their critical limitations, which work only with a fixed number of classes. Zero-shot learning is as a joint embedding problem of domain specific and side information, which includes ontology, wikis, dictionary, and blogs.

To be certain, ZSL overcomes few-shot learning and one-shot learning limitation and promised to yield a result better than FSL and OSL. The goal is to classify unseen samples of different classes without having a training dataset. This is possible once you have proper information about the domain and classes, properties, and most importantly the functionality of the problem. The ZSL process is a journey from feature space to a vector space in which it leverages feature extraction and dimension reduction algorithms technically. The feature vector describes shared features among classes. Reed et al. [36] applied neural language model to overcome supervised learning limitation. ZSL has been accomplished for visual recognition [36], music classification [40], and image classification [32]. More recent methods have been proposed by Kodirov et al. [38] using auto-encoders for ZSL, Nagabandi et al. [23] to deploy MTL for online Learning and by Finn et al. [44] for online MTL.

6.5   Discussion

Last 5 years researchers have worked on meta-learning with different promises. We evaluate important research studies and examine them based on different criteria like meta-learning promises, proposed method, meta-learning models, followed by conferences or journal where the papers have been published and domain of the study in Table 6.2. This table illustrates the importance of meta-learning and image classification challenges. Choosing the appropriate type of data for machine learning

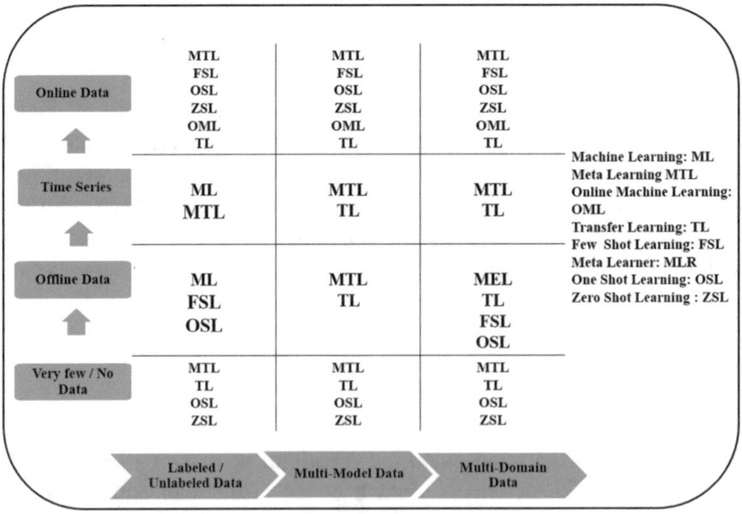

Fig. 6.8 Machine learning: ML, meta-learning: MTL, online machine learning: OML, transfer learning: TL, few-shot learning: FSL, meta-learner: MLR, one-shot learning: OSL

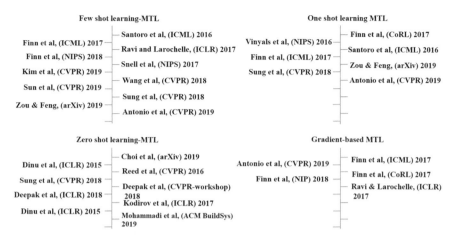

Fig. 6.9 A brief studies over promises of meta-learning

algorithms is an important yet challenging task. According to [46], it is crucial to select an optimal algorithm to solve each specific problem to ensure optimal decision making. They combined experimental result and interviewed with domain experts. It is essential to know where we are, what are the challenges, and what kind of data we have now. Further, what is the relationship among emerging data with respect to traditional and modern machine learning algorithms. Figure 6.8 presents the information to choose which algorithms are suitable, compatible, and applicable given the specific type of data.

In Fig. 6.9, we have identified some of the publications in top venues. According to our investigation, few-shot learning is one of the most promising areas.

6.6   Conclusion

Optimizing algorithms to work with offline data is almost ubiquitous in each domain, such as engineering applications. The majority of studies have determined an optimal way to deal with large-scale problems. Advancing technologies have people provided data available wherever they have access to internet. Thus, it is critical to process continues data which is online and introduce an advance learning algorithm to help scientists to predict future properly. In this chapter, we addressed this problems and investigated an advanced machine learning algorithm to solve them optimally using MTL. Majority of research studies in few shot learning have accomplished recently are categorized in MTL's important promises. One model based, one metric based, one gradient decent based, which also known as optimization method. Further, MTL has three critical extension for emerging data and large-scale problems. The first one, few-shot learning which is practically worked on k-shots of training classes. The second extension is special type of few-shot learning which here we have only one-shot for each training classes. The last one but not the least one is zero-shot learning. Although decent work has been done using FSL and OSL, but ZSL is the promising extension of meta-learning where researchers have no idea about the new classes and no enough data available.

References

1. J. Schmidhuber, Evolutionary principles in self-referential learning, Diploma thesis, 1987
2. F.G. Mohammadi, M.H. Amini, H.R. Arabnia, Evolutionary computation, optimization and learning algorithms for data science (2019). arXiv preprint arXiv: 1908.08006
3. F.G. Mohammadi, M.H. Amini, H.R. Arabnia, Applications of nature-inspired algorithms for dimension reduction: enabling efficient data analytics (2019). arXiv preprint arXiv: 1908.08563
4. M.H. Amini, Distributed computational methods for control and optimization of power distribution networks, PhD Dissertation, Carnegie Mellon University, 2019
5. M.H. Amini, J. Mohammadi, S. Kar, Distributed holistic framework for smart city infrastructures: tale of interdependent electrified transportation network and power grid. IEEE Access **7**, 157535–157554 (2019)
6. M.H. Amini, J. Mohammadi, S. Kar, Distributed intelligent algorithm for interdependent electrified transportation and power networks, *Proceedings of the 9th ACM Symposium on Design and Analysis of Intelligent Vehicular Networks and Applications* (ACM, New York, 2019)
7. A. Imteaj, M.H. Amini, J. Mohammadi, Leveraging decentralized artificial intelligence to enhance resilience of energy networks. arXiv preprint arXiv:1911.07690 (2019)
8. M.H. Amini, B. Nabi, M.R. Haghifam, Load management using multi-agent systems in smart distribution network, in *2013 IEEE Power and Energy Society General Meeting* (IEEE, New York, 2013)
9. M.H. Amini, S. Bahrami, F. Kamyab, S. Mishra, R. Jaddivada, K. Boroojeni, P. Weng, Y. Xu, Decomposition methods for distributed optimal power flow: panorama and case studies of the DC model, in *Classical and recent aspects of power system optimization* (Academic Press, Cambridge, 2018), pp. 137–155
10. A. Imteaj, M.H. Amini, Distributed sensing using smart end-user devices: pathway to federated learning for autonomous IoT, in *2019 International Conference on Computational Science and Computational Intelligence* (Las Vegas, 2019)
11. S. Ravi, H. Larochelle, Optimization as a model for few-shot learning, in *International Conference on Learning Representations (ICLR)* (2017), pp. 281–288
12. R. Vilalta, Y. Drissi, A perspective view and survey of meta-learning. Artif. Intell. Rev. **18**(2), 77–95 (2002)
13. J. Hannan, Approximation to Bayes risk in repeated play. Contrib. Theory Games **3**, 97–139 (1957)
14. N. Cesa-Bianchi, G. Lugosi, *Prediction, Learning, and Games* (Cambridge University Press, Cambridge, 2006)
15. S.M. Weiss, N. Indurkhya, Rule-based machine learning methods for functional prediction. J. Artif. Intell. Res. **3**, 383–403 (1995)
16. J.M. Banda, M. Seneviratne, T. Hernandez-Boussard, N.H. Shah, Advances in electronic phenotyping: from rule-based definitions to machine learning models. Annu. Rev. Biomed. Data Sci. **1**, 53–68 (2018)
17. M. Chen, U. Challita, W. Saad, C. Yin, M. Debbah, Artificial neural networks-based machine learning for wireless networks: a tutorial. IEEE Commun. Surv. Tutorials (2019). https://ieeexplore.ieee.org/abstract/document/8755300/

18. A. Iranmehr, H. Masnadi-Shirazi, N. Vasconcelos, Cost-sensitive support vector machines. Neurocomputing **343**, 50–64 (2019)
19. R. Agrawal, Integrated parallel k-nearest neighbor algorithm, in *Smart Intelligent Computing and Applications* (Springer, Berlin, 2019), pp. 479–486
20. D. Poland, S. Rychkov, A. Vichi, The conformal bootstrap: theory, numerical techniques, and applications. Rev. Mod. Phys. **91**(1), 015002 (2019)
21. Y. Bengio, S. Bengio, J. Cloutier, *Learning a Synaptic Learning Rule* (Université de Montréal, Département d'informatique et de recherche . . . , Montréal, 1990)
22. S. Hochreiter, J. Schmidhuber, Long short-term memory. Neural Comput. **9**(8), 1735–1780 (1997)
23. A. Nagabandi, C. Finn, S. Levine, Deep online learning via meta-learning: continual adaptation for model-based RL (2018). arXiv preprint arXiv:1812.07671
24. A. Santoro, S. Bartunov, M. Botvinick, D. Wierstra, T. Lillicrap, Meta-learning with memory-augmented neural networks, in *International Conference on Machine Learning* (2016), pp. 1842–1850
25. G. Koch, R. Zemel, R. Salakhutdinov, Siamese neural networks for one-shot image recognition, in *ICML Deep Learning Workshop*, vol. 2 (2015)
26. O. Vinyals, C. Blundell, T. Lillicrap, D. Wierstra et al., Matching networks for one shot learning, in *Advances in Neural Information Processing Systems* (2016), pp. 3630–3638
27. C. Finn, P. Abbeel, S. Levine, Model-agnostic meta-learning for fast adaptation of deep networks, in *Proceedings of the 34th International Conference on Machine Learning-Volume 70*, JMLR. org (2017), pp. 1126–1135
28. A. Antoniou, H. Edwards, A. Storkey, How to train your MAML (2018). arXiv preprint arXiv:1810.09502
29. C. Finn, K. Xu, S. Levine, Probabilistic model-agnostic meta-learning, in *Advances in Neural Information Processing Systems* (2018), pp. 9516–9527
30. J. Snell, K. Swersky, R. Zemel, Prototypical networks for few-shot learning, in *Advances in Neural Information Processing Systems* (2017), pp. 4077–4087
31. Y. Zou, J. Feng, Hierarchical meta learning (2019). arXiv preprint arXiv:1904.09081
32. F. Sung, Y. Yang, L. Zhang, T. Xiang, P.H.S. Torr, T.M. Hospedales, Learning to compare: relation network for few-shot learning, in *Proceedings of the IEEE Conference on Computer Vision and Pattern Recognition* (2018), pp. 1199–1208
33. J. Kim, T. Kim, S. Kim, C.D. Yoo, Edge-labeling graph neural network for few-shot learning, in *Proceedings of the IEEE Conference on Computer Vision and Pattern Recognition* (2019), pp. 11–20
34. Y.-X. Wang, R. Girshick, M. Hebert, B. Hariharan, Low-shot learning from imaginary data, in *The IEEE Conference on Computer Vision and Pattern Recognition (CVPR)* (June 2018)
35. C. Finn, T. Yu, T. Zhang, P. Abbeel, S. Levine, One-shot visual imitation learning via meta-learning, in *1st Conference on Robot Learning (CoRL)* (2017)
36. S. Reed, Z. Akata, H. Lee, B. Schiele, Learning deep representations of fine-grained visual descriptions, in *Proceedings of the IEEE Conference on Computer Vision and Pattern Recognition* (2016), pp. 49–58
37. A.L.G. Dinu, M. Baroni, Improving zero-shot learning by mitigating the hubness problem, in *Proceedings of the 3rd International Conference on Learning Representations (ICLR 2015)*, workshop track (2015)
38. E. Kodirov, T. Xiang, S. Gong, Semantic autoencoder for zero-shot learning, in *Proceedings of the IEEE Conference on Computer Vision and Pattern Recognition* (2017), pp. 3174–3183
39. F.G. Mohammadi, M.H. Amini, Promises of meta-learning for device-free human sensing: learn to sense, in *Proceedings of the 1st ACM International Workshop on Device-Free Human Sensing* (DFHS'19) (ACM, New York, 2019), pp. 44–47. https://doi.org/10.1145/3360773.3360884
40. J. Choi, J. Lee, J. Park, J. Nam, Zero-shot learning and knowledge transfer in music classification and tagging (2019). arXiv preprint arXiv:1906.08615

41. D. Pathak, P. Mahmoudieh, G. Luo, P. Agrawal, D. Chen, Y. Shentu, E. Shelhamer, J. Malik, A.A. Efros, T. Darrell, Zero-shot visual imitation, in *Proceedings of the IEEE Conference on Computer Vision and Pattern Recognition Workshops* (2018), pp. 2050–2053
42. D. Pathak, P. Mahmoudieh, G. Luo, P. Agrawal, D. Chen, Y. Shentu, E. Shelhamer, J. Malik, A.A. Efros, T. Darrell, Zero-shot visual imitation, in *International Conference on Learning Representations (ICLR)* (2018)
43. B.M. Lake, R. Salakhutdinov, J.B. Tenenbaum, Human-level concept learning through probabilistic program induction. Science **350**(6266), 1332–1338 (2015)
44. C. Finn, A. Rajeswaran, S. Kakade, S. Levine, Online meta-learning (2019). arXiv preprint arXiv:1902.08438
45. Q. Sun, Y. Liu, T.-S. Chua, B. Schiele, Meta-transfer learning for few-shot learning, in *Proceedings of the IEEE Conference on Computer Vision and Pattern Recognition*, pp. 403–412 (2019)
46. Q. Wang, Y. Ming, Z. Jin, Q. Shen, D. Liu, M.J. Smith, K. Veeramachaneni, H. Qu, Atmseer: increasing transparency and controllability in automated machine learning, in *Proceedings of the 2019 CHI Conference on Human Factors in Computing Systems* (ACM, New York, 2019), p. 681

Part II
Application of Optimization, Learning and Control in Interdependent Complex Networks

Chapter 7
Predictive Analytics in Future Power Systems: A Panorama and State-Of-The-Art of Deep Learning Applications

Sakshi Mishra, Andrew Glaws, and Praveen Palanisamy

Abstract The challenges surrounding the optimal operation of power systems are growing in various dimensions, due in part to increasingly distributed energy resources and a progression towards large-scale transportation electrification. Currently, the increasing uncertainties associated with both renewable energy generation and demand are largely being managed by increasing operational reserves—potentially at the cost of suboptimal economic conditions—in order to maintain the reliability of the system. This chapter looks at the big picture role of forecasting in power systems from generation to consumption and provides a comprehensive review of traditional approaches for forecasting generation and load in various contexts. This chapter then takes a deep dive into the state-of-the-art machine learning and deep learning approaches for power systems forecasting. Furthermore, a case study of multi-time-horizon solar irradiance forecasting using deep learning is discussed in detail. Smart grids form the backbone of the future interdependent networks. For addressing the challenges associated with the operations of smart grid, development and wide adoption of machine learning and deep learning algorithms capable of producing better forecasting accuracies is urgently needed. Along with exploring the implementation and benefits of these approaches, this chapter also considers the strengths and limitations of deep learning algorithms for power

S. Mishra (✉)
Integrated Applications Center, National Renewable Energy Laboratory, Golden, CO, USA
e-mail: sakshi.mishra@nrel.gov

A. Glaws
Computational Science Center, National Renewable Energy Laboratory, Golden, CO, USA
e-mail: andrew.glaws@nrel.gov

P. Palanisamy
Perception Planning and Decision Systems, General Motors, Warren, MI, USA
e-mail: praveen.palanisamy@gm.com

© Springer Nature Switzerland AG 2020
M. H. Amini (ed.), *Optimization, Learning, and Control for Interdependent Complex Networks*, Advances in Intelligent Systems and Computing 1123,
https://doi.org/10.1007/978-3-030-34094-0_7

systems forecasting applications. This chapter, thus, provides a panoramic view of state-of-the-art of predictive analytics in power systems in the context of future smart grid operations.

Keywords Smart grid · Deep learning · Predictive analytic · Machine learning · Time series · Energy forecast · Power systems

7.1 Introduction

Overview The challenges surrounding the optimal operation of power systems are growing in various dimensions, due in part to increasingly distributed energy resources and a progression towards large-scale transportation electrification. Currently, the increasing uncertainties associated with both renewable energy generation and demand are largely being managed by increasing operational reserves— potentially at the cost of suboptimal economic conditions—in order to maintain the reliability of the system. This chapter looks at the big picture role of forecasting in power systems from generation to consumption and provides a comprehensive review of traditional approaches for forecasting generation and load in various contexts. This chapter then takes a deep dive into the state-of-the-art machine learning and deep learning approaches for power systems forecasting. Furthermore, a case study of multi-time-horizon solar irradiance forecasting using deep learning is discussed in detail. Smart grids form the backbone of the future interdependent networks. For addressing the challenges associated with the operations of smart grid, development and wide adoption of machine learning and deep learning algorithms capable of producing better forecasting accuracies is urgently needed. Along with exploring the implementation and benefits of these approaches, this chapter also considers the strengths and limitations of deep learning algorithms for power systems forecasting applications. This chapter, thus, provides a panoramic view of state-of-the-art of predictive analytics in power systems in the context of future smart grid operations.

Forecasting has long played an essential role in power systems planning and operations. With the introduction of deregulated markets, forecasting has emerged as a critical component of electricity markets as well. Reliable forecasting models allow electrical utilities and independent systems operations (ISOs) to make optimal capacity building and dispatch decisions by understanding their economic implications while still maintaining a reliable energy supply. Forecasting models are also used by the market participants to place strategic bids. The significance of forecasting has dramatically increased because of the rapidly changing landscape of traditional power systems. Some of the main drivers of this change are (1) increasing penetration of intermittent renewable energy resources on utility scale as well as distributed energy resources (DERs), (2) deployment of various smart grid technologies such as advanced metering infrastructure, (3) deregulation of electricity markets, (4) demand response programs turning static loads into dynamic loads, (5) forthcoming electrification of the transportation fleet, (6) greenhouse gas reduction targets, and (7) declining costs of energy storage technologies, among others.

The forecasted data in power systems include meteorological variables such as solar irradiance, wind speed, and wind direction; energy production from renewable energy sources such as photovoltaic plants, wind farms, and hydroelectric dams; load or demand; price of electricity or locational marginal prices; price of fossil fuels such as coal, oil, and natural gas; electric vehicle (EV) charging loads, and so on. These quantities are forecasted for different timescales as well as different spatial resolutions. Long-term forecasts are useful for power systems infrastructure building decisions while short-term forecasts are utilized to inform optimal decision-making by system operators dispatching energy on the grid and market participants trading energy in the markets.

7.1.1 Motivation

Traditionally, in the regulated electricity sector, mostly vertically integrated utilities had a monopoly. The reliability of supply was primarily the utilities' responsibility and was maintained using short-term load forecasts. The fossil fuel-based generation sources were dispatchable so that variability associated with the demand was the primary source of uncertainty in the system. Electricity users were passive consumers; that is, there was neither a bidirectional flow of energy from the distribution grid end nor any provision of demand response. Planning and investment in new capacity were based on long-term demand forecasts and utilities were responsible for building the transmission capacity to serve their customers. Traditional forecasting methodologies served well in this regulated business scenario.

Competitive electricity markets have been introduced since the last decade of the twentieth century as a part of the deregulation of the electricity sector [1]. Consequently, energy is now traded in competitive markets, making electricity price and demand forecasts fundamental inputs to the day-to-day decision-making process of the various energy-selling entities, including the utilities, independent power producers, large industrial customers with significant amounts of distribution generation production, and so on.

Moreover, building new transmission capacity is not a straightforward decision made by a single utility anymore. FERC Order No. 1000 [2] established new rules regarding the transmission planning and cost allocation requirement for public utility transmission providers, which have made capacity expansion a competitive process as well. As a result, accurate long-term load forecasting for different geographic areas has become even more important for maintaining the reliability of the system and economically expanding the network to accommodate future demand growth as well as distributed generation penetration on the grid.

The shift toward a digital and electrified economy is causing increased research and planning for networks of electrified transportation and a smart grid, operating interdependently. Forecasting will play an essential role in the transition to this new system as well. These future interdependent networks require reliable long-term EV growth forecasts for the planning of EV charging infrastructure as well as distribution network enhancements to accommodate the high penetrations of dynamic EV charging loads. In the operations domain, short-term forecasts of EV charging or discharging are required to get accurate load forecasts. Additional complexity is added when daily EV charging profiles are optimized using intelligent controls. The operational schedule of EV charging responds to the market price (even without the initiation of demand response events from the utility), making it even more challenging for traditional forecasting approaches to predict the dynamically changing demand.

The macrogrid in the USA (as well as many other industrialized countries) is a century old. Various components of generation, transmission, and distribution systems are reaching the end of their useful life and need to be refurbished. Though there are significant capital costs necessary to renovate the thousands of miles of distribution infrastructure, the reliability threats are even more dire. Several recent wildfires in California can be attributed to the aging power systems infrastructure of Pacific Gas and Electric [3]. The remaining useful life of the assets can be assessed to strategically plan the renovation of aging power systems infrastructure by leveraging advanced machine learning and deep learning-based predictive analytics. Accurate remaining useful life predictions for distribution grid components can inform economic investment such that the components with the highest risk of failure are replaced first.

Growing uncertainty in energy consumption, increasing penetration of inter-mittent renewable energy generation sources (at both utility scale and for small DERs), the burgeoning share of microgrid deployment, smart grid technologies enabling the internet of things (IoT), aging grid infrastructure, and the forthcoming revolution of electrified transportation are rapidly changing the landscape of power systems. Advanced and innovative predictive analytics approaches are urgently needed to enable more accurate forecasts to improve decision-making and provide the foundation for a smart, resilient, and sustainable grid of the future.

7.1.2 Classification of Power Systems Forecasting Models

Power systems forecasts may be done for various timescales based on the application of the predicted data. These forecasts can also be classified based on their application domain and their role in the power systems generation, transmission, distribution, and consumption areas. When a quantity is forecasted for different time horizons, the input variables used for producing the forecast also change. For example, when forecasting load for a long-term horizon in a given geographical area, inputs such as macroeconomic uncertainty, population growth, climate change patterns (for predicting the extreme loads), and distributed generation penetration projections are considered. For short-term load forecasting like day-ahead forecasting, input variables such as day of week, time of day, and past load consumption are relied upon instead. The effectiveness of machine learning and deep learning algorithms varies based on the timescale and type of input variables. The classification of forecasting models is discussed in detail in the following sections.

7.1.2.1   Classification Based on the Domain of Application in Power Systems

There are three functionally different parts of power systems studies and management, which make it possible to provide reliable and economical electricity to consumers in the present and in the future. These three parts—planning, operations, and market—are described in the following sections. Different types of predictive models are used in these three parts for obtaining forecasts for different quantities, such as load, resource, production, and so on, as shown in Fig. 7.1.

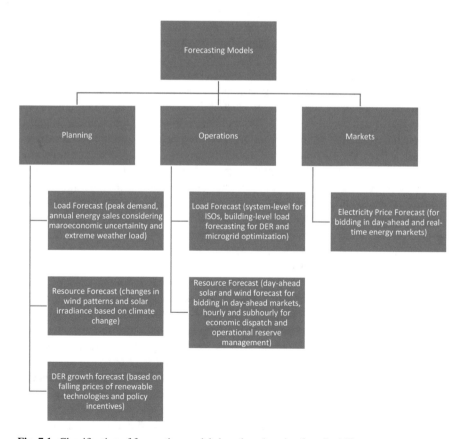

Fig. 7.1 Classification of forecasting models based on domain of applicability

7.1.2.1.1 Planning

The process of power systems planning is ever evolving and has the largest strategic impact on the future of power systems. Future power systems are planned using a set of forecasts, including *load forecasting* (which in turn depends on macroeconomic uncertainty, extreme weather, and climate changes for a given geographic location), *distributed generation technologies growth predictions* (which depend on the rate of decrease in the cost of these technologies and energy policy that provides incentives), and *resource forecasting* (which includes short-term production forecasts as well as long-term changes in solar irradiance and wind patterns for a given geographic area).

Smart and clean energy technologies form the foundation of the future smart grid. The key to enabling the adoption of clean energy technologies lies in how well power systems enhancements are planned to accommodate new technologies, enabling their smooth integration with the existing power systems. The goal of planning is to build and modify the generation, transmission, and distribution infrastructure that are needed to meet predicted future needs. Therefore, power systems planning has traditionally been divided into centralized generation planning, transmission planning, and distribution planning. The outcome of planning studies is to address what to build (more generation or transmission/distribution), how much to build, and where to build.

Traditionally, generation planning begins with load forecasting. Reliability evaluation is then conducted to determine if and when additional generation is needed. The remaining useful life of existing base load plants, which are largely powered by fossil fuels, is also accounted for in the next step. This is followed by capacity expansion studies based on economic considerations [4]. Nowadays, however, generation planning is not a solitary process. DER penetration forecasts, including behind-the-meter distributed generation, need to be accounted for in the process. Also, economical siting of utility-scale renewable generation plants depends on availability of solar and wind resources, which may or may not coincide with the demand pockets and existing transmission infrastructure. High penetration of utility-scale renewable energy resources, given their intermittent and variable nature, adds increased complexity to generation planning studies that depend on renewable resource forecasts [5, 6].

Transmission planning is aimed at optimizing the use of a generation portfolio by supplying loads from the most economical sources of power and improving the reliability of overall systems by operating generation stations flexibly [7]. Generation and transmission planning are closely related because the powerflows through the transmission system are a direct result of generation dispatch [8, 9]. Distribution system planning, on the other hand, is optimized for the lowest cost operation that meets the desired reliability of the electricity service. However, the introduction and increased adoption of DERs has changed the process of distribution system planning drastically. This is because components of distribution and transmission systems are not designed to handle the bidirectional flow of power from the DERs, so additional measures must be taken to refurbish the distribution grid with this capability [10, 11].

For generation and transmission planning, load forecasting is done for a long time horizon—often between 2 and 10 years. This is because system capacity expansion projects require a long lead time. Peak annual demand/load (in kilowatts) and total annual energy sales (in kilowatt-hours) are calculated for long-term load forecasts [12]. Peak load is highly correlated with weather. Therefore, peak load forecast is normalized based on extreme weather predictions. Projected EV and DER growth in the future has led to researching and employing methodologies that explicitly consider DERs as well as EV load along with its charging patterns [13, 14]. Load forecasting also needs to be specific to geographical locations, along with maintaining reasonable accuracies of the predicted magnitude.

7.1.2.1.2 Operations

Power systems operations are associated with making decisions regarding the use of existing equipment and infrastructure to generate, transmit, and deliver energy. It is primarily aimed at doing so safely, reliably, and efficiently. The operations domain deals with three different time horizons: (1) operations planning (a few weeks to months), (2) near real time (a few hours to days), and (3) real time (typically 5–10 min) [15].

Operations planning ensures that sufficient resources are available to meet demand for the next few months. It takes load forecasts (and associated errors), utility-scale renewable generation forecasts, and generation and transmission outages into account. Operations planning also defines the reserve capacity requirements to mitigate the risk imbalances because of forecast errors and unplanned outages of generation or transmission components [16]. The aim of near real-time operations is to select the most economic generation portfolio for the next few days using a process called unit commitment. Real-time operations are aimed at ensuring system reliability and supply sufficiency by revising the near real-time schedule on an as-needed basis.

Load forecasting is the first step of all three time horizons of power systems operations, making it a critical component. For the operations planning and near real-time applications, hourly load forecasts are used. For real-time applications, however, subhourly (minute-level) resolution is typically required. Once the magnitude and geographic location of demand are obtained using load forecasts, least-cost generation is scheduled to meet that demand. The production forecasts of utility-scale renewable generation plants are also considered while scheduling the generation. In the regions with high DER penetration, their production is also considered; behind-the-meter DERs are typically considered negative load.

7.1.2.1.3 Markets

The landscape of the power sector has substantially changed after the introduction of competitive markets coupled with the deregulation of the industry. This has led to the trading of electricity under market rules using spot and derivative contracts. But the price dynamics of this unique commodity is different from any other commodity because of its unique properties, requirements, and dependencies. For example, energy typically experiences a constant balance between production and consumption because large quantities are not economically storable. Additionally, power demand can depend on weather factors, such as temperature, precipitation, and wind speeds, and on the magnitude of activity in different sectors (i.e., holidays vs. workdays, weekdays vs. weekends, on-peak vs. off-peak hours).

Electricity prices in the wholesale market, therefore, exhibit seasonality at various timescales (daily, weekly, annually) as well as abrupt and brief price spikes. According to [17], "[t]he costs of over-/under-contracting and then selling/buying power in the balancing (or real-time) market are typically so high that they can lead to huge financial losses or even bankruptcy. Extreme price volatility, which can be up to two orders of magnitude higher than that of any other commodity or financial asset, has forced market participants to hedge not only against volume risk but also against price movements."

Short-term electricity price forecasting is done for the day-ahead market, where the bids are submitted for the delivery of electricity during each load period, which can be hourly or subhourly. Medium-term time horizons are used for risk management and derivative pricing. These forecasts can either be point-forecasts or probability distributions of the prices. Long-term electricity price forecasts are done for planning and economic feasibility analysis of future power plants, establishing long-term power purchase agreements, forward capacity markets, seasonal capacity markets, financial transmission rights auctions, and so on. The time horizon can vary from months to years for such applications.

Renewable generation forecasts in the short term are also required for owners to bid in the market. ISOs need the production forecasts of intermittent energy sources to schedule the generation with sufficient reserves to minimize the risk of underproduction. To avoid financial losses associated with underbidding or overbidding, renewable generation plant owners need reasonably accurate forecasts of solar and wind resources [18].

For each time horizon, the choice of input variables plays a significant role in the effectiveness of the model for both traditional forecasting approaches as well as deep learning methods. For short-term forecasts, the daily and hourly variability must be considered. On the other hand, medium-term forecasting favors annual variations more than weekly ones. For long-term price forecasts, seasonality itself becomes irrelevant. Instead, long-term trends such as load-growth in a certain geographic area, large penetration of cheap renewable energy resources in close proximity, and EV load demand play the major role.

7.1.2.2 Classification Based on Timescale

In the previous section, various power systems forecasting models were discussed in the context of their applicability to planning, operation, and market domains.

Table 7.1 Types of forecasting models based on timescale

Forecasting type	Time horizon	Applications	Methods
Nowcasting	5 min–6 h	Load frequency control, battery-use optimization, real-time market participation, economic dispatch	Satellite-based physical models (use cloud motion vector-based method); sky imagery-based physical models; statistical/machine learning models based on historical data (e.g., persistence, ARMA, SVRs, deep learning)
Short-term forecasting	6 h–1 week	Unit commitment, switching source, rescheduling means of production, day-ahead market participation	Hybrid NWP/statistical/machine learning models
Medium-term forecasting	1 week–2 year	Scheduling maintenance, capacity markets bidding, and pricing	Statistical models based on predicted growth/change (e.g., end-use, econometric models)
Long-term forecasting	2 year–5 year	Long-term purchase agreements, forward capacity market, management of multiyear reservoirs, nuclear fuel management	Statistical models based on predicted growth/change (e.g., end-use, econometric models)
Very long-term forecasting	25+ year	Capacity expansion, infrastructure retirement, policymaking	Statistical models based on predicted growth/change (e.g., end-use, econometric models); general circulation models (climate models)

Another way of classifying the forecasting models in power systems is based on the timescale for which the quantities are being forecasted. These timescales can mainly be classified into five types, as given in Table 7.1.

7.1.3 Organization of the Chapter

The introduction section first lays out the motivation behind exploring newer approaches such as deep learning for power systems predictive analytics. The power systems forecasting problems are then classified in broad categories based on timescale of forecasting as well as their application in power systems planning, operations, and market domains. The second section, *forecasting power systems using classical approaches*, takes a deeper look at the widely used statistical times-series forecasting methods as well as traditional machine learning-based approaches. The third section then introduces state-of-the-art deep learning algorithms and explores their recent applications in the power systems forecasting literature. A solar irradiance forecasting case study is discussed in detail in the fourth section. The fifth section identifies future work areas in this domain and concludes the chapter.

7.2 Forecasting in Power Systems Using Classical Approaches

The power systems forecasting problems discussed in the previous section most closely align with the mathematical framework of the time series forecasting problem. This section introduces this general mathematical framework and provides a broad overview of several statistical and machine learning approaches to time series forecasting. Note that deep learning methods are left to Sect. 7.3 to be explored in more detail.

7.2.1 Time Series Data

A general time series dataset can be written as

$$\{x_1, x_2, x_3, \ldots\}, \tag{7.1}$$

where each x_t for $t = 1, 2, 3, \ldots$ represents the realization of some random variable. A common additive modeling approach to characterizing Eq. (7.1) is to partition the time series into a trend, seasonality, and stochastic term,

$$x_t = T_t + S_t + Z_t. \tag{7.2}$$

The trend term T_t represents the long-term, nonperiodic changes in the data, the seasonality term S_t describes any periodic behavior of the time series, and the stochastic term Z_t is stationary process (defined later) that models the random noise in the data.

Note that Eq. (7.1) frames the time series data in terms of scalar-valued quantities. This is done to simplify the discussion in this section in order to provide a clear and broad overview of traditional approaches to time series forecasting. The extension of this perspective to the multivariate case is relatively straightforward. One feature of multivariate time series data that is important to power systems modeling is the concept of exogenous variables. In time series forecasting, exogenous variables are causally independent of other factors in the system. In the case of solar irradiance forecasting, examples of exogenous variables may include factors like wind speed and cloud cover. Including exogenous variables in the forecasting process may improve performance.

Recall that each x_t in Eq. (7.1) is a realization of some random variable. The complete time series is then fully characterized by the joint distribution of these random variables. However, such a perspective is typically impractical or impossible for real-world applications. A more reasonable approach is to characterize the time series in terms of secondary properties, such as the mean and covariance functions of the series,

$$\mu_t = \mathbb{E}[x_t] \quad \text{and} \quad \sigma_{t,s} = \mathbb{Cov}[x_t, x_s] = \mathbb{E}[(x_t - \mu_t)(x_s - \mu_s)]. \tag{7.3}$$

The dependence of the value of x_t on previous terms is characterized by the autocovariance function $\gamma_t(h) = \sigma_{t,t+h}$ and the autocorrelation function $\rho_t(h) = \gamma_t(h)/\gamma_t(0)$, where h is the lag parameter.

A key property of time series data is the idea of stationarity. A given time series is said to be strictly stationary if any two subseries,

$$\{x_t, x_{t+1}, x_{t+2}, \ldots, x_{t+n}\} \quad \text{and} \quad \{x_s, x_{s+1}, x_{s+2}, \ldots, x_{s+n}\} \quad \text{for } t, s, n \in \mathbb{N}, \tag{7.4}$$

have the same joint distribution. Notice that if each x_t in a given time series is independent and identically distributed (iid), then the time series is strictly stationary. Such a sequence drawn from a distribution with mean 0 and variance σ^2 is typically referred to as white noise.

As discussed earlier, characterizing the full joint distribution of a time series is not realistic for most real-world applications, making the identification of a time series as strictly stationary infeasible. Alternatively, a time series is said to be weakly (or wide-sense) stationary if any two subseries have the same mean and covariance functions, μ_t and $\sigma_{t,s}$, respectively. Equivalently, a weakly stationary time series has mean and covariance functions that are independent of t. That is, $\mu_t = \mu$ and $\sigma_{t,s} = \sigma$. Notice that this also implies that the autocovariance and autocorrelation functions only depend on the lag parameter, $\gamma_t(h) = \gamma(h)$ and $\rho_t(h) = \rho(h)$. Because this definition is of more practical use, it is common to use the term stationary to refer to weakly stationary and specifically refer to a time series as strictly stationary when the stricter definition is meant.

The next two sections explore various statistical and machine learning approaches to time series forecasting. In general, the goal of forecasting is to predict values of future datapoints $\{\hat{x}_{n+1}, \hat{x}_{n+2}, \ldots\}$ given a finite set of observed data $\{x_1, x_2, x_3, \ldots, x_n\}$. Generally, time series forecasting is classified according to the horizon out to which the forecast is made, as illustrated in Table 7.1.

The differences between short-, medium-, and long-term forecasts are highly dependent on the problem under consideration. However, short- and medium-

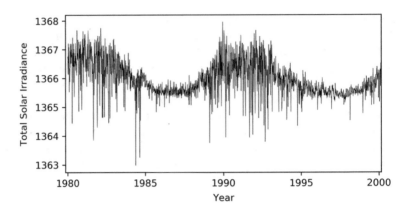

Fig. 7.2 Total solar irradiance measured daily from 1980 to 2000. These data come from the NOAA's National Centers for Environment Information database [19]

term forecasts typically are more dependent on autocorrelation factors and shorter seasonality behaviors. These prediction horizons tend to be more amenable to the types of data-driven methods covered here. Long-term forecasting seeks to model trends in the data and often depends on the additional models of the relevant systems to help predict changes in these trends.

Power systems forecasting is a particularly difficult problem. Figure 7.2 shows an example time series data of total solar irradiance over a 20-year range [19]. It is immediately obvious that this dataset is nonstationary (as is the case of many time series data arising from power systems). The data show seasonal cycles of increased and decreased solar irradiance that have a period of approximately 11 years. In addition to fluctuations in the mean of the data, the seasonality also changes the variance of the data. The time series varies more significantly during periods of high solar irradiance and less significantly during periods of low solar irradiance. Lastly, the example data in Fig. 7.2 highlight the differences in short-, medium-, and long-term forecasting. Short-term forecasts are focused on accurately capturing the high-frequency fluctuations in the data. Medium- and long-term predictions cannot hope to perfectly predict these behaviors and instead focus on the large-scale trends and seasonal characteristics in the time series.

7.2.2 Statistical Forecasting Approaches

7.2.2.1 Naïve Model Approach

The naïve model approach to time series forecasting simply predicts that the next value in the sequence is the same as the current value,

$$\hat{x}_{t+1} = x_t. \tag{7.5}$$

This approach produces the optimal prediction for random walk data and is therefore also known as the random walk model for time series forecasting. The main purpose of this model is to serve as a simple baseline to compare with more sophisticated models. This naïve model approach is also called the persistence model [20].

7.2.2.2 Exponential Smoothing

Exponential smoothing is a relatively simple approach to modeling time series that predicts new values in time series using a weighted moving average that more heavily favors recent datapoints [21]. Given time series data $\{x_1, x_2, x_3, \ldots, x_n\}$, the simple exponential smoothing model computes the smoothed approximation of $\hat{x}_{n+1}$ as

$$\hat{x}_{n+1} = \alpha x_n + (1 - \alpha) \hat{x}_n, \tag{7.6}$$

where $\alpha \in (0, 1)$ is the smoothing factor. Notice that this method computes a weighted average of the current true value and the current predicted (or smoothed) value. The current smoothed value was computed similarly. Thus, previous terms contribute to the current predict value with exponentially decreasing importance. The rate of this decay is controlled by the smoothing factor α. Extensions to simple exponential smoothing incorporate trends and seasonality [22, 23].

Simple exponential smoothing is among the earliest forecasting techniques applied to load forecasting [24]. In particular, this work explores the application of exponential smoothing to short-term forecasting at hourly intervals. More recently, several studies have explored the application of double seasonality exponential smoothing to short-term load forecasting and found this approach to be robust despite its relative simplicity [25, 26].

7.2.2.3 Autoregressive Moving Average (ARMA) Models

The autoregressive moving average (ARMA) model and its variations are powerful forecasting tools that are among the most popular statistical methods for power systems analysis. The ARMA model has long been used for power-related problems, such as solar irradiance and load forecasting [27, 28]. More recently, an ARMA variant called ARIMA (covered in the next section) has been applied to short-term solar forecasting [29, 30], next-day electricity pricing [31], and hourly load predictions [32].

As the name suggests, the ARMA model makes two key assumptions on the time series. The first is that the time series data can be modeled by an autoregressive process. An autoregressive process of order p, denoted by $AR(p)$, assumes a linear dependence of the current timestep on the previous p timesteps,

$$\hat{x}_t = \sum_{i=1}^{p} \omega_i x_{t-i} + z_t, \qquad (7.7)$$

where ω_i are constants and z_t is a white noise term. The second assumption of the ARMA model is that of a moving average model. The moving average model of order q, denoted by $MA(q)$, represents the sequence as a linear relationship of some other white noise sequence,

$$\hat{x}_t = \sum_{i=1}^{q} \theta_i z_{t-i} + z_t, \qquad (7.8)$$

where θ_i are constants and each z_t is an iid white noise term. The ARMA model of orders p and q, denoted by $ARMA(p, q)$, combines Eqs. (7.7) and (7.8) to form

$$\hat{x}_t = \sum_{i=1}^{p} \omega_i x_{t-i} + \sum_{i=1}^{q} \theta_i z_{t-i} + z_t. \qquad (7.9)$$

The ARMA model is typically solved using the Box-Jenkins method [33]. This is an iterative process of specifying the model, fitting the parameters, and verifying the process. Specifying the model involves the order of the $ARMA(p, q)$ model (i.e., selecting the appropriate values of p and q). Heuristically, this can be accomplished by examining the autocorrelation function $\rho_t(h)$ and the partial autocorrelation function. Recall that the autocorrelation function explains the relationship between two terms with lag h. However, because this relationship can have a recursive structure, it may be difficult to distinguish between a time series that is dependent on the previous n points and one that is highly dependent only on the previous one. The partial autocorrelation addresses this concern by filtering out the influence of the intermediate terms $\{x_{t-1}, x_{t-2}, \ldots, x_{t-h+1}\}$. This is computed by solving the linear system

$$\boldsymbol{\Sigma\alpha} = \boldsymbol{\gamma}, \qquad (7.10)$$

where $(\boldsymbol{\Sigma})_{i,j} = \gamma_t(i - j)$ and $(\boldsymbol{\gamma})_i = \gamma_t(i)$ for $i, j = 1, 2, \ldots, h$. The partial autocorrelation with lag h is $\alpha_t(h) = (\boldsymbol{\alpha})_h$. Reasonable guesses of p and q for ARMA can be made from examining plots of the autocorrelation and partial autocorrelation functions. If the autocorrelation plot slowly decays to zero and the partial autocorrelation plot abruptly decays to zero after a lag of h, then the model is likely $ARMA(h, 0)$, or equivalently $AR(h)$. Alternatively, if the partial autocorrelation plot slowly decays to zero and the autocorrelation plot abruptly decays to zero after a lag of h, then the model is likely $ARMA(0, h)$ or $MA(h)$. If both values slowly decay to zero, then the model is likely $ARMA(p, q)$ where the orders are taken to be a lag after which the plots have sufficiently decayed. Selecting the appropriate value of p and q can be difficult and take some trial and error.

Once the order of the ARMA model has been determined, the parameters ω_i and θ_i must be fit. This is accomplished using any preferred numerical optimization method to solve for the maximum likelihood estimate of these parameters. Once ω_i and θ_i have been computed, the model is examined for errors and overfitting. If necessary, the process is repeated with a new model selection.

7.2.2.4 Autoregressive Moving Integrated Average (ARIMA) Models

The success and popularity of the ARMA model have led to multiple variations and extensions of the method. The autoregressive integrated moving average (ARIMA) model was introduced to address the stationarity assumption on the time series data. ARIMA has been used recently for predicting the EV charging demand for stochastic power systems operation [34]. It incorporates differencing of the time series data to attempt to remove any nonstationary behavior. The number of differencing steps d is treated as another modeling parameter so that the model is written $ARIMA(p, d, q)$.

Notice that the discussion of the Box–Jenkins method discussed in the previous section appears to assume that both the autocorrelation and the partial autocorrelation functions will eventually decay to zero (whether slowly or rapidly). If this is not the case, then differencing may be applied to the data to remove the nonstationarity. Differencing is a common approach to producing a stationary time series. One differencing iteration produces a new time series with

$$y_t = x_{t+1} - x_t. \tag{7.11}$$

The Box–Jenkins method determines d by differencing on the time series until the autocorrelation and partial autocorrelation plots decay appropriately.

7.2.3 Machine Learning Forecasting Approaches

Supervised machine learning seeks to construct a predictive model $f_\Theta(\mathbf{x})$, based on a given training set of data $\{\mathbf{x}_i, y_i\}_{i=1}^N$, where $\mathbf{x}_i$ and y_i represent the feature vector and the target value [35]. For time series forecasting, the feature vectors are typically constructed by a moving window over the given data, $\mathbf{x}_i = [x_i, x_{i+1}, x_{i+2}, \ldots, x_{i+n}]^\top$, and the target value is the first datapoint after this window, $y_i = x_{i+n+1}$. The subscript Θ in the model denotes the collection of parameters that are tuned to best fit the data. Machine learning methods fit the model parameters from the data through iterative updates to reduce some loss function, such as the squared-error loss $L = \sum_{i=1}^N (y_i - f_\Theta(\mathbf{x}_i))^2$ or the absolute-error loss $L = \sum_{i=1}^N |y_i - f_\Theta(\mathbf{x}_i)|$.

This section introduces two popular machine learning methods for time series forecasting: the support vector regression (SVR) and the Gaussian process regression (GPR). There are many more approaches that may be appropriate depending on the specific problem at hand, such as k-nearest neighbor regression or regression trees. A more comprehensive overview of these methods can be found in [20, 36, 37]. Deep learning (or neural network) methods tend to fall within the realm of machine learning as well. However, their discussion is reserved for Sect. 7.3 so that they may be explored more in depth.

7.2.3.1 Support Vector Regression

Support vector regression (SVR) is a form of the popular machine learning approach known as support vector machine (SVM) [38]. The linear SVR attempts to fit the model

$$y = \boldsymbol{\theta}^\top \mathbf{x} + \theta_0 \tag{7.12}$$

to the data while minimizing $\|\boldsymbol{\theta}\|$. A linear model may be insufficient to describe the complex relationships underlying real-world datasets. Nonlinear or kernel SVR reformulates the model as

$$y = \sum_{i=1}^{N} \theta_i k\left(\mathbf{x}_i, \mathbf{x}\right) + \theta_0, \tag{7.13}$$

where $k(\cdot, \cdot)$ is a kernel function such as the radial basis function, or squared-exponential kernel $k(\mathbf{x}_i, \mathbf{x}_j) = \exp\left(-\gamma(\mathbf{x}_i - \mathbf{x}_j)^2\right)$, where γ is a hyperparameter that can be tuned using a grid search with cross-validation. The use of the kernel function implicitly defines a nonlinear mapping of the feature vector to some higher-dimensional space where a linear model is applied. This nonlinear mapping provides greater flexibility than simply applying the linear model directly to the features as in Eq. (7.12). Such a mapping is guaranteed to exist, provided that the kernel satisfies the so-called Mercer condition [39].

As mentioned previously, training any machine learning model requires the formulation of some loss function that informs the optimal set of model parameters. For SVR, it is common to use the ϵ-insensitive loss function. This loss ignores any points within $\pm\epsilon$ of the model prediction and is equal to the absolute error in the model for datapoints outside this range. Using the ϵ-insensitive loss, the SVR learning problem can be stated as

$$\min_{\Theta} \|\Theta\| + c \sum_{i=1}^{N} \xi_i \text{ subject to } \left| y_i - \sum_{i=1}^{N} \theta_i k\left(\mathbf{x}_i, \mathbf{x}\right) + \theta_0 \right| \leq \epsilon + \xi_i \tag{7.14}$$

where ξ_i are slack variables that penalize deviation outside the ϵ-insensitive region of the loss function.

SVR is a popular machine learning approach to load forecasting [40–42]. SVRs have also been combined with other approaches to enhance performance on the short-term load forecasting problem. For example, an SVR can be combined with a locally weighted regression method that more heavily favors nearby points when making predictions [43]. Another approach combines SVRs with the empirical mode decomposition that separates out the high- and low-frequency components of a time series [44]. Both of these hybrid approaches were found to outperform the classical SVR method.

7.2.3.2 Gaussian Process Regression

Gaussian process regression (GPR) approaches time series forecasting from a Bayesian perspective by assuming that the underlying model for the data is drawn from prior distribution of functions [45]. For GPR, this prior is assumed to be a mixture of multivariate Gaussian random variables, or a Gaussian process,

$$f\left(\mathbf{x}\right) \sim \mathcal{GP}\left(m\left(\mathbf{x}\right), k\left(\mathbf{x}, \mathbf{x}'\right)\right), \tag{7.15}$$

where $m(\mathbf{x})$ and $k(\mathbf{x}, \mathbf{x}')$ are the mean and variance function, respectively. Often, the problem is formulated with mean zero and the kernel function equal to the squared-exponential from SVR.

Conditioning on the given dataset generates the posterior distribution of f, which is also a Gaussian process with mean and variance

$$\mathbb{E}\left[f\left(\mathbf{x}\right) | \{\mathbf{x}_i, y_i\}_{i=1}^N\right] = \mathbf{k(x)}^\top \mathbf{K}^{-1}\mathbf{y},$$

$$\mathbb{V}\mathrm{ar}\left[f\left(\mathbf{x}\right) | \{\mathbf{x}_i, y_i\}_{i=1}^N\right] = k\left(\mathbf{x}, \mathbf{x}\right) - \mathbf{k(x)}^\top \mathbf{K}^{-1}\mathbf{k}\left(\mathbf{x}\right), \tag{7.16}$$

where $(\mathbf{y})_i = y_i$, $(\mathbf{K})_{i,j} = k(x_i, x_j)$ and $(\mathbf{k(x)})_i = k(x_i, \mathbf{x})$ for any $\mathbf{x}$. Using the decaying exponential kernel, it can be shown that the model interpolates the data without any variance. By assuming the data are corrupted by Gaussian noise with variance σ^2, the posterior distribution then has mean and variance

$$\mathbb{E}\left[f\left(\mathbf{x}\right) | \{\mathbf{x}_i, y_i\}_{i=1}^N\right] = \mathbf{k(x)}^\top \left(\mathbf{K} + \sigma^2 \mathbf{I}\right)^{-1}\mathbf{y},$$

$$\mathbb{V}\mathrm{ar}\left[f\left(\mathbf{x}\right) | \{\mathbf{x}_i, y_i\}_{i=1}^N\right] = k\left(\mathbf{x}, \mathbf{x}\right) - \mathbf{k(x)}^\top \left(\mathbf{K} + \sigma^2 \mathbf{I}\right)^{-1}\mathbf{k}\left(\mathbf{x}\right). \tag{7.17}$$

Gaussian processes have been applied to the load forecasting problem with promising success [46]. Additionally, GPR has been used for renewable energy forecasting relating to solar radiation [47] and wind power [48]. One study used GPRs with time-based composite covariance to handle seasonality in solar radiation data [49].

7.2.4 Shortcomings of Classical Approaches

Statistical and machine learning approaches to time series forecasting are powerful tools for understanding and modeling power systems forecasts. These methods have been performing reasonably well for short- and medium-term forecasting with traditionally acceptable level of accuracies. However, these methods can require significant data preprocessing that is not explored deeply here. For example, most of the statistical approaches assume stationary time series data with variable-independence and normality assumptions. Extensions to these methods that effectively deal with nonstationary data require manual tuning of various meta-parameters that essentially transform the data to be stationary.

Additionally, with the increasing dynamism in the future power systems, there is a need to obtain forecasts with higher accuracy than what is being achieved with traditional statistical and machine learning methods. The operations of power systems are getting more dynamic in nature with bidirectional flow of power through distributed energy resources, prosumer participation with demand-response, and other smart grid technologies. Increasing renewable energy penetration and decreasing synchronous generation resources are reducing the overall inertia of the grid [50]. This requires a finer temporal resolution of the forecasts in order to maintain reliable real-time operations of the grid. Furthermore, price forecasting for electricity markets can benefit greatly from a small percentage gain in the prediction accuracy, and better renewable energy forecasts are required by ISOs to lower the amount of the costly operational reserves [51].

The next section examines the history and current state-of-the-art in deep learning methods for power systems forecasting. With their ability to represent complex nonlinear behaviors in nonstationary, high-frequency, and high-dimensional time series data, these methods have been shown to be more robust to some of the abovementioned pitfalls of traditional approaches, but at the expense of some new hurdles.

7.3 Forecasting in Power Systems Using Deep Learning

7.3.1 Deep Learning

Artificial neural networks (ANN) are universal function approximators [52]; that is, it is possible to represent complex nonlinear behavior in a high-dimensional, high-frequency, and nonstationary time series using ANNs. A deep neural network is an ANN with multiple hidden layers and nodes cascaded between input and output layers. Deep neural networks are sophisticated neural networks that have been successfully applied to analyze data in many disciplines in the past several years such as computer vision, image recognition, automatic speech recognition, bioinformatics, finance, and nature language processing [53].

In general, supervised machine learning algorithms are particularly task specific. However, deep learning networks are capable of learning intricate structures in large datasets, allowing them to generalize better to scenarios not present in the training data. Because of their capability to learn nonlinear relationships between input features, these networks can identify and ignore features that do not impact the target variable by minimizing the appropriate weights. Consequently, deep learning algorithms typically do not require the type of extensive data preprocessing and feature engineering that is required of other traditional machine learning methods. Additionally, deep learning algorithms are also capable of managing high-dimensional datasets better than traditional machine learning algorithms.

Recurrent neural networks (RNN), long short-term memory networks (LSTM), convolutional neural networks (CNN), autoencoders, restricted Boltzmann machines, deep belief networks, and deep Boltzmann machines are all common types of deep learning algorithms. The following sections introduce deep learning algorithms that are often applied to power systems forecasting problems and describe their mathematic framework briefly.

7.3.1.1 Recurrent Neural Network

Unlike traditional feedforward neural networks in which information flows from each layer to the next, RNNs allow the output from a layer to flow back into itself. This allows RNNs to process sequential data without assuming the independence among the time series samples or the datapoints [54]. Feedforward networks lose any knowledge of the system state after processing each time series sample, thereby failing to account for the relationship between exogenous variables along the temporal dimension. The recurrent edges in an RNN introduce temporal coupling into the model. The internal memory, formed by the feedback connections of the neurons in the hidden-layer nodes, updates the states of each neuron in the network with the previous input. The addition of this temporal coupling, which unfolds over time, allows RNNs to learn and exhibit complex system dynamics, making them efficient at time series forecasting problems.

The input to an RNN is a sequence of real-valued datapoints $\{\mathbf{x}_1, \mathbf{x}_2, \mathbf{x}_3, \ldots\}$, where $\mathbf{x}_t$ represents the value of time series variables timestep t. Given a finite input subsequence of length n, the target output for the RNN is the next value $\mathbf{y}_{n+1}$. Note that the target $\mathbf{y}$ values may contain the same variables as the input $\mathbf{x}$ values but at future timesteps, or they may be different if the input includes exogenous variables. The network output (i.e., the predictions from the network) is denoted

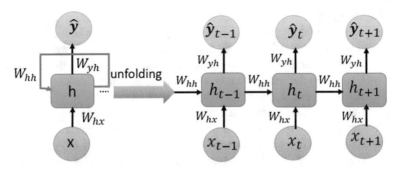

Fig. 7.3 Unfolding of an RNN over the temporal dimension

by $\hat{\mathbf{y}}_t$. Figure 7.3 shows how the network unfolds the data along the temporal dimension. Mathematically, this unfolding is written as

$$\mathbf{h}_t = f_h \left(\mathbf{W}_{hx}\mathbf{x}_t + \mathbf{W}_{hh}\mathbf{h}_{t-1} + \mathbf{b}_h \right),$$

$$\hat{\mathbf{y}}_{t+1} = f_o \left(\mathbf{W}_{yh}\mathbf{h}_t + \mathbf{b}_y \right), \tag{7.18}$$

where the current input datapoint $\mathbf{x}_t$ is fed into the network along with the output of the hidden layer from the previous timestep $\mathbf{h}_{t-1}$, and the output from the hidden layer is used to generate the prediction $\hat{\mathbf{y}}_{t+1}$. The remaining terms in Eq. (7.18) include the activation functions f_o and f_h, the weight matrices $\mathbf{W}_{hx}$, $\dot{\mathbf{W}}_{hh}$, and $\mathbf{W}_{yh}$, and the biases for each layer $\mathbf{b}_h$ and $\mathbf{b}_y$.

7.3.1.2 Long Short-Term Memory Network

In theory, RNNs should be capable of handling long-term temporal relationships because of their ability to retain information from previous timesteps. In practice, vanishing gradients make it difficult for them to learn long-term dependencies. Long short-term memory networks (LSTM) are a variation on the traditional RNN that are more effective at learning long-term trends in data, making them efficient at time series forecasting problems.

The key difference between RNNs and LSTMs is that the latter replaces hidden nodes with a more complex memory cell that handles the recurrent transfer of information (see Fig. 7.4). Four layers of neural connections, which exchange information in a particular special way, form the foundation of these memory cells. LSTMs are capable of learning long-term dependencies because the memory cells retain the existing information and append the unit with the new information; in RNNs, the content of the hidden node is replaced with the new value calculated from the current input.

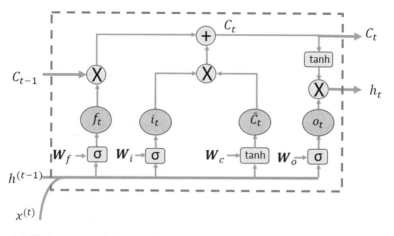

Fig. 7.4 LSTM memory cell diagram (from [55])

The mathematical formulation governing the flow of information in a LSTM cell is

$$\mathbf{f}_t = \sigma \left(\mathbf{W}_f \left[\mathbf{h}_{t-1}^\top \ \mathbf{x}_t^\top \right]^\top + \mathbf{b}_f \right),$$

$$\mathbf{i}_t = \sigma \left(\mathbf{W}_i \left[\mathbf{h}_{t-1}^\top \ \mathbf{x}_t^\top \right]^\top + \mathbf{b}_i \right),$$

$$\tilde{\mathbf{C}}_t = \tanh \left(\mathbf{W}_c \left[\mathbf{h}_{t-1}^\top \ \mathbf{x}_t^\top \right]^\top + \mathbf{b}_c \right),$$

$$\mathbf{C}_t = \mathbf{f}_t \circ \mathbf{C}_{t-1} + \mathbf{i}_t \circ \tilde{\mathbf{C}}_t,$$

$$\mathbf{o}_t = \sigma \left(\mathbf{W}_o \left[\mathbf{h}_{t-1}^\top \ \mathbf{x}_t^\top \right]^\top + \mathbf{b}_o \right),$$

$$\mathbf{h}_t = \mathbf{o}_t \circ \tanh (\mathbf{C}_t) . \tag{7.19}$$

Note that the nodes in the cell operate on the concatenated vector $\left[\mathbf{h}_{t-1}^\top \ \mathbf{x}_t^\top \right]^\top$ where $\mathbf{x}_t$ is the current input vector and $\mathbf{h}_{t-1}$ denotes the output from the cell at the previous timestep. The value $\mathbf{C}_t$ is the current state of the cell and is defined by a combination of the information from the forget gate $\mathbf{f}_t$ and the input gate $\mathbf{i}_t$ (where $\circ$ denotes the element-wise Hadamard product). The output gate $\mathbf{o}_t$ is acted on by the cell state to produce the output of the cell $\mathbf{h}_t$. The various

W's and **b**'s represent the weights and biases in the cell, while σ and tanh are the sigmoid and hyperbolic tangent activation functions, respectively.

Once the model is chosen, there are two main iterative phases in the learning algorithm: (1) forward propagation and (2) weight update. For an RNN or LSTM, the architecture first unfolds the time series input along the temporal dimension, making the network similar to a traditional feedforward neural network. In the forward propagation phase, the input vector propagates through the hidden layers (using randomly initialized values for the weight matrices and biases) to compute the output vector. The mismatch between the interim prediction output and the actual target is calculated as a loss function (e.g., the mean squared-error loss). The weights are updated using gradient descent with the gradients with respect to the loss function computed using the backpropagation through time algorithm.

7.3.1.3 Other Relevant Models

WaveNet deep learning models were recently introduced that apply deep learning techniques from audio signal processing and computer vision models to time series (sequential) data [56]. Convolutional neural networks (CNN) are a type of deep feedforward ANN that have been used to analyze visual imagery on a large scale. A deep convolutional WaveNet architecture, which is variation of CNN, has been successfully used for conditional time series forecasting [57].

7.3.2 Deep Learning Applications

Deep learning has been applied to a variety of power systems prediction problems recently, including solar forecasting, building load forecasting, system load forecasting, wind forecasting, and electricity price forecasting. RNN and LSTMs are the most popular architectures published in the literature for power systems forecasting problems. The following sections discuss the recent literature of power systems predictive analytics using deep learning.

7.3.2.1 Load Forecasting

Load forecasting may be done at either a systems level or building level and for different time horizons. Deep neural networks have been used for building energy load forecasting using an LSTM and an LSTM-based sequence to sequence modeling approach [58]. Short-term residential load forecasting is done using an LSTM in [59]. Shi et al. [60] propose a pooling-based RNN architecture, which outperforms traditional RNNs, along with other traditional machine learning algorithms in residential load forecasting. Another variation of RNN, called the gated recurrent unit network, is used in [61] for daily peak load forecasting. CNNs with k-means clustering have also been applied for short-term load forecasting for smart grids [62].

7.3.2.2   Generation Forecasting

In the field of renewable energy, deep learning has been applied for wind and solar forecasting problems. A short-term wind forecasting problem is addressed using RNNs with a so-called infinite feature selection method in [63] and using CNNs in [64]. A hybrid deep learning approach is proposed for day-ahead wind power forecasting in [65]. Wind and solar irradiance forecasting are done using CNNs with input data obtained from numerical weather prediction in [66].

Solar forecasting methodologies vary widely based on the type of inputs being used for the process. For example, a standard time series forecasting problem may only make use of previous solar irradiance measurement (endogenous variables). Alternatively, one might use ground-based meteorological parameters (exogenous variables) or sky imagery/video for predicting solar irradiance. Siddiqui et al. propose a deep learning-based approach for solar irradiance forecasting using sky videos [67]. LSTMs are used for solar power forecasting by Gensler et al. in [68] and RNNs are used in [69] for solar irradiance forecasting. Section 7.4 in this chapter examines a case study in multi-time-horizon solar irradiance forecasting using RNNs and LSTMs [55, 70].

7.3.2.3   Electricity Price Forecasting and Electric Vehicle Charging

Electricity price forecasting in competitive energy markets is a challenging prediction problem because of the rare characteristics of electricity. Electricity cannot be treated like other commodities because trading requires a balance between supply and demand at every point in time. The failure to maintain this balance results in blackouts and brownouts that are hugely detrimental to the society as a whole. The research around deep learning-based approaches to electricity price forecasting is growing. There have been a few articles exploring this topic [71–75]. Deep learning has also been applied for demand-side management for smart charging of EVs [76].

7.3.3   Deep Learning Strengths and Shortcomings

Deep learning has shown promising results in the field of predictive analysis, because of its ability to model complex, nonlinear relationships between various exogenous input variables and the associated output. It is capable of uncovering trends in the historical dataset, providing highly accurate forecasts. For power systems forecasting problems, deep learning algorithms are increasingly outpacing traditional approaches for nowcasting and short-term forecasting. In order to produce such accurate predictions for these time horizons, deep learning algorithms require a relatively significant amount of training data. The next two sections summarize the strengths and weaknesses of deep learning approaches in the context of power systems forecasting problems.

7.3.3.1 Strengths

Deep learning algorithms with recurrent connections (e.g., RNN and LSTM) are capable of capturing short- and long-term trends in time series data. When trained using exogenous variables, these algorithms are effective at finding and modeling the complex temporal relationships between various input variables. Deep learning also has the rather unique capability of performing in situ feature engineering; that is, extensive manual feature engineering is not required for deep learning algorithms like traditional machine learning algorithms. The data availability in power systems has exploded in recent years, creating a natural environment for the emergence of deep learning algorithms. For this reason, it is reasonable to assume that deep learning has yet to reach its full potential in revolutionizing power systems predictive analytics field.

7.3.3.2 Shortcomings

Deep learning models have traditionally been difficult to train because of their expensive computational costs. This limitation has been overcome in recent years with technical advances in GPUs, network architectures, and development of performance optimization techniques. While ANNs act as universal function approximators, they are also often a black-box approach to modeling. They lack interpretability and are prone to overfitting because of the high capacity to learn (especially deep neural networks). Also, deep learning algorithms require significant amounts of data for training. For cases where data are limited, deep learning algorithms may not be the optimal method to use. Lastly, for long-term forecasting horizons (5+ years), statistical methods still provide reasonably good predictions, given the limited data availability.

7.4 Case Study: Multi-Timescale Solar Irradiance Forecasting Using Deep Learning

This section reviews an example of using deep learning for real-time forecasting of solar irradiance [55, 70], where a unified architecture is proposed for predicting multi-time-horizon solar irradiance. This work uses both RNNs and LSTMs to make

Fig. 7.5 The seven SURFRAD research stations distributed across the continental USA

predictions of global horizontal irradiance (GHI), also referred to as the total solar irradiance. Recall from Sect. 7.3 that these deep learning architectures use data from previous timesteps to inform the current one. This allows the models to learn the underlying dynamics of system in order to enhance their predictive capabilities.

7.4.1 Data

The data for this study come from the seven Surface Radiation Budget Network (SURFRAD) measurement stations, scattered across the continental USA (see Fig. 7.5) that measure various meteorological parameters, including solar radiation. The distribution of these research stations across various climate zones demonstrates the robustness of the constructed networks in predicting GHI.[1] Minute-by-minute meteorological data for 2009–2011 from this database is used in this study. The data are averaged over each hour to obtain mean hourly GHI values for forecasting.

---

[1]Because the supervised training approach relies on the data from the given geographic location to learn the relationships between the various meteorological inputs and the GHI, it is difficult to transport/reuse this model for a climatically different geographical location. Thus, the model will need to be trained with a location-specific dataset for using it in various geographical locations. Therefore, the algorithm itself is robust for various locations, but the model needs to be retrained for different locations.

Data from 2010 and 2011 at each location are used for training while corresponding data from 2009 are used for measuring performance.

7.4.1.1 Global Horizontal Irradiance

Global horizontal irradiance (GHI) refers to the total solar power per unit area that is incident on some surface (e.g., a photovoltaic solar panel) and is typically measured in W/m^2. This value has two main components: (1) direct normal irradiance (DNI) and (2) diffuse horizontal irradiance (DHI). The GHI at a particular time t can be expressed as

$$\text{GHI}_t = \text{DNI}_t \times \cos(\theta) + \text{DHI}_t, \tag{7.20}$$

where θ denotes the solar zenith angle, which is the angle between the zenith (overhead) and the sun. This value is important to understanding the availability of solar energy on the grid.

The constructed networks (discussed in Sect. 7.4.2) directly predict a value known as the clear-sky index K_t. This value is a ratio of the true GHI to the expected GHI in a cloud-free scenario,

$$K_t = \text{GHI}_t \Big/ \text{GHI}_t^{\text{clear}}. \tag{7.21}$$

The clear-sky index is a dimensionless value that describes the total solar irradiance relative to a theoretical upper limit, which occurs in cloud-free situations. This acts as a type of normalization for the model that can increase robustness to location or seasonality. The clear-sky GHI ($\text{GHI}_t^{\text{clear}}$) in Eq. (7.21) is calculated using the Bird clear-sky model [77] based on latitude, longitude, elevation, and atmospheric parameters, such as column water vapor, ozone optical thickness, and aerosol optical depth. Based on this calculation and the predicted clear-sky index from the deep learning model, one can easily obtain the predicted GHI.

7.4.1.2 Exogenous Input Variables

The input to the deep learning model is a vector of 20 exogenous variables for each timestep:

- downwelling global solar (W/m^2),
- upwelling global solar (W/m^2),
- direct-normal solar (W/m^2),
- downwelling diffuse solar (W/m^2),
- downwelling thermal infrared (W/m^2),
- downwelling infrared case temperature (K),
- downwelling infrared dome temperature (K),

- upwelling thermal infrared (W/m^2),
- upwelling infrared case temperature (K),
- upwelling infrared dome temperature (K),
- global UVB (mW/m^2),
- photosynthetically active radiation (W/m^2),
- net solar (W/m^2),
- net infrared (W/m^2),
- net radiation (W/m^2),
- 10-mean air temperature (C),
- relative humidity (%),
- wind speed (*m/s*),
- wind direction ($^\circ$),
- station pressure (mb).

Not all of the variables listed here are necessarily important to the solar irradiance forecast, but they have been used in this case study. As an extension to this work, further experiments can be conducted to understand the relevance of the individual input variables and accordingly reduce the dimensionality of the dataset.

7.4.1.3 Data Preprocessing and Postprocessing

The algorithmic approach in this case study begins by preprocessing the data. This includes removing extreme outliers (values which are $+/-$ 4 standard deviation away from the mean) as well as nighttime values, filling in missing data with the mean value of surrounding points, and normalizing the input data vectors. The clear-sky GHI is computed using the Bird model (see Sect. 7.4.1.1) and used to transform target GHI values to the clear-sky index K_t. Postprocessing the data includes recovering the predicted GHI from the clear-sky index and computing the performance of the network using the mean squared error.

7.4.2 Model Architecture and Training

This case study examines two scenarios: (1) a fixed-time horizon that is similar to other statistical and machine learning forecasting approaches (such as those discussed in Sects. 7.2.2 and 7.2.3) and (2) a multi-time-horizon that is better suited for the flexibility of a deep learning model. In the fixed-time case, separate models are trained for each desired time horizon (1, 2, 3, and 4 h) while a single model is used to predict all of the time horizons in the multi-time case. In both scenarios, separate models are trained for each of the seven SURFRAD locations.

For the fixed-time-horizon problem, this study only considers traditional RNN models and compares the performance of this deep learning method to standard machine learning approaches. The network is constructed using rectified linear units (ReLU) activation functions for all hidden layers and a linear activation on the output layer. The output is scalar-valued because the goal is to predict GHI for a single time horizon.

The multi-time-horizon networks predict GHI at 1-, 2-, 3-, and 4-h time horizons simultaneously, producing a four-dimensional output vector. This work also proposes an extension to the unified architecture for predicting multi-time-scale solar irradiance, which covers 5-min, 15-min, and other such intrahour time horizons. This work compares LSTMs and RNNs; however, no comparison is made to traditional machine learning methods because these approaches are unable to perform multi-time-horizon predictions. The RNNs have similar activation architectures to the fixed-time-horizon case. The LSTM networks use sigmoid and hyperbolic tangent activations within the memory cells.

For training, the deep learning models have access to the target GHI values so that the mean squared loss can be computed. The models are trained using stochastic gradient descent where the gradients with respect to this loss are computed using backpropagation through time. The training minibatch sizes are $n = 100$, and the networks are trained for 1000 epochs.

7.4.3 Results

7.4.3.1 Single Time Horizon Model

Table 7.2 shows the comparison between the RNN performance and the performance of other machine learning forecasting approaches. The values in the ML column are those presented in [78] where the authors perform the same fixed-time-horizon study using several traditional machine learning algorithms (SVRs, random forests, and gradient boosting) as well as a traditional feedforward neural network. The listed performance is the optimal performance across all testing algorithms for each horizon/location combination. In each case, the RNN approach significantly outperforms the others.

7.4.3.2 Multi-Time-Horizon Model

Table 7.3 contains the results of the RNN/LSTM comparison study for the multi-time-horizon problem.[2] Recall that for this study, a single RNN or LSTM network is trained for each location that forecasts GHI out to all four time horizons. Neither network architecture outperformed the other across all seven locations. However, within each location there is a significant increase in error from the 3-h to the 4-h

---

[2]Note that Table 7.2 compares the GHI W/m^2 values between RNN and traditional ML approaches while Table 7.3 is comparing the performance of RNN and LSTM algorithms based on clear-sky index (which is a dimensionless parameter).

Table 7.2 Root mean squared error of predicted GHI (W/m^2) for fixed-time-horizon RNNs and machine learning (ML) approaches

	Bondville, IL		Boulder, CO		Desert Rock, NV		Fort Peck, MT		Goodwin Creek, MS		Penn State, PA		Sioux Falls, SD	
	RNN	ML	RNN	ML	RNN	ML	RNN	ML	RNN	ML	RNN	ML	RNN	ML
1-h	16.8	62	17	74	41.7	52	21.2	56	24.8	71	8.64	67	27.2	52
2-h	20.73	98	20.7	108	57.23	72	29.7	81	25.2	103	10.5	97	32.1	81
3-h	18.78	116	21.2	123	60.54	83	25.5	94	26.9	125	11.8	114	30.6	96
4-h	17.98	121	22.9	125	49.71	82	29.4	93	22	120	10.7	117	35.3	103
Mean	18.57	99.25	20.45	107.5	52.29	72.25	26.45	81	24.73	104.8	10.41	98.75	31.3	83

Table 7.3 Root mean squared error of the predicted clear-sky index for multi-time-horizon RNN and LSTM models

	Bondville, IL		Boulder, CO		Desert Rock, NV		Fort Peck, MT		Goodwin Creek, MS		Penn State, PA		Sioux Falls, SD	
	RNN	LSTM	RNN	LSTM	RNN	LSTM	RNN	LSTM	RNN	LSTM	RNN	LSTM	RNN	LSTM
1-h	0.71	0.65	0.59	0.58	0.47	0.46	0.71	0.72	0.65	0.67	0.68	0.69	0.69	0.71
2-h	1.12	1.18	1.06	1.06	0.98	0.96	1.35	1.34	1.17	1.20	1.13	1.11	1.25	1.23
3-h	4.52	4.30	4.01	4.45	4.53	4.71	6.31	6.18	4.59	4.58	3.16	3.12	5.06	4.99
4-h	52.13	43.65	49.45	58.88	153.27	126.87	69.99	77.99	58.98	54.48	28.40	24.70	73.41	77.16
Mean	14.62	12.45	13.78	16.24	39.81	33.25	19.59	21.56	16.35	15.23	8.34	7.40	20.10	21.02

forecasting horizon. This could indicate that autocorrelation in the time series data decays after 3 h.

7.5 Summary and Future Work

Given the rapidly occurring technological changes in power systems and their forthcoming transformation into the smart grid, which operates as a part of a complex amalgamation of interdependent transportation, communication, and IoT networks, there is an urgent need for developing and deploying better algorithms for forecasting the various power systems quantities. Moreover, operational uncertainties continue to increase with the burgeoning share of utility-scale as well as DER-scale renewable energy generation on the grid, calling for better short-term forecasts. These play a significant role in the optimization of the operational efficiency of power systems, both economically and in terms of reliability.

Forecasting accuracies for electricity price prediction play a major role in maintaining the economic viability of energy producers' businesses in the market. Resource and load forecasting also have a major role to play in large-scale deployment of microgrids because these quantities are the main inputs to the optimization algorithms aimed at operating the microgrids intelligently (i.e., maximizing economic benefit while maintaining the reliability of the local supply).

Deep learning algorithms (e.g., RNNs and LSTMs) have been applied to power systems forecasting problems with promising results in the recent literature. They also offer the potential to continue improving as they are trained further on the continuous stream of newly generated data. As with any research area, the goal is to ultimately move these algorithms to the industry deployment phase. Because of their low forward inference time (on the order of milliseconds), these algorithms and architectures can provide forecasts in the near real-time horizon. The performance of the deployed systems can be further improved by implementing sophisticated hyperparameter tuning mechanisms.

The following two sections briefly note some areas where there is plenty of scope as well as a need for further development of deep learning applications for power systems.

7.5.1 Deterministic Versus Probabilistic Forecasting

The forecasted values from the deep learning models discussed in Sect. 7.3 are deterministic in nature. That is, given the same sequence of inputs, the networks will always produce the same output. Furthermore, there is no measure of confidence related to the predictions. Recall the Gaussian process regression (GPR) from Sect. 7.2.3.2. This probabilistic forecasting approach naturally produces a measure of confidence based on the variance in the GP. As the predicted values get further from any given data, the variance grows and the confidence in the predicted value decreases. Such understanding of uncertainties in power systems forecasting is critical because of the highly variable nature of the data and the large cost of grid blackouts and brownouts. Some work has considered how deep learning can be recast as a probabilistic model [79, 80], but continued research into the topic is critical.

7.5.2 Other Potential Applications

Anomaly detection in smart grids is a timely and relevant topic as the distribution grid infrastructure in industrialized countries like the USA has aged and needs refurbishment and replacement to maintain the reliability of the supply. There has been relatively less progress in applying deep learning algorithms for anomaly detection in power systems prognostics and fault prediction problems [81–83]. The application areas include anomaly detection for predicting the remaining useful life of the components of power systems, predicting impending fault on power systems, and predicting building level faults based on the data from building sensors.

Acknowledgments The authors wish to thank Kate Anderson and Adam Warren (National Renewable Energy Laboratory) for the encouragement to pursue this research work. The authors would also like to thank Manajit Sengupta and Ryan King (National Renewable Energy Laboratory) for providing useful suggestions to refine the manuscript. This research did not receive any specific grant from funding agencies in the public, commercial, or not-for-profit sectors. The views expressed in the article do not necessarily represent the views of the DOE or the U.S. Government.

References

1. M. Shahidhpour, H. Yamin, Z. Li, *Market Operations in Electric Power Systems: Forecasting, Scheduling, and Risk Management* (Wiley-IEEE Press, Hoboken, 2002)
2. Federal Energy Regulatory Commission, Order no. 1000 - Transmission planning and cost allocation, 21 July 2011. (Online). https://www.ferc.gov/whats-new/comm-meet/2011/072111/E-6.pdf. Accessed 27 June 2019
3. I. Penn, P. Eavis, J. Glanz, How PG&E ignored fire risks in favor of profits, New York Times, 2019
4. H. Chen, Y. Zhang, H. Ngan, Power System Planning, in Power System Optimization: Large-Scale Complex Systems Approaches (Wiley, Singapore, 2016), pp. 76–130
5. A.S. Malik, C. Kuba, Power generation expansion planning including large scale wind integration: a case study of Oman. J. Wind Energy **2013**, 1 (2013)
6. M. Child, C. Kemfert, D. Bogdanov, C. Breyer, Flexible electricity generation, grid exchange and storage for the transition to a 100% renewable energy system in Europe. Renew. Energy **139**, 80–101 (2019)
7. I.F. Abdin, E. Zio, An integrated framework for operational flexibility assessment in multi-period power system planning with renewable energy production. Appl. Energy **222**, 898–914 (2018)
8. O.J. Guerra, D.A. Tejada, G.V. Reklaitis, An optimization framework for the integrated planning of generation and transmission expansion in interconnected power systems. Appl. Energy **170**, 1–21 (2016)

9. H. Kim, S. Lee, S. Han, W. Kim, K. Ok, S. Cho, *Integrated generation and transmission expansion planning using generalized Bender's decomposition method*, in IEEE International Conference on Computational Intelligence & Communication Technology, Ghaziabad, India, 2015
10. S. Hasanvand, M. Nayeripour, H. Fallahzadeh-Abarghouei, A new distribution power system planning approach for distributed generations with respect to reliability assessment. J. Renewable Sustainable Energy **8**, 045501 (2016)
11. B. Singh, C. Pal, V. Mukherjee, P. Tiwari, M.K. Yadav, Distributed generation planning from power system performances viewpoints: a taxonomical survey. Renew. Sust. Energ. Rev. **75**, 1472–1492 (2017)
12. M.R. AlRashidi, K.M. EL-Naggar, Long term electric load forecasting based on particle swarm optimization. Appl. Energy **87**(1), 320–326 (2010)
13. A. Poghosyan, D.V. Greetham, S. Haben, T. Lee, Long term individual load forecast under different electrical vehicles uptake scenarios. Appl. Energy **157**(1), 699–709 (2015)
14. M. Zhao, W. Liu, J. Su, L. Zhao, X. Dong, W. Liu, *Medium and long term load forecasting method for distribution network with high penetration DGs*, in 2014 China International Conference on Electricity Distribution (CICED), Shenzhen, China, 2014
15. E. Vaahedi, *Practical power system operation* (Wiley, Hoboken, 2014)
16. M. Matos, R. Bessa, A. Botterud, Z. Zhou, Forecasting and setting power system operating reserves, in *Renewable Energy Forecasting - From Models to Applications*, ed. by G. Kariniotakis (Woodhead Publishing Elsevier, Duxford, 2017), pp. 279–308
17. R. Weron, Electricity price forecasting: a review of the state-of-the-art with a look into the future. Int. J. Forecast. **30**, 1030–1081 (2014)
18. N. Mazzi, P. Pinson, Wind power in electricity markets and the value of forecasting, in *Renewable Energy Forecasting - From Models to Applications*, Woodhead Publishing Series in Energy (Woodhead Publishing Elsevier, Duxford, 2017), pp. 259–278
19. C. Frohlich, J. Lean, Total Solar Irradiance (TSI) composite database, National Oceanic and Atmospheric Administration (Online). https://www.ngdc.noaa.gov/stp/solar/solarirrad.html. Accessed July 2019
20. C. Voyant, G. Notton, S. Kalogirou, M.-L. Nivet, C. Paoli, F. Motte, A. Fouilloy, Machine learning methods for solar radiation forecasting: a review. Renew. Energy **105**, 569–582 (2017)
21. R.G. Brown, Exponential smoothing for predicting demand. Oper. Res. **5**(1), 145 (1957)
22. C.C. Holt, *Forecasting Seasonals and Trends by Exponentially Weighted Moving Averages* (Office of Naval Research, Arlington, 1957)
23. P.R. Winters, Forecasting sales by exponentially weighted moving averages. Manag. Sci. **6**(3), 324–342 (1960)
24. W.R. Christiaanse, Short-term load forecasting using general exponential smoothing. IEEE Trans. Power Syst. **PAS-90**(2), 900–911 (1971)
25. J.W. Taylor, L.M. de Menezes, P.E. McSharry, A comparison of univariate methods for forecasting electricity demand up to a day ahead. Int. J. Forecast. **22**, 1–16 (2006)
26. J.W. Taylor, P.E. McSharry, Short-term load forecasting methods: an evaluation based on European data. IEEE Trans. Power Syst. **22**(4), 2213–2219 (2007)
27. T.N. Goh, K.J. Tan, Stochastic modeling and forecasting of solar radiation data. Sol. Energy **19**, 755–757 (1977)
28. S. Vemuri, W.L. Huang, D.J. Nelson, On-line algorithms for forecasting hourly loads of an electric utility. IEEE Trans. Power Syst. **PAS-100**(9), 3775–3784 (1981)
29. G. Reikard, Predicting solar radiation at high resolutions: a comparison of time series forecasts. Sol. Energy **83**(3), 342–349 (2009)
30. A. Moreno-Munoz, J. J. G. de la Rosa, R. Posadillo, F. Bellido, *Very short term forecasting of solar radiation*, in 2008 33rd IEEE Photovoltaic Specialists Conference, 2008
31. J. Contreras, R. Espinola, F.J. Nogales, A.J. Conejo, ARIMA models to predict next-day electricity prices. IEEE Trans. Power Syst. **18**(3), 1014–1020 (2003)
32. N. Amjady, Short-term hourly load forecasting using time-series modeling with peak load estimation capability. IEEE Trans. Power Syst. **16**(3), 498–505 (2001)

33. G.E.P. Box, G.M. Jenkins, *Time Series Analysis: Forecasting and Control* (Prentice Hall, Englewood Cliffs, 1994)
34. H. Amini, A. Kargarian, O. Karabasoglu, ARIMA-based decoupled time series forecasting of electric vehicle charging demand for stochastic power system operation. Electr. Power Syst. Res. **140**, 379–390 (2016)
35. T. Hastie, R. Tibshirani, J. Friedman, *The Elements of Statistical Learning* (Springer, New York, 2001)
36. N.K. Ahmed, A.F. Atiya, N.E. Gayar, H. El-Shishiny, An empirical comparison of machine learning models for time series forecasting. Econ. Rev. **29**(5–6), 594–621 (2010)
37. R.H. Inman, H.T.C. Pedro, C.F.M. Coimbra, Solar forecasting methods for renewable energy integration. Prog. Energy Combust. Sci. **39**(6), 535–576 (2013)
38. V. Vapnik, *The Nature of Statistical Learning Theory* (Springer, New York, 1995)
39. N. Cristianini, J. Shawe-Taylor, *An Introduction to Support Vector Machines and Other Kernel-Based Learning Methods* (Cambridge University Press, Cambridge, 2000)
40. E. Ceperic, V. Ceperic, A. Baric, A strategy for short-term load forecasting by support vector regression machines. IEEE Trans. Power Syst. **28**(4), 4356–4364 (2013)
41. J. Dhillon, S.A. Rahman, S.U. Ahmad, M.D.J. Hossain, *Peak electricity load forecasting using online support vector regression*, in IEEE Canadian Conference on Electrical and Computer Engineering, 2016
42. J. Jose, V. Margaret, K.U. Rao, *Impact of demand response contracts on short-term load forecasting in smart grid using SVR optimized by GA*, in Innovations in Power and Advanced Computing Technologies, 2017
43. E.E. Elattat, J. Goulermas, Q.H. Wu, Electric load forecasting based on locally weighted support vector regression. IEEE Trans. Syst. Man Cybern. Part C Appl. Rev. **40**(4), 438–447 (2010)
44. L. Ghelardoni, A. Ghio, D. Anguita, Energy load forecasting using empirical mode decomposition and support vector regression. IEEE Trans. Smart Grid **4**(1), 549–556 (2013)
45. C.E. Rasmussen, C.K.I. Williams, *Gaussian Processes for Machine Learning* (The MIT Press, Cambridge, 2006)
46. H. Mori, M. Ohmi, *Probabilistic short-term load forecasting with Gaussian processes*, in Proceedings of the 13th International Conference on Intelligent Systems Application to Power Systems, 2005
47. A. Rohani, M. Taki, M. Abdollahpour, A novel soft computing model (Gaussian process regression with K-fold cross validation) for daily and monthly solar radiation forecasting (part: I). Renew. Energy **115**, 411–422 (2018)
48. N. Chen, Z. Qian, I.T. Nabney, X. Meng, Wind power forecasts using Gaussian processes and numerical weather prediction. IEEE Trans. Power Syst. **29**(2), 656–665 (2014)
49. S. Salcedo-Sanz, C. Casanova-Mateo, J. Munoz-Mari, G. Camps-Valls, Prediction of daily global solar irradiation using temporal Gaussian processes. IEEE Geosci. Remote Sens. Lett. **11**(11), 1936–1940 (2014)
50. A. Ulbig, T. Borsche, G. Andersson, Impact of low rotational inertia on power system stability and operation. IFAC Proc. Vol. **47**(3), 7290–7297 (2014)
51. E. Bitar, P.P. Khargonekar, K. Poolla, Systems and control opportunities in the integration of renewable energy into the smart grid. IFAC Proc. Vol. **44**(1), 4927–4932 (2011)
52. K. Hornik, M. Stinchombe, H. White, Multilayer feedforward networks are universal approximators. Neural Netw. **2**(5), 359–366 (1989)
53. Y. LeCun, Y. Bengio, G. Hilton, Deep learning. Nature **521**, 436–444 (2015)
54. A. Graves, *Supervised sequence labelling with recurrent neural networks* (Springer-Verlag, Berlin, 2012)
55. S. Mishra, P. Palanisamy, An integrated multi-time-scale modeling for solar irradiance forecasting using deep learning, arXiv:1905.02616, 2019
56. A.v.d. Oord, S. Dieleman, H. Zen, K. Simonyan, O. Vinyals, A. Graves, N. Kalchbrenner, A. Senior, K. Kavukcuoglu, *WaveNet: a generative model for raw audio*, in DeepMind, 2016

57. A. Borovykh, S. Bohte, C.W. Oosterlee, Conditional time series forecasting with convolutional neural networks, arXiv:1703.04691, 2018
58. D.L. Marino, K. Amarasinghe, M. Manic, *Building energy load forecasting using deep neural networks*, in IECON 2016 - 42nd Annual Conference of the IEEE Industrial Electronics Society, Florence, Italy, 2016
59. W. Kong, Z.Y. Dong, Y. Jia, D.J. Hill, Y. Xu, Y. Zhang, Short-term residential load forecasting based on LSTM recurrent neural network. IEEE Trans. Smart Grid 10(1), 841–851 (2019)
60. H. Shi, M. Xu, R. Li, Deep learning for household load forecasting - a novel pooling deep RNN. IEEE Trans. Smart Grid 9(5), 5271–5280 (2018)
61. Z. Yu, Z. Niu, W. Tang, Q. Wu, Deep learning for daily peak load forecasting–a novel gated recurrent neural network combining dynamic time warping. IEEE Access 7, 17184–17194 (2019)
62. X. Dong, L. Qian, L. Huang, *Short-term load forecasting in smart grid: a combined CNN and K-means clustering approach*, in 2017 IEEE International Conference on Big Data and Smart Computing, Jeju, South Korea, 2017
63. H. Shao, X. Deng, Y. Jiang, A novel deep learning approach for short-term wind power forecasting based on infinite feature selection and recurrent neural network. J. Renewable Sustainable Energy 10, 043303 (2018)
64. C.-J. Huang, P.-H. Kuo, A short-term wind speed forecasting model by using artificial neural networks with stochastic optimization for renewable energy systems. Energies 11, 2777 (2018)
65. Y.-Y. Hong, C.L.P.P. Rioflorido, A hybrid deep learning-based neural network for 24-h ahead wind power forecasting. Appl. Energy 250, 530–539 (2019)
66. D. Diaz-Vico, A. Torres-Barran, A. Omari, J.R. Dorronsoro, Deep neural networks for wind and solar energy prediction. Neural Process. Lett. 46, 829 (2017)
67. T.A. Siddiqui, S. Bharadwaj, S. Kalyanaraman, *A deep learning approach to solar-irradiance forecasting in sky-videos, in 2019 IEEE Winter Conference on Applications of Computer Visions (WACV),* Waikoloa Village, HI, USA, 2019
68. A. Gensler, J. Henze, B. Sick, N. Raabe, *Deep learning for solar power forecasting - an approach using AutoEncoder and LSTM neural networks*, in IEEE International Conference on Systems, Man, and Cybernetics (SMC), Budapest, Hungary, 2016
69. A. Alzahrani, P. Shamsi, C. Dagli, M. Ferdowsi, Solar irradiance forecasting using deep neural networks. Proc. Comput. Sci. 114, 304–313 (2017)
70. S. Mishra, P. Palanisamy, *Multi-time-horizon solar forecasting using recurrent neural network*, in 2018 IEEE Energy Conversion Congress and Exposition (ECCE), Portland, OR, 2018
71. W. Zhang, F. Cheema, D. Srinivasan, *Forecasting of electricity prices using deep learning networks*, in 2018 IEEE PES Asia-Pacific Power and Energy Engineering Conference (APPEEC), Kota Kinabalu, Malaysia, 2018
72. R.A. Chinnathambi, S.J. Plathottam, T. Hossen, A.S. Nair, P. Ranganathan, *Deep neural networks (DNN) for day-ahead electricity price markets*, in 2018 IEEE Electrical Power and Energy Conference (EPEC), Toronto, ON, Canada, 2018
73. J. Lago, F.D. Ridder, B.D. Schutter, Forecasting spot electricity prices: deep learning approaches and empirical comparison of traditional algorithms. Appl. Energy 221, 386–405 (2018)
74. A. Brusaferri, M. Matteucci, P. Portlani, A. Vitali, Bayesian deep learning based method for probabilistic forecast of day-ahead electricity prices. Appl. Energy 250, 1158–1175 (2019)
75. S. Atef, A.B. Eltawil, *A comparative study using deep learning and support vector regression for electricity price forecasting in smart grids*, in 2019 IEEE 6th International Conference on Industrial Engineering and Applications (ICIEA), Tokyo, Japan, 2019
76. K.L. Lopez, C. Gagne, M.-A. Gardner, Demand-side management using deep learning for smart charging of electric vehicles. IEEE Trans. Smart Grid 10(3), 2683–2691 (2019)
77. R.E. Bird, R.L. Hulstrom, *Simplified Clear Sky Model for Direct and Diffuse Insolation on Horizontal Surfaces* (Solar Energy Research Institute, Golden, CO, 1981)

78. A. Dobbs, T. Elgindy, B.-M. Hodge, A. Florita, J. Novacheck, *Short-Term Solar Forecasting Performance of Popular Machine Learning Algorithms* (National Renewable Energy Laboratory, Golden, CO, 2017)
79. G. Li, J. Shi, J. Zhou, Bayesian adaptive combination of short-term wind speed forecasts from neural network models. Renew. Energy **36**(1), 352–359 (2011)
80. N. Bassamzadeh, R. Ghanem, Multiscale stochastic prediction of electricity demand in smart grids using Bayesian networks. Appl. Energy **193**, 369–380 (2017)
81. Z. Fengming, L. Shufang, G. Zhimin, W. Bo, T. Shiming, P. Mingming, Anomaly detection in smart grid based on encoder-decoder framework with recurrent neural network. J. China Univ. Posts Telecommun. **24**(6), 67–73 (2017)
82. L. Banjanovic-Mehmedovic, A. Hajdarevic, M. Kantardzic, F. Mehmedovic, I. Dzananovic, Neural network-based data-driven modelling of anomaly detection in thermal power plant. J. Control Meas. Electron. Comput. Commun. **58**(1), 69 (2017)
83. C. Zhang, D. Song, Y. Chen, X. Feng, C. Lumezanu, W. Cheng, J. Ni, B. Zong, H. Chen, N.V. Chawla, *A deep neural network for unsupervised anomaly detection and diagnosis in multivariate time series data,* in Proceedings of the AAAI Conference on Artificial Intelligences, 2019

Chapter 8
Bi-level Adversary-Operator Cyberattack Framework and Algorithms for Transmission Networks in Smart Grids

M. Hadi Amini, Javad Khazaei, Darius Khezrimotlagh, and Arash Asrari

Abstract Transmission system is one of the most important assets in secure power delivery. Recent advancements toward automation of smart grids and application of supervisory control and data acquisition (SCADA) systems have increased vulnerability of power grids to cyberattacks. Cyberattacks on transmission network, specifically the power transmission lines, are among crucial emerging challenges for the operators. If not identified properly and in a timely fashion, they can cause cascading failures leading to blackouts. This chapter tackles false data injection modeling from the attacker's perspective. It further develops an algorithm for detection of false data injections in transmission lines. To this end, first, a bi-level mixed integer programming problem is introduced to model the attack scenario, where the attacker can target a transmission line in the system and inject false data in load measurements on targeted buses in the system to overflow the targeted line. Second, the problem is analyzed from the operator's viewpoint and a detection

M. H. Amini (✉)
School of Computing and Information Sciences, Florida International University, Miami, FL, USA

Sustainability, Optimization, and Learning for InterDependent Networks Laboratory (solid lab), Florida International University, Miami, FL, USA
e-mail: moamini@fiu.edu; hadi.amini@ieee.org; www.solidlab.network

J. Khazaei
Electrical Engineering, Architectural Engineering, Penn State University Park, State College, PA, USA
e-mail: jxk792@psu.edu

D. Khezrimotlagh
Mathematics, Penn State Harrisburg, Middletown, PA, USA
e-mail: dzk349@psu.edu

A. Asrari
Electrical and Computer Engineering, Southern Illinois University, Carbondale, IL, USA
e-mail: arash.asrari@siu.edu

© Springer Nature Switzerland AG 2020
M. H. Amini (ed.), *Optimization, Learning, and Control for Interdependent Complex Networks*, Advances in Intelligent Systems and Computing 1123,
https://doi.org/10.1007/978-3-030-34094-0_8

algorithm is proposed using l_1 norm minimization approach to identify the bad measurement vector in data readings. In order to evaluate the effectiveness of the proposed attack model, case studies have been conducted on IEEE 57-bus test system.

Keywords Cyberphysical security · Optimal attacker strategy · Attack detection · Sparsity-based decomposition · Energy systems · Smart grid

8.1 Introduction

8.1.1 Overview

Transmission system is one of the most important assets in secure power delivery. Recent advancements toward automation of smart grids and application of supervisory control and data acquisition (SCADA) systems have increased vulnerability of power grids to cyberattacks. Cyberattacks on transmission network, specifically the power transmission lines, are among crucial emerging challenges for the operators. If not identified properly and in a timely fashion, they can cause cascading failures leading to blackouts. This chapter tackles false data injection modeling from the attacker's perspective. It further develops an algorithm for detection of false data injections in transmission lines. To this end, first, a bi-level mixed integer programming problem is introduced to model the attack scenario, where the attacker can target a transmission line in the system and inject false data in load measurements on targeted buses in the system to overflow the targeted line. Second, the problem is analyzed from the operator's viewpoint and a detection algorithm is proposed using l_1 norm minimization approach to identify the bad measurement vector in data readings. In order to evaluate the effectiveness of the proposed attack model, case studies have been conducted on IEEE 57-bus test system.

In recent years, as online monitoring devices have widely being developed and implemented in smart grids, cybersecurity has also become a more serious issue to be tackled. In addition, interconnection of power systems in different areas and development of advanced communication technologies to automate smart grid assets have made the grid more vulnerable to cyber-physical attacks. A cyberattacker can therefore inject computer viruses or anomalies to endanger the security and resiliency of the smart grid system [1–4]. Example of such attach includes Russian's cyberattack on obtaining detailed data on nuclear power plants and water facilities in the USA in 2018 [6]. Another real-world example is the successful cyberattack by Russian hackers in December 2015 on the Ukraine power grid. In this cyberattack, 30 substations were switched off by hackers which resulted in a power outage of 1–6 h for almost 230,000 people [7].

Transmission lines are the most important assets in power delivery in smart grids, and if failed, can cause serious cascading problems leading to blackouts. One example of such cascading failure was a blackout in Italy that happened in 2004

[8]. Cyberattacks on transmission lines are normally designed to overflow a line or series of lines with the aim of cascading failures. Such transmission line congestion can be achieved if an attacker injects false data on load measurements without being detected by bad data detection algorithms in state estimation process. Recent studies show that by having information on the topology of the system, these false data injections can be designed in a way to bypass state estimation methods without being detected [9–11]. Thus, to protect the smart grid against these vulnerabilities and increase the resiliency of the system, it is crucial to understand the problem from attacker's point of view and develop models that account for various attack scenarios in transmission line congestion.

A few studies developed models for false data injection attacks with the aim of bypassing state estimation in smart grids [12–17]. For instance, [12] extensively modeled false data injection attacks which could not be detected by DC state estimation algorithms. It was shown that if the injected false values in load buses follow system's admittance matrix (B), the attack can be successful. To identify the worst attacking strategy in false data injection attacks, a heuristic algorithm was developed in [14]. Furthermore, an attack model was formulated in [16] that would allow the attacker to make profits in real-time markets. Moreover, a comparison between few bad data detection algorithms were reported in [17]. Although these studies provided a full insight to false data injection problems in smart grids, they did not focus on transmission line congestion and also would require the attacker to have complete access to the system, which was not realistic.

There are also several studies which aimed to generalize false load data injection attacks by focusing on incomplete power system models and limited access by the attackers [18–22]. For example, a practical model on false data injection attack was introduced in [19]. Physical constraints of the smart grid system were considered to formulate the model. The model was developed based on the fact that the attackers could not alter the generator output powers. As a result, the power balance needed to be met continuously. Furthermore, an attack model was developed in [18], which only relied on the data and the topology of a local targeted region that could bypass the bad data detection algorithm. Recent studies showed that mixed integer linear programming (MILPs) modeling of the cyberattacks is the most suitable and practical modeling procedures for false data injection attacks [19, 21]. Several studies introduced MILP attack models to initially maximize the system costs [20, 22]. Nevertheless, the main focus of these studies were on attacks which maximized the system costs, and attacks on transmission lines' congestion were not considered.

In [23], in order to overflow transmission lines, a tri-level MILP solution was developed. This model took into account the security constrained economic dispatch problem to find an optimal strategy to overflow multiple targeted transmissions lines through data falsification. Although this model is effective for the scenario in which attacker has access to all lines, in realistic scenarios attacker does not have access to all buses. Due to complexity of tri-level model, it increases the complexity of protection schemes as well. In order to tackle the complexity issues regarding the model in [23], a bi-level MILP model was proposed in [24]. This model also ignored some practical limits. Optimal injection on load buses was not modeled in this study, i.e., there is no specific strategy to choose the most vulnerable bus from attacker's perspective. In practice, this may not be possible due to limited access to a part of network.

This study addresses two cybersecurity problems from two completely different perspectives: (1) optimal attack strategy from the attacker's viewpoint, (2) state estimation to identify bad data injection considering presence of cyberattacks from operator's viewpoint. The main advantages of the presented MILP approach for the attacker are:

- Optimizing false data injections rather than number of target buses to achieve a more realistic attack model to overflow the transmission lines.
- Restricting transmission line overflow to a certain upper limit to prevent unrealistic spike in line flow, e.g., two times as compared with normal operating scenario in some cases [24].
- Making the realistic assumption that all lines/buses may not be accessible to the attacker. To this end, our model assumes attacking to a target load bus without being detected by the operator.
- Developing a bi-level MILP model as an alliterative to tri-level methods. This simplifies protection scheme in resilient power system studies. Further, we have included load injection limits to outperform the bi-level model in [24].

More importantly, this chapter proposes a detection framework that identifies the false data injections in smart grids, this has never been studied in the existing literature.

The organization of this chapter is as follows. Section 8.2 provides the preliminaries regarding DC power flow model. Section 8.3 explains the false data injection attacks and their effect on smart grid state estimation. Section 8.4 explores a bi-level MILP model to find the optimal strategy of an attacker in transmission networks. Section 8.5 is devoted to the operator's strategy for bad data injection identification using sparsity-based decomposition. Section 8.6 includes the case studies on IEEE 57-bus benchmark model, followed by Sect. 8.7 that concludes the paper. The overall structure of this study is provided in Fig. 8.1.

8.2 DC Power Flow Model

Here, we provide the preliminaries for DC power flow formulation. This notation is used both for finding the optimal cyberattack strategy for the attacker and developing a bad data detection method for the operator.

The following assumptions are made due to the physical characteristics of transmission networks to obtain linear DC power flow formulation as opposed to nonlinear AC power flow model:

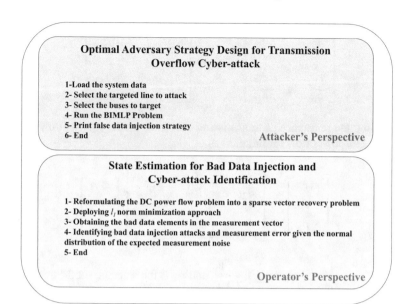

Optimal Adversary Strategy Design for Transmission Overflow Cyber-attack

1-Load the system data
2- Select the targeted line to attack
3- Select the buses to target
4- Run the BIMLP Problem
5- Print false data injection strategy
6- End

Attacker's Perspective

State Estimation for Bad Data Injection and Cyber-attack Identification

1- Reformulating the DC power flow problem into a sparse vector recovery problem
2- Deploying l_j norm minimization approach
3- Obtaining the bad data elements in the measurement vector
4- Identifying bad data injection attacks and measurement error given the normal distribution of the expected measurement noise
5- End

Operator's Perspective

Fig. 8.1 Overall structure of this study: exploring the cyberattack from attacker's and operator's perspectives

1. Due to the high X/R ratio in transmission networks, only the inductive component of impedance is considered [25], i.e., $R_{ij} = 0$ for all lines.
2. Voltage mismatch among neighboring (connected) buses is negligible, i.e., $\delta_i - \delta_j \approx 0$; hence, we can make the following approximations $\cos(\delta_i - \delta_j) \approx 1$ and $\sin(\delta_i - \delta_j) \approx 0$.
3. Due to the low voltage deviation in transmission networks, all voltage magnitudes are set to 1 p.u.

Newton–Raphson power flow method is the most suitable approach to solve the power flow equations due to its quadratic convergence. DC power flow is a simplified version of decoupled power flow by further dropping the $Q - |V|$ equations and assuming a constant voltage profile for all the buses in the system. Therefore, it is assumed that $|V_i| = 1$ p.u. for all the buses. Assuming the system has N buses and bus number 1 is the slack bus (e.g., $V_1 = 1\angle 0$ p.u.), the decoupled power flow can be modeled by

$$\begin{bmatrix} \Delta P \\ \Delta Q \end{bmatrix} = \begin{bmatrix} J_1 & 0 \\ 0 & J_4 \end{bmatrix} \begin{bmatrix} \Delta\theta \\ \Delta|V| \end{bmatrix} \tag{8.1}$$

where $\Delta P = [\Delta P_2, \Delta P_3, \ldots, \Delta P_N]$ and $\Delta Q = [\Delta Q_2, \Delta Q_3, \ldots, \Delta Q_N]$ are vectors representing the mismatch between the scheduled and calculated active/reactive powers (e.g., $\Delta P_i = P_i^{\text{sch}} - P_i^{(k)}$, $\Delta Q_i = Q_i^{\text{sch}} - Q_i^{(k)}$), J_1 and J_4 are elements of Jacobean matrix represented by

$$
J_1 = \begin{bmatrix} \dfrac{\partial P_2}{\partial \theta_2} & \cdots & \dfrac{\partial P_2}{\partial \theta_N} \\ \vdots & \ddots & \vdots \\ \dfrac{\partial P_N}{\partial \theta_2} & \cdots & \dfrac{\partial P_N}{\partial \theta_N} \end{bmatrix}, \quad J_4 = \begin{bmatrix} \dfrac{\partial Q_2}{\partial |V|_2} & \cdots & \dfrac{\partial Q_2}{\partial |V|_N} \\ \vdots & \ddots & \vdots \\ \dfrac{\partial Q_N}{\partial |V|_2} & \cdots & \dfrac{\partial Q_N}{\partial |V|_N} \end{bmatrix} \tag{8.2}
$$

Disregarding the resistance of transmission lines, this model can be further simplified to the DC power flow formulation as,

$$
\begin{bmatrix} \Delta P_2 \\ \vdots \\ \Delta P_N \end{bmatrix} = \begin{bmatrix} B_{11} & B_{12} & B_{13} & \dots & B_{1N} \\ B_{21} & B_{22} & B_{23} & \dots & B_{2N} \\ \vdots & \vdots & \vdots & \ddots & \vdots \\ B_{N1} & B_{N2} & B_{N3} & \dots & B_{NN} \end{bmatrix} \begin{bmatrix} \Delta \theta_2 \\ \vdots \\ \Delta \theta_N \end{bmatrix} \tag{8.3}
$$

where $B_{ik} = -\dfrac{1}{x_{ik}}$ and $B_{ii} = \sum_{i=1}^{N} \dfrac{1}{x_{ik}}$, and x_{ik} is the reactance of the transmission line between bus i and bus k. In a matrix form, the DC power flow is formulated as

$$
P - D = B\theta \tag{8.4}
$$

where P is the vector of generated active powers and D is the demand vector in the system. The power flow in transmission lines between bus i and k using DC power flow is formulated by

$$
P_{ik} = \frac{1}{x_{ik}}(\theta_i - \theta_k) \tag{8.5}
$$

8.3 False Data Injection Attacks Based on DC State Estimation

In smart grids, the supervisory control and data acquisition is in charge of estimating the parameters of the system based on received phasor measurement unit (PMU) measurements. This is mainly to ensure an error-free measurement by running a bad data detection algorithm in state estimation process. State estimation is the process of using sample measurements to calculate the values of state variables in power systems. In DC power flow, since the only variables are bus voltage angles, the objective of state estimation is to estimate the bus voltage angles using measured values. The maximum likelihood criterion is normally used to estimate the parameters of the system in this case. The objective of maximum likelihood criterion is to maximize the probability that estimated value $(\hat{x})$ is the true value of the state variable x. This can be formulated as

$$P(\hat{x}) = x \qquad (8.6)$$

In the maximum likelihood criterion, it is assumed that the probability density function (PDF) of the random measurement errors is known. However, if the PDF of sample measurements is assumed to follow a normal (Gaussian) distribution function, the least square estimates can be used to estimate the states of the system [26]. Therefore, if a single parameter, θ, is to be estimated using N_m measurements, the objective is to [26]

$$\min \sum_{i=1}^{N_m} \frac{\left[\theta_i^{\text{meas}} - f_i(\theta)\right]^2}{\sigma_i^2} \qquad (8.7)$$

where $f_i(\theta)$ is the function used to calculate the bus voltage angles, which is equal to $H\theta$ in DC power flow. It is noted that H is an $N_m \times N_s$ matrix of the coefficients of linear function $f_i(\theta)$ and is related to transmission line reactances. Furthermore, σ_i^2 is the variance of ith measurement, θ_i^{meas} is the measured bus voltage angle, and N_m is the total number of measurements in the system. By converting (8.7) to matrix form, the problem can be written as [26]

$$\min J(\theta) = [\theta^{\text{meas}} - H\theta]^T W^{-1} [\theta^{\text{meas}} - H\theta] \qquad (8.8)$$

where $J(\theta)$ is the measurement residual, and θ^{meas} and W are defined as

$$\theta^{\text{meas}} = \begin{bmatrix} \theta_1^{\text{meas}} \\ \theta_2^{\text{meas}} \\ \vdots \\ \theta_{N_m}^{\text{meas}} \end{bmatrix}, W = \begin{bmatrix} \sigma_1^2 & & & \\ & \sigma_2^2 & & \\ & & \ddots & \\ & & & \sigma_{N_m}^2 \end{bmatrix} \qquad (8.9)$$

To find the minimum of $J(\theta)$ in (8.8), the gradient of measurement residual must be zero (e.g., $\nabla J(\theta) = 0$), this will result in a solution of estimated values

$$\hat{\theta} = [H^T W^{-1} H]^{-1} H^T W^{-1} \theta^{\text{meas}} \qquad (8.10)$$

After deriving the estimated values, the SCADA system normally runs a bad data check to calculate the two-norm value of the mismatch between the estimated values and measured values. If the error is greater than the threshold, the bad data is detected. It was proved in [23] that an attacker can bypass the bad data detection algorithm if the false data is designed to satisfy

$$\Delta z = H\Delta\theta \qquad (8.11)$$

where $\Delta\theta$ is the change in the bus voltage angles, and Δz is the change in the measurement vector due to the false data injection.

It should be noted that the successful attack might not be guaranteed, this is because of the fact that after the false data injections, the control center should adjust the generation powers for an optimal power flow that results in a lower system cost. The economic dispatch problem will then be run in presence of false data injection, which results in a new transmission flow that deviates from normal load flow results. Since the economic dispatch might have multiple solutions, the success of attack cannot be guaranteed [23]. The main assumptions for false data injection attacks due to system limitations can be listed as [27]:

- The synchronous generator readings cannot be altered
- The measurement tampering on each load is limited within its nominal rating
- Power balance, which is a mismatch between the generation and load, should always be met

These limitations will mathematically be modeled and included in the cyberattack problem to be formulated in the next section. It is also assumed that the attacker has limited access to the system buses for false data injection. Therefore, a subset of all system buses (F) that the attacker can access is defined to highlight the fact that the attack can only be done on the subset. In addition, to avoid supply–demand violation, sum of injected powers by the attacker has to be zero, this can be formulated as

$$\sum_{i \in F} \Delta D_i = 0 \tag{8.12}$$

where ΔD_i is false active power injection at bus i. To account for limited injection on each measurement, a new constraint has to be defined,

$$-\tau D_i \leq \Delta D_i \leq \tau D_i \tag{8.13}$$

where τ is the limit on the maximum injection and is considered as 15% in this study, and D_i is the nominal load at bus i.

8.4 Attacker's Problem: Finding the Optimal Set of Target Transmission Lines using MILP

In this section, the attacker's problem is modeled as a bi-level mixed integer linear programming (MILP) optimization model, where an attacker can target a transmission line and overflow the targeted line by injecting false data on targeted buses. The assumption is that the attackers have enough information on topology of the system to conduct attacks; however, they might not have access to all buses in the system. The attacker's problem is designed in a way that the transmission flow of a targeted line always exceeds the maximum thermal limit of the line after running the security constraint economic dispatch problem. As a result, regardless of the economic dispatch solution (it might have multiple solutions), the targeted line will always be overflowed. Another assumption of this problem is that all generating units are online in the period of false data injection. Thus, instead of unit commitment, economic dispatch is formulated.

$$\min \sum_{i \in F} \Delta D_i \tag{8.14}$$

$$s.t. \ P_{ij}^t + U_i M \geq \alpha P_{ij}^{\max} \tag{8.15}$$

$$- P_{ij}^t + (1 - U_i)M \geq \alpha P_{ij}^{\max} \tag{8.16}$$

$$- 1.2 \leq \frac{P_{ij}^t}{P_{ij}^{\max}} \leq 1.2 \tag{8.17}$$

$$\sum_{i \in F} \Delta D_i = 0 \tag{8.18}$$

$$- \tau D_i \leq \Delta D_i \leq \tau D_i \tag{8.19}$$

$$P - (D + \Delta D) = B\theta^t \tag{8.20}$$

$$P_{ij}^t = \frac{\theta_i^t - \theta_j^t}{x_{ij}} \tag{8.21}$$

$$- 2\pi \leq \theta_i^t \leq 2\pi \tag{8.22}$$

$$\min \sum_{i=1}^{n_G} C_{g,i}(P_{G,i}) \tag{8.23}$$

$$s.t. \ C_{g,i}(P_{G,i}) = a_i + b_i P_{G,i} + c_i P_{G,i}^2 \tag{8.24}$$

$$P - (D + \Delta D) = B\theta^f \tag{8.25}$$

$$P_{G,i}^{\min} \leq P_{G,i} \leq P_{G,i}^{\max} \tag{8.26}$$

$$- P_{ij}^{\max} \leq \frac{\theta_i^f - \theta_j^f}{x_{ij}} \leq P_{ij}^{\max} \tag{8.27}$$

The developed MILP problem is separated into two sub-problems known as upper level and lower level problems. The upper level problem formulates the attack using a given power flow results and outputs the injection vector on targeted lines to ensure bypassing the bad data detection method in DC optimal power flow. The lower level problem formulates the DC economic dispatch problem with false data injections to retain the operation of power system within the desired limits. The problem is shown in (8.14)–(8.27).

The upper level objective function (8.14) is designed to find the minimum injections needed in a subset of targeted buses (F). The main target of the attacker is to overflow a targeted transmission line, it is also noted that the transmission line flow can be bi-directional, this can be formulated as

$$|P_{ij}^t| \geq \alpha_l P_{ij}^{max} \tag{8.28}$$

where $|P_{ij}^t|$ is the transmission flow between bus i and j, the absolute value is used to reflect the fact the flow can be bi-directional, P_{ij}^{max} is the maximum thermal limit of the line between bus i and j, and α is a number greater than 1 to ensure the transmission line flow will be greater than the limit. Due to the existence of absolute value in the constraint (8.28), the problem becomes nonlinear. To solver the issue, this constraint can be linearized by introducing a binary variable U_i and a large enough constant M. The linearization will result in two constraints shown in (8.15) and (8.16). The readers are encouraged to refer to [28] for more information on linearizing the constraints with absolute value using the big M method.

Constraint (8.17) enforces the overflow to be within 20% of maximum power flow, although the main target is to overflow a line, to avoid being detected by the operator, the max flow should be limited. Constraint (8.18) is designed to ensure the power balance is always met; therefore, summation of all the injections should be zero at any moment. Constraint (8.19) refers to the fact that an attacker cannot inject any amount of data at any load bus; therefore, the false injected power at any bus is limited to $\tau\%$ of the nominal load at that bus. Constraint (8.20) ensures bypassing the bad data detection. As it was mentioned in (8.11), the attacker can bypass the DC state estimation without being detected if the injected data is designed based on (8.11). By rearranging (8.20),

$$(P - D) + \Delta D = B\theta + B\Delta\theta \tag{8.29}$$

where $\Delta\theta$ is the change in the bus voltage angles due to false data injection. Knowing the fact that $P - D = B\theta$, (8.29) can be simplified to (8.11). Constraints (8.21) and (8.22) relate the transmission flow to maximum allowable limits of bus voltage angles. The lower level problem is the economic dispatch problem based on DC power flow, which is formulated through (8.23)–(8.27). In the lower level problem, the objective is to minimize the generation cost for all n_G generators in the system. It is assumed that the generating units do not contain several control valves. Hence, the "convex" cost function of the generators is represented by $a_i + b_i P_{G,i} + c_i P_{G,i}^2$, where $P_{G,i}$ is the active power generated by generator i, and a, b, and c are cost function constants (constraint (8.24)). Constraint (8.25) represents the power balance equation in presence of false data injection, constraints (8.26) and (8.27) ensure the generator powers and transmission line flows are within the limit after injection, where x_{ij} is the reactance of the transmission line between bus i and j.

8.4.1   Identifying Feasible Attacks

Realistically, physical limitations of power systems and solution of economic dispatch problem limit the feasibility of the attacks from attacker's point of view. In other words, it is impractical to target all the transmission lines in the system and only a few lines might be practically targeted at any moment. Therefore, the attacker's first step is to recognize the feasible attacks and then inject data on selected buses (targeted buses) to overflow those lines that can result in a feasible solution. An algorithm is defined to identify the feasible solution of overflowing transmission lines in the system. The model requires the power flow data from previous step to result in a successful subset of transmission line numbers that can be targeted without violating the bad data detection algorithm in state estimation procedure. The flowchart of the proposed algorithm that results in feasible solutions of the line overflows is shown in Fig. 8.2.

8.5   Operator's Problem: Bad Data Detection to Prevent Outages Caused by Cyberattack

In the previous section, we explained the algorithm for finding the optimal attack strategy from attacker's perspective. Successful attack will affect the result of DC power flow problem. Power system operator is responsible for maintaining situational awareness, as well as ensuring resilient and secure energy delivery by identifying these attacks and deploying preventive measures [29]. In this section, we focus on a specific algorithm to identify bad data injection attacks, i.e., attacks that manipulate the measurement vector to falsify the real values used for DC power flow calculation in transmission networks. The main idea is to design a state estimator which is robust to error/attack in the measurement vector while obtaining the power flows. This estimator leverages the structure of admittance matrix to efficiently identify bad data injection attacks.

The following section is devoted to the details of this state estimator. To this end, we build on the proposed sparsity-based error detection algorithm proposed in [30]. This algorithm leverages the singularity of B matrix due to the geographically dispersed transmission networks, as well as the sparse nature of estimation error vector. Sparsity-based cyberattack detection algorithm is based on the following steps:

1. Reformulating the DC power flow problem into a sparse vector recovery problem
2. Deploying l_1-norm minimization approach [31]
3. Obtaining the bad data elements in the measurement vector
4. Identifying bad data injection attacks and measurement error given the normal distribution of measurement noise

Fig. 8.2 Flowchart of the
proposed BMILP program for
line overflow feasibility in

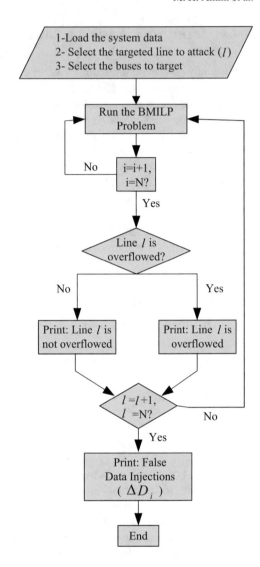

There is a low dimensional structure in most of the collected data and measurements in real-world applications, including DC power flow problem. This has been leveraged by some of the prior studies to improve sparse vector recovery [31, 32], e.g., they have explored sparse vector recovery in case of having limited available measurements. In order to recover an N-dimensional sparse vector, these algorithms do not necessarily require N data points [32, 33].

Candes et al. [32] proposed a sparse vector recovery algorithm that only needs a minor portion of random orthogonal projection [30]. Let $\mathbf{v} \in \mathbb{R}^N$ denote an arbitrary sparse vector with its number of non-zero components defined as l_0-norm, i.e., $\|\mathbf{v}\|_0$. We define set of orthonormal basis matrix as $\mathbf{O} \in \mathbb{R}^{N \times N}$. We have

$$\mathbf{O}^T \mathbf{O} = \mathbf{I}, \tag{8.30}$$

where $\mathbf{I}$ represents identity matrix. We represent a set of random columns from $\mathbf{O}$ as $\mathbf{M} \in \mathbb{R}^{N \times m}$. We further let μ denote the lowest value that meets the following inequality:

$$\max_i \|\mathbf{M}^T \mathbf{e}_i\|_2 \leq \frac{\mu m}{N}, \tag{8.31}$$

where e_i denotes a standard basis. Note that the dimension of space is $N \times m$. Small values of μ refers to the fact that the subspace corresponding to columns of $\mathbf{M}$ is not in the same direction as standard basis. The orthogonal matrix $\mathbf{M}$ is used to measure the sparse vector. As we use this matrix to measure sparse vector, it should not be sparse.

According to [32], if

$$m \geq c \|\mathbf{v}\|_0 \mu \log \frac{N}{\delta} \tag{8.32}$$

where c is a constant, then, solution of following optimization problem

$$\min_{\hat{\mathbf{z}}} \quad \|\hat{\mathbf{z}}\|_1$$
$$\text{subject to} \quad \mathbf{M}^T \hat{\mathbf{z}} = \mathbf{M}^T \mathbf{v} \tag{8.33}$$

is the same as $\mathbf{v}$ with a lower probability bound of $(1 - \delta)$, i.e., we can reconstruct the N-dimensional vector given a limited set of random measurements [34].

Given the above-mentioned preliminaries on sparse vector recovery, we now explain the sparsity-based decomposition algorithm for bad data injection in DC power flow calculation [30]. In the B matrix, due to the row corresponding to slack bus, there is at least one row which can be obtained as linear combination of the other rows, i.e., B matrix is not full rank. Let r_B denote rank of B matrix. Hence, we can reformulate DC power flow problem as $\mathbf{p} = \mathbf{Q}\theta + \epsilon$, where $\mathbf{Q} \in \mathbb{R}^{\aleph \times r_B}$ represents orthonormal basis for column subspace of B matrix. Conventionally, in order to estimate the coefficient vector, which is equivalent to the voltage angles vector in DC power flow problem (i.e., θ) least square approach is used as follows:

$$\min_{\hat{\theta}} \|\mathbf{p} - \mathbf{Q}\hat{\theta}\|_2 \tag{8.34}$$

The optimization problem in (8.34) basically projects $\mathbf{p}$ on the columns subspace of $\mathbf{Q}$. Hence, the effectiveness of the estimator depends on the noise vector ϵ and its projection on column subspace of B matrix. Note that noise vector aims at modeling the natural measurement noise in normal situation. However, in presence of attackers (e.g., the attack scenario that has been introduced in previous section) or measurement anomalies (e.g., communication failure or defective equipment), some elements of this vector will have abnormal value.

We assume that the error/attack vector ϵ is a sparse vector, i.e., attackers cannot manipulate all measurements at the same time. As opposed to l_2-minimization, l_1-minimization methods are adaptive in presence of sparse error/attack vector [34, 35], i.e., if ϵ is sufficiently sparse and $\mathbf{Q}$ meets incoherency criterion [31, 35], solution of

$$\min_{\hat{\theta}} \|\mathbf{p} - \mathbf{Q}\hat{\theta}\|_1 \tag{8.35}$$

is θ.

In the proposed sparsity-based decomposition algorithm in [30], based on the sparsity assumption for B matrix, $r_B < \aleph$. If $\mathbf{Q}^\perp \in \mathbb{R}^{\aleph \times (\aleph - r_B)}$ represent the matrix that complements the column subspace of $\mathbf{Q}$, according to [31, 35], we can rewrite (8.35) as

$$\begin{aligned} \min_{\hat{\epsilon}} \quad & \|\hat{\epsilon}\|_1 \\ \text{subject to} \quad & (\mathbf{Q}^\perp)^T \hat{\epsilon} = (\mathbf{Q}^\perp)^T \mathbf{p} \end{aligned} \tag{8.36}$$

If the following inequality holds,

$$(\aleph - r_B) \geq c \|\epsilon\|_0 \mu_B \log \frac{N}{\delta} \tag{8.37}$$

then, with a lower bound probability of $(1 - \delta)$, the optimal solution of (8.36) is θ, where μ_B is obtained using the following:

$$\max_i \|(\mathbf{Q}^\perp)_i^T\|_2 \leq \frac{\mu_B(\aleph - r_B)}{\aleph} \tag{8.38}$$

Consequently, given the assumption regarding low rank of B matrix ($(\aleph - r_B)$ is fairly large), the optimization problem in (8.35) can find the exact estimation.

8.6 Case Studies

The proposed BMILP cyberattack model is validated in IEEE 57-bus benchmark model using MATLAB. The parameters of the system are derived from MAT-POWER toolbox, which is an open access MATLAB toolbox used for load flow studies [36]. The maximum flow limits (P_{ij}^{max}) for IEEE 57-bus system were not provided in the toolbox, and it was considered to be 120 MW for each transmission line. The structure of the system is illustrated in Fig. 8.3, the system is composed of 7 generators, 80 transmission lines, and 42 loads. The upper level problem for best scenario false data injection attacks represented in (8.14)–(8.22) is solved using ("intlinprog") function of MATLAB, which is designed to solve mixed integer linear

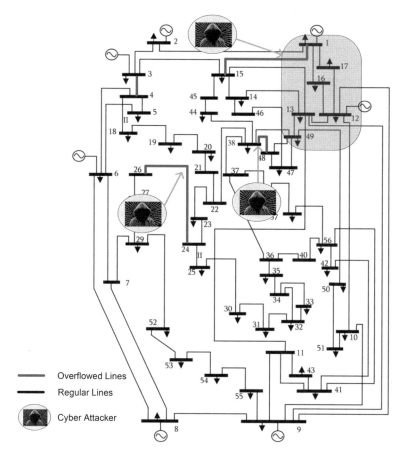

Fig. 8.3 Schematic of the IEEE 57-bus system studied in this work; transmission lines highlighted by red color indicate the potential lines that can be targeted by attackers. The highlighted area is used to illustrate the effect of gaining access to a complete area in the system

programming problems. The lower problem represented in (8.23)–(8.27) involves a nonlinear objective function and therefore, a nonlinear solver of MATLAB named ("fmincon") is used to solve the lower level problem. A few case studies are carried out in the following sections.

8.6.1   Feasibility of Line Overflow

Due to the physical limitations imposed on cyberattacks and false data injections, the attackers cannot target any transmission line in the system. For example, the factor τ limits the injection on each load to 15% in this study. Furthermore, the

Table 8.1 Feasibility of line overflow in IEEE 57 case

Line number	From bus	To bus	U_i
3	3	4	0
15	1	15	1
37	24	26	0
79	38	48	1

flow of a line cannot exceed 120% of the maximum limit (1.2 × 120 MW). These limitations will leave the attackers with few options to overflow. The proposed attack model in (8.14)–(8.27) can be run for all the buses in the system in order to identify a list of transmission lines that can be overflowed successfully. In this case study, the algorithm was run for all 80 transmission lines in the IEEE 57-bus system to identify a list of lines to be targeted for overflowing. To solve this problem, the objective function in (8.14) is run for all buses in the system instead of a subset F of system buses. This means, the subset F is considered as $i = 1, 2, \ldots, 57$.

Test results for the list of feasible attacks on targeted transmission lines to overflow in IEEE 57-bus system are illustrated in Table 8.1. It is seen that four transmission lines can be targeted in this system by attackers, which includes transmission lines number 3, 15, 37, and 79. The second and third column represent that each line number is assigned based on the MATPOWER data from one bus to another, and the last column shows the value of the binary variable for overflowing the line (U_i). These transmission lines are highlighted by red color in IEEE 57-bus system depicted in Fig. 8.3. It is worth mentioning that the results in Table 8.1 are only valid for the conditions considered in this case (e.g., $\tau = 15\%$, $\alpha = 1.01$). Therefore, the results might change if the security constraints in (8.15), (8.16), and (8.19) are relaxed.

8.6.2 Targeted Attack on Line 15

This case investigates targeted attacks to overflow transmission line number 15, which connects bus 1 to bus 15 in IEEE 57-bus system as shown in Fig. 8.3. Since the attacks are targeted, the attacker can target any set of buses to inject false data and overflow this line. Ideally, the attacker would access buses close to the targeted line and inject data. A few set of feasible solutions to overflow line number 15 is shown in Table 8.2. Three different combinations are considered to overflow line 15, in the first scenario, the attacker targets buses 12, 44, 47, and 49 that are very close to the targeted transmission line. In the second case, the attacker targets buses 8, 9, and 12, and finally, it is shown that if the attacker only targets buses 9 and 12, the attack can successfully be done. Table 8.2 shows the attack results for these three test scenarios. It is noted that the attackers target specific buses in the system that they can access and the program outputs the minimum injections needed to

Table 8.2 False data injection on buses to overflow line 15

(Case 1 buses)	ΔD	(Case 2 buses)	ΔD	(Case 3 buses)	ΔD
12	−11.94 MW	8	−5.62 MW	9	24.2 MW
44	2.4 MW	9	24.2 MW	12	−24.2 MW
47	5.94 MW	12	−18.58 MW		
49	3.6 MW				
P_{15}	121.49 MW	P_{15}	121.2 MW	P_{15}	121.5 MW
Cost (no attack)	$45,342	Cost (no attack)	$45,342	Cost (no attack)	$45,342
Cost (with attack)	$45,938	Cost (with attack)	$45,108	Cost (with attack)	$44,902

Table 8.3 False data injection on buses to overflow one area

Targeted buses	ΔD	Targeted buses	ΔD
1	−11 MW	12	28 MW
13	−3.6 MW	16	−8.4 MW
17	−8.6 MW	49	3.6 MW
$P_{15} = 125.9$ MW			
Cost (no attack) = $ 45,342			
Cost (with attack) = $ 45,685			

overflow the targeted transmission line. It is observed that the targeted line 15 is overflowed in all three cases as its power flow (P_{15}) is more than 120 MW limit. The last two columns show the cost of the system with no attack and after attack, it can be inferred from the Table 8.2 that in case 2 and 3, the system cost is less with data injections, compared to the no data injection case, which might trick the operator to choose this scenario as an acceptable economic dispatch case study.

8.6.3 Severe Attack on an Area

This case study considers a severe false data injection attack, where the attacker(s) can access an area in the power system and can inject false data in any bus within that area. This concept is illustrated in the highlighted area shown in Fig. 8.3. The attacker targets line 15 and injects data on buses 1, 12, 13, 16, 17, and 49 within the area. Results of false data injection on the whole area are illustrated in Table 8.3. The false data injections are also shown in second and fourth column for each bus. It is observed that the attack can successfully overflow line 15 and the cost of operation has also increased after the attack.

8.7 Conclusion

In this chapter, a framework was proposed to initially model the false data injection attacks on smart grids with the aim of transmission line congestion on targeted buses and eventually proposed a detection framework to detect such injections. A bi-level mixed integer programming problem was considered for the attacker's problem that would allow attackers to target a transmission line in the system and inject false data on targeted buses in vicinity of the targeted transmission line to cause congestion without being detected by DC state estimation algorithm. Through case studies, it was shown that the proposed attack model results in a list of available transmission lines to overflow. Furthermore, it was shown that the attacker can inject false data on selected buses or buses in a hacked area in the system to overflow a targeted line. To detect these attacks, a detection framework from operator's point of view is also developed that uses l_1 norm minimization to identify the bad measurement vector. The proposed model can easily be integrated to the security constraint economic dispatch problem in order to protect the smart grid against transmission lines congestion cyberattacks. Future research will focus on (1) validating the detection algorithm in IEEE benchmarks, (2) proposing a framework that would protect the system against congestion attacks on multiple lines at the same time, and (3) introducing market frameworks in which market participants can gain profit by contributing to mitigation of system congestion caused by cyberattacks.

Acknowledgements This work was under support from Penn State's Center for Security Research and Education (CSRE) seed grant 2019.

References

1. X. Yu, Y. Xue, Smart grids: a cyberphysical systems perspective. Proc. IEEE **104**(5), 1058–1070 (2016)
2. M.H. Cintuglu, O.A. Mohammed, K. Akkaya, A.S. Uluagac, A survey on smart grid cyber-physical system testbeds. IEEE Commun. Surv. Tutorials **19**(1), 446–464 (2017)
3. H. Chung, W. Li, C. Yuen, W. Chung, Y. Zhang, C. Wen, Local cyber-physical attack for masking line outage and topology attack in smart grid. IEEE Trans. Smart Grid **10**, 4577–4588 (2019)
4. A. Imteaj, M.H. Amini, J. Mohammadi, Leveraging decentralized artificial intelligence to enhance resilience of energy networks (2019). arXiv preprint arXiv:1911.07690
5. Y. Tang, Q. Chen, M. Li, Q. Wang, M. Ni, and X. Fu, Challenge and evolution of cyber attacks in cyber physical power system, in *2016 IEEE PES Asia-Pacific Power and Energy Engineering Conference (APPEEC)* (IEEE, Xi'an, 2016), pp. 857–862
6. N. Perlroth, D.E. Sanger, Cyberattacks put Russian fingers on the switch at power plants, US says. New York Times **15** (2018)
7. T. Maurer, *Cyber Mercenaries* (Cambridge University Press, Cambridge, 2018)
8. Y. Cai, Y. Cao, Y. Li, T. Huang, B. Zhou, Cascading failure analysis considering interaction between power grids and communication networks. IEEE Trans. Smart Grid **7**(1), 530–538 (2015)
9. X. Liu, Z. Li, X. Liu, and Z. Li, Masking transmission line outages via false data injection attacks. IEEE Trans. Inf. Forensics Secur. **11**(7), 1592–1602 (2016)
10. L. Wei, D. Gao, C. Luo, False data injection attacks detection with deep belief networks in smart grid, in *2018 Chinese Automation Congress (CAC)* (2018), pp. 2621–2625
11. Y. Xiang, Z. Ding, Y. Zhang, L. Wang, Power system reliability evaluation considering load redistribution attacks. IEEE Trans. Smart Grid **8**(2), 889–901 (2016)
12. Y. Liu, P. Ning, M.K. Reiter, False data injection attacks against state estimation in electric power grids. ACM Trans. Inf. Syst. Secur. **14**(1), 13 (2011)
13. L. Liu, M. Esmalifalak, Q. Ding, V.A. Emesih, Z. Han, Detecting false data injection attacks on power grid by sparse optimization. IEEE Trans. Smart Grid **5**(2), 612–621 (2014)
14. O. Kosut, L. Jia, R.J. Thomas, L. Tong, Malicious data attacks on the smart grid. IEEE Trans. Smart Grid **2**(4), 645–658 (2011)
15. G. Chaojun, P. Jirutitijaroen, M. Motani, Detecting false data injection attacks in AC state estimation. IEEE Trans. Smart Grid **6**(5), 2476–2483 (2015)
16. L. Xie, Y. Mo, B. Sinopoli, Integrity data attacks in power market operations. IEEE Trans. Smart Grid **2**(4), 659–666 (2011)

17. S. Barreto, M. Pignati, G. Dán, J.-Y. Le Boudec, M. Paolone, Undetectable timing-attack on linear state-estimation by using rank-1 approximation. IEEE Trans. Smart Grid **9**(4), 3530–3542 (2018)
18. X. Liu, Z. Li, Local load redistribution attacks in power systems with incomplete network information. IEEE Trans. Smart Grid **5**(4), 1665–1676 (2014)
19. Y. Yuan, Z. Li, K. Ren, Quantitative analysis of load redistribution attacks in power systems. IEEE Trans. Parallel Distrib. Syst. **23**(9), 1731–1738 (2012)
20. Z. Li, M. Shahidehpour, A. Alabdulwahab, A. Abusorrah, Bilevel model for analyzing coordinated cyber-physical attacks on power systems. IEEE Trans. Smart Grid **7**(5), 2260–2272 (2016)
21. M. Tian, M. Cui, Z. Dong, X. Wang, S. Yin, L. Zhao, Multilevel programming-based coordinated cyber physical attacks and countermeasures in smart grid. IEEE Access **7**, 9836–9847 (2019)
22. J. Liang, L. Sankar, O. Kosut, Vulnerability analysis and consequences of false data injection attack on power system state estimation. IEEE Trans. Power Syst. **31**(5), 3864–3872 (2016)
23. X. Liu, Z. Li, Trilevel modeling of cyber attacks on transmission lines. IEEE Trans. Smart Grid **8**(2), 720–729 (2017)
24. Y. Tan, Y. Li, Y. Cao, M. Shahidehpour, Cyber-attack on overloading multiple lines: a bilevel mixed-integer linear programming model. IEEE Trans. Smart Grid **9**(2), 1534–1536 (2018)
25. G. Giannakis, V. Kekatos, N. Gatsis, S.-J. Kim, H. Zhu, B. Wollenberg, Monitoring and optimization for power grids: a signal processing perspective. IEEE Signal Process. Mag. **30**(5), 107–128 (2013)
26. A.J. Wood, B.F. Wollenberg, G.B. Sheblé, *Power Generation, Operation, and Control* (Wiley, New York, 2013)
27. Y. Yuan, Z. Li, K. Ren, Modeling load redistribution attacks in power systems. IEEE Trans. Smart Grid **2**(2), 382–390 (2011)
28. G. Dantzig, *Linear Programming and Extensions* (Princeton University Press, Princeton, 2016)
29. A. Gholami, T. Shekari, M.H. Amirioun, F. Aminifar, M.H. Amini, A. Sargolzaei, Toward a consensus on the definition and taxonomy of power system resilience. IEEE Access **6**, 32035–32053 (2018)
30. M.H. Amini, M. Rahmani, K.G. Boroojeni, G. Atia, S.S. Iyengar, O. Karabasoglu, Sparsity-based error detection in dc power flow state estimation, in *2016 IEEE International Conference on Electro Information Technology (EIT)* (IEEE, Grand Forks, 2016), pp. 0263–0268
31. M. Rahmani, G.K. Atia, High dimensional low rank plus sparse matrix decomposition. IEEE Trans. Signal Process. **65**(8), 2004–2019 (2017)
32. E. Candes, J. Romberg, Sparsity and incoherence in compressive sampling. Inverse Probl. **23**(3), 969 (2007)
33. E.J. Candès, J. Romberg, T. Tao, Robust uncertainty principles: exact signal reconstruction from highly incomplete frequency information. IEEE Trans. Inf. Theory **52**(2), 489–509 (2006)
34. E.J. Candes, T. Tao, Near-optimal signal recovery from random projections: universal encoding strategies? IEEE Trans. Inf. Theory **52**(12), 5406–5425 (2006)
35. E.J. Candes, T. Tao, Decoding by linear programming. IEEE Trans. Inf. Theory **51**(12), 4203–4215 (2005)
36. R.D. Zimmerman, C.E. Murillo-Sanchez, R.J. Thomas, MATPOWER: steady-state operations, planning, and analysis tools for power systems research and education. IEEE Trans. Power Syst. **26**(1), 12–19 (2011)

Chapter 9
Toward Operational Resilience of Smart Energy Networks in Complex Infrastructures

Babak Taheri, Ali Jalilian, Amir Safdarian, Moein Moeini-Aghtaie, and Matti Lehtonen

Abstract Smart energy systems can mitigate electric interruption costs provoked by manifold disruptive events via making efforts toward proper pre-disturbance preparation and optimal post-disturbance restoration. In this context, effective contingency management in power distribution networks calls for contemplating disparate parameters from interconnected electric and transportation systems. This chapter, while considering transportation issues in power networks' field operations, presents a navigation system for pre-positioning resources such as field crews and reconfiguring the network to acquire a more robust configuration in advance of the imminent catastrophe. Also, after the occurrence of the calamity, this navigator optimally allocates the resources to recover the devastating system. So, providing a coordination framework for manual field operation and automation system, this navigator takes a step from traditionally operated systems accommodation toward smart networks. During the contingency management process, there might be modifications in initial data due to the dynamic and time-varying condition of electric and transportation systems. Therefore, the mentioned navigator copes with a real-time problem of data-driven decision making in which, the decisions need to track online changes to the input data. Decision making by the navigation system in this environment is based on a mixed integer linear programming (MILP) optimization which is described in this chapter in detail.

Keywords Electric power distribution system · Complex infrastructure · System resilience · Reconfiguration · Proactive actions · Crew dispatching · Manual switch · Remote controlled switch

B. Taheri · A. Jalilian · A. Safdarian (✉) · M. Moeini-Aghtaie
Electrical Engineering Department, Sharif University of Technology, Tehran, Iran
e-mail: safdarian@sharif.edu

M. Lehtonen
Department of Electrical Engineering and Automation, Aalto University, Finland

© Springer Nature Switzerland AG 2020
M. H. Amini (ed.), *Optimization, Learning, and Control for Interdependent Complex Networks*, Advances in Intelligent Systems and Computing 1123,
https://doi.org/10.1007/978-3-030-34094-0_9

203

9.1 Introduction

9.1.1 Overview

Smart energy systems can mitigate electric interruption costs provoked by manifold disruptive events via making efforts toward proper pre-disturbance preparation and optimal post-disturbance restoration. In this context, effective contingency management in power distribution networks calls for contemplating disparate parameters from interconnected electric and transportation systems. This chapter, while considering transportation issues in power networks' field operations, presents a navigation system for pre-positioning resources such as field crews and reconfiguring the network to acquire a more robust configuration in advance of the imminent catastrophe. Also, after the occurrence of the calamity, this navigator optimally allocates the resources to recover the devastating system. So, providing a coordination framework for manual field operation and automation system, this navigator takes a step from traditionally operated systems accommodation toward smart networks. During the contingency management process, there might be modifications in initial data due to the dynamic and time-varying condition of electric and transportation systems. Therefore, the mentioned navigator copes with a real-time problem of data-driven decision making in which, the decisions need to track online changes to the input data. Decision making by the navigation system in this environment is based on a mixed integer linear programming (MILP) optimization which is described in this chapter in detail.

The continuous supply of energy has a direct influence on customers' satisfaction, hence, has always been one of the most important issues for power distribution utilities. In this regard, the occurrence of a fault caused by equipment failure or due to events such as adverse weather conditions may cause a blackout in the grid and jeopardize the reliability of electric energy services. In this situation, the restoration of the power supply to interrupted customers should be done as quickly as possible and with the lowest cost. To this end, researchers have always been trying to describe the optimal operation required in post-fault situations [1–8]. In the fault management process, the use of remote controlled switches (RCSs) due to their ability to promptly restore a significant share of the network's disconnected load has been noticed [8–17]. However, manual switching in the network and repair operations require the presence of field operation crews at the network level. The dispersed manual switches (MSs) at the network level and the limited crew teams necessitate the optimal use of the transportation system in implementing the load restoration process.

Meanwhile, optimal fault management can have a more important role in service restoration after natural disasters. Severe weather events can have a significant impact on power system infrastructures. In recent years, the incidence and severity of these events have been increasing due to climate changes [18]. In 2016, based on Energy Information Administration (EIA) reports, West Virginia recorded the highest SAIDI[1] of 6 h in the USA when major events were excluded. While, including major events, highest reliability index was more than 20 h per customer because of hurricane Matthew in South Carolina. Given such incidents, the development of methods to increase the resilience of power systems against natural disasters, especially climatic events, is becoming more important. Power systems as a critical infrastructure should be reliable under normal circumstances and predictable faults [19]. In this regard, power systems are designed and exploited based on the key features of reliability, namely, security and adequacy [20, 21]. However, due to the increase in natural disasters in recent years, the power system infrastructures must be resilient to high impact low probability (HILP) incidents too. Therefore, in this chapter, measures to increase resilience of the power system before and after predicted events are described. Using meteorological forecasts, these measures take the distribution network to the most robust configuration to avoid likely failures. Also, the transportation system issues are considered in performing field operations by the crew teams.

9.2 Resilience Enhancement Scheme

Nowadays, due to remarkable advances in weather forecasting, natural disasters and their severity can be predicted much more precisely than before. For example, today's weather forecasting systems are able to predict tornados for 2 h and hurricanes and tropical storms from 24 to 72 h before their occurrence [21]. Therefore, in this chapter, the concept of proactive planning of electric power distribution networks has been harnessed to increase the resilience of these networks to natural disasters, which is related to the operations performed before the incident [22–25]. In this regard, in the pre-storm stage, distribution network operators try to change the network configuration to acquire a storm-resistant structure. In fact, by calculating the failure probability of the distribution system branches, the system operator tends to supply the customers especially critical ones via available redundant paths rather than exploiting the branches with higher failure probability. The proposed model thus provides an optimal switching sequence for RCSs and MSs. In addition, with the aim of speeding up the likely post-event actions, the system operator, after reconfiguring the system, dispatches the available crew teams to the available staging locations; thus, in the post-event stage, he/she can recover the system more quickly. This placement will be such that post-event operations are carried out as quickly as possible, and customers of the distribution network experience the least possible interruption duration. It is worthwhile to note that all operations must be carried out within a specific time period (i.e., the remaining time to the incident) and crews must arrive at the prepared sites before the event occurs.

---

[1]System Average Interruption Duration Index.

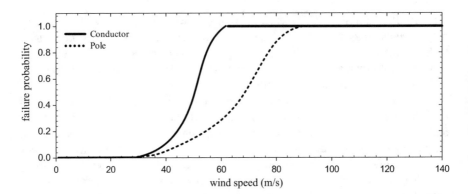

Fig. 9.1 Typical fragility curves against wind speed [14]

All of these pre-event preparation scheduling is obtained as the result of identifying the likely damages of the distribution system branches. In fact, there has to be a prediction of damages caused by the forthcoming event. For this purpose, the fragility curves of distribution network equipment are used. A fragility curve is a statistical tool representing the probability of exceeding a given damage state (or performance) as a function of an engineering demand parameter. There are sample fragility curves for lines and poles in Fig. 9.1 [26–28]. In this study, fragility curves associated with poles and lines are utilized in order to obtain their failure probability. Thus, aggregating the related data to characteristics of the upcoming disaster (e.g., tornado) such as intensity and approaching angel, and mapping the obtained intensity to the associated fragility curve, the failure probability of each component could be calculated.

The failure probabilities can be calculated as follows:

$$\rho_l(w) = \rho_{l,c}(w) + \rho_{l,p}(w) - \rho_{l,c}(w)\rho_{l,p}(w) \tag{9.1}$$

$$\rho_{l,p}(w) = 1 - \left(1 - \rho_{p-\mathrm{ind}}(w)\right)^{N_{p,l}} \tag{9.2}$$

where $\rho_l(w)$ is the failure probability of line l associated with wind speed w. In addition, parameters $\rho_{l,c}(w)$ and $\rho_{l,p}(w)$ are the failure probability of conductors and poles, respectively, and $\rho_{p-\mathrm{ind}}(w)$ is the failure probability of an individual pole. Also, $N_{p,l}$ is the number of poles holding line l. Therefore, exerting the above relations, failure probability of distribution lines can be calculated.

After applying the proactive actions, the system operator monitors the system during the natural calamity and aggregates the data related to the operating condition of the distribution system's components. Then, after the occurrence of the event, he/she tends to recover the system based on the initial configuration of the system which is the result of the pre-event actions, and the status of the distribution system branches after confronting the disaster. To do so, he/she reconfigures the system optimally via co-optimizing the operation of MSs, RCSs, and various distributed energy sources such as photovoltaics (PVs), wind turbines (WTs), and energy storages (ESs).

9.3 Real-Time Decision Making Process

Optimal fault management in power distribution networks not only depends on technical electrical parameters but is also dependent on other interconnected infrastructures. Field operations are performed in the context of the transportation system, and meteorological forecasts provide necessary information for pre-event preventive measures. In this regard, the method presented in this chapter can be used as the core of the decision making algorithm to develop a navigator system for fault management procedure. Data preparation procedure for the navigation system is shown in Fig. 9.2. This navigator utilizes locational and technical data from different toolboxes such as travel time estimator, load estimator, and failure estimator for data-directed decision making to produce information about the optimal schedule of switching actions and routing for field crews. The failure estimator toolbox as the connection interface of the electrical system with the meteorological system uses event characteristics data and components fragility curves to produce probable

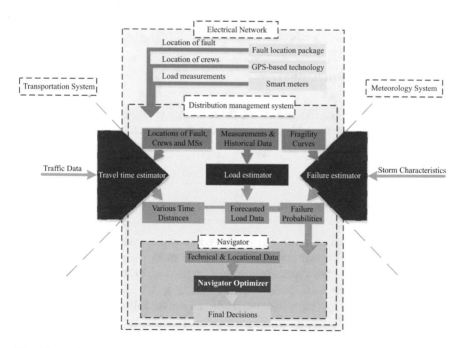

Fig. 9.2 Navigator data flow in complex infrastructures

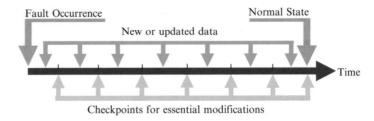

Fig. 9.3 Periodic decision making

post-incident damages. Also, travel time estimator toolbox as the link between the electrical system and transportation system processes the locational data of crews, faults, and MSs locations and also traffic data to calculate mutual distances. Fortunately, effective approaches to the online estimation of travel time have been widely investigated [29, 30]. It should be noted that fault location data in the pre-disaster situation are probabilistic data from the failure estimator toolbox.

During the restoration process, the initial data, due to network and area dynamic conditions might be corrected or changed. In this regard, one-time decision making for the entire future operations may make the expected outcome vulnerable to many unexpected incidents. Therefore, the presented navigator periodically gathers new or updated data and decides about the best possible decisions in the future horizon. Then, comparing new decisions with previous ones, essential modifications can be made (see Fig. 9.3).

9.4   Optimization Model

As discussed before, in order to enhance the resilience of electric power distribution systems, a three-stage model is presented in this chapter each of which is described in below.

9.4.1   Pre-event Preparation Strategy

As was discussed before, it is assumed that before the approaching event hits the system, distribution system operators tend to use more reliable redundant paths in the system rather than most likely fragile paths (i.e., lines with higher failure probability); thus, the total demand of the system will be supplied with highly reliable lines. In this regard, system operators can use the following model to attain the mentioned desire.

$$\text{Min} \sum_{t,l} \rho_l \times S_{t,l} \times \overbrace{\left| P^f_{t,l} \right|}^{\vartheta_{t,l}} \tag{9.3}$$

$$y_1 - y_2 = P^f_{t,l} \tag{9.4}$$

$$\left| P^f_{t,l} \right| . w = y_1 \tag{9.5}$$

$$\left| P^f_{t,l} \right| . (1 - w) = y_2 \tag{9.6}$$

$$-S_{t,l} \times \text{Cap} \le \vartheta_{t,l} \le S_{t,l} \times \text{Cap} \tag{9.7}$$

$$\left| P^f_{t,l} \right| - (1 - S_{t,l}) \times \text{Cap} \le \vartheta_{t,l} \le \left| P^f_{t,l} \right| + (1 - S_{t,l}) \times \text{Cap} \tag{9.8}$$

where, $S_{t,l}$ is a binary variable indicating whether line l at time t is in a closed state or not. Also, $P^f_{t,l}$ is a variable which indicates the power flow of line l at time t. $\left| P^f_{t,l} \right|$ is the absolute value of $P^f_{t,l}$. y_1 and y_2 are positive auxiliary variables. Also, w is an auxiliary binary variable. Cap_l represents the capacity of line l, and M is a satisfactorily big positive number. The parameter ρ_l stands for failure probability of line l. Here, using objective function (9.3), the system operator trips the lines with higher power flow and failure probability out as much as possible subject to supplying the total demand. It should be noted that the set of Eqs. (9.4–9.8) stand for linearization of the objective function.

Power Balance Equations The following expressions ensure that active and reactive power balance is satisfied in each bus in the time horizon.

$$P^D_{m,t} - P^{ES}_{m,t} - P^{DG}_{m,t} - P^{WT}_{m,t} - P^{PV}_{m,t} + \sum_t \sum_l P^f_{t,l} = 0, \forall m, t \tag{9.9}$$

$$Q^D_{m,t} - Q^{ES}_{m,t} - Q^{DG}_{m,t} - Q^{WT}_{m,t} - Q^{PV}_{m,t} + \sum_t \sum_l Q^f_{t,l} = 0, \forall m, t \tag{9.10}$$

where, $P^D_{m,t}$ and $Q^D_{m,t}$ are the active and reactive demand at bus m at time t. Also, $P^{DG}_{m,t}$ and $Q^{DG}_{m,t}$ are the active and reactive output power of the distributed generation

(DG) at bus m at time t. In addition, $P_{m,t}^{ES}$, $P_{m,t}^{WT}$, and $P_{m,t}^{PV}$ are the active power of ES, WT, and PV systems at bus m at time t.

Power Flow Equations Here, the linear AC power flow equations [31] and expressions associated with the allowed range for the capacity of each line of the system are provided.

$$-S_{t,l}.\text{Cap}_l \leq P_{t,l}^f \leq S_{t,l}.\text{Cap}_l, \forall t, l \tag{9.11}$$

$$v_{t,n} - v_{t,m} + \frac{r_l.P_{t,l}^f + x_l.Q_{t,l}^f}{V_{t,1}} \geq (\alpha_{t,l} - 1) M, \forall t, l \tag{9.12}$$

$$v_{t,n} - v_{t,m} + \frac{r_l.P_{t,l}^f + x_l.Q_{t,l}^f}{V_{t,1}} \leq (1 - \alpha_{t,l}) M, \forall t, l \tag{9.13}$$

$$\underline{V} \leq v_{m,t} \leq \overline{V}, \forall m, t \tag{9.14}$$

$$A^2 \geq \left(\sum_l P_{t,l}^f\right)^2 + \left(\sum_l Q_{t,l}^f\right)^2, \forall t, l \in (m, n), m = 1 \tag{9.15}$$

$$\alpha_{m,n,t} + \alpha_{n,m,t} = S_{t,l}, \forall l \in (m, n), n \in \varphi_m, t = \text{EP} \tag{9.16}$$

$$\sum_{n \in \varphi_m} \alpha_{m,n,t} = 1, \forall m \notin \varphi_{\text{root}}, t = \text{EP} \tag{9.17}$$

$$\alpha_{1,n,t} = 0, \forall m \in \varphi_{\text{root}}, t = \text{EP} \tag{9.18}$$

where, M is a satisfactorily big positive number. In addition, parameters r_l and x_l are the resistance and reactance of line l, respectively, and variable $v_{m,t}$ represents the voltage magnitude at bus m at time t. Needless to mention, line l is between buses m and n. Finally, parameters $\overline{V}$ and $\underline{V}$ stand for the maximum and minimum allowable voltage range, and parameter A is the maximum apparent power of the main grid. Also, $\alpha_{m,n,t}$ is a binary variable which is equal to 1 if bus n is the parent of bus m. φ_m is a binary parameter which indicates the set of buses connected to bus m via a line. In addition, parameter EP represents the event's predictability, which means that how much early the system operator can foresee the upcoming event. Note that at each time interval, the power flowing through a given line cannot exceed the line's capacity, which is assured by (9.11). As can be observed, (9.12) and (9.13) are the linear AC power flow equations, and (9.14) implies that the voltage magnitude at the buses cannot violate the acceptable range. Finally, (9.15) is used to restrict the apparent power imported from the main grid. Note that, (9.15) is linearized using the special-ordered-sets-of-type 2 (SOS2) method [32]. Set of constraints (9.16)–(9.18) are considered for maintaining the radial structure of the distribution system [33].

Output Power of Distributed Energy Resources Here, the allowed output power ranges for various distributed energy resources are defined as follows:

$$\gamma_{t,m} \cdot \underline{P_m^{DG}} \le P_{t,m}^{DG} \le \gamma_{t,m} \cdot \overline{P_m^{DG}}, \quad \forall t, m \tag{9.19}$$

$$\gamma_{t,m} \cdot \underline{Q_m^{DG}} \le Q_{t,m}^{DG} \le \gamma_{t,m} \cdot \overline{Q_m^{DG}}, \forall t, m \tag{9.20}$$

$$0 \le P_{m,t}^{PV} \le \overline{P_{m,t}^{PV}}, \quad \forall t, m \tag{9.21}$$

$$0 \le P_{m,t}^{WT} \le \overline{P_{m,t}^{WT}}, \forall t, m \tag{9.22}$$

$$0 \le P_{m,t}^{ES} \le \overline{P_{m,t}^{ES}}, \forall t, m \tag{9.23}$$

where $\gamma_{t,m}$ indicates whether the DG at bus m at time t is scheduled or not. Also, parameters $\overline{P_m^{DG}}/\underline{P_m^{DG}}$ and $\overline{Q_m^{DG}}/\underline{Q_m^{DG}}$ represent the max/min active and reactive output powers of DGs, respectively. Furthermore, parameters $\overline{P_{m,t}^{PV}}$, $\overline{P_{m,t}^{WT}}$, and $\overline{P_{m,t}^{ES}}$ are the maximum active power of PV system, WT, and ES at bus m at time t, respectively. The set of constraints (9.19–9.23) represent the limitation of various types of energy sources, including DGs, PVs, WTs, and ESs. It is assumed that the PVs, WTs, and ESs are constant power sources with a constant power factor.

Crew Dispatching in the Transportation System Since changing the configuration of the system requires crew dispatching to the MS locations, the movement of the crew teams in the transportation system is modeled as follows:

$$-\psi_{t,l} \le S_{t,l} - S_{t-1,l} \le \psi_{t,l}, \forall t, l \tag{9.24}$$

$$\psi_{t,l} \le S_{t,l} + S_{t-1,l} \le 2 - \psi_{t,l}, \forall t, l \tag{9.25}$$

where binary variable $\psi_{t,l}$ indicates whether the status of the switch of line l at time t is changed or not. In this regard, as delineated in (9.24) and (9.25) whenever the status of a switch changes, it is set to 1. The following constraints are considered for modeling the travel of the crew teams in the transportation system and manual switching actions.

$$\tau_{c,i} + TT_{i,j} + SCT_j - \tau_{c,i} \leq M \left(1 - \beta_{c,i,j}^{\text{route}}\right), \forall c, i, j \tag{9.26}$$

$$\tau_{c,i} + TT_{i,j} + SCT_j - \tau_{c,i} \geq -M \left(1 - \beta_{c,i,j}^{\text{route}}\right), \forall c, i, j \tag{9.27}$$

$$0 \leq \tau_{c,i} \leq M \sum_i \beta_{c,i,j}^{\text{route}}, \forall c, j \tag{9.28}$$

$$\sum_l \sum_{dp} \beta_{c,dp,l}^{\text{route}} \leq 1, \forall c \tag{9.29}$$

$$\sum_c \sum_i \beta_{c,i,l}^{\text{route}} = \sum_t \psi_{t,l}, \quad \forall l \in x_l^m \tag{9.30}$$

$$\sum_i \beta_{c,i,l}^{\text{route}} - \sum_j \beta_{c,l,j}^{\text{route}} = 0, \forall c, l \in x_l^m \tag{9.31}$$

$$\sum_c \sum_i \beta_{c,i,l}^{\text{route}} \leq 1, \forall l \in x_l^m \tag{9.32}$$

$$\sum_{sl} \sum_i \beta_{c,i,st}^{\text{route}} = 1, \forall c \tag{9.33}$$

$$\sum_j \beta_{c,st,j}^{\text{route}} = 0, \forall c, st \tag{9.34}$$

$$\sum_t t.\psi_{t,l} \geq \sum_c \left(\tau_{c,l} + SCT_l \sum_i \beta_{c,i,l}^{\text{route}}\right), \forall l \in x_l^m \tag{9.35}$$

$$\sum_t t \cdot \psi_{t,l} \leq \sum_c \left(\tau_{c,l} + \text{SCT}_l \sum_i \beta_{c,i,l}^{\text{route}} \right) + 1 - \varepsilon, \forall l \in x_l^m \tag{9.36}$$

$$\sum_i \beta_{c,i,j}^{\text{route}} - \sum_i \beta_{c,j,i}^{\text{route}} \geq 0, \forall c, j / \{dp\} \tag{9.37}$$

$$\sum_c \sum_i \beta_{c,i,l}^{\text{route}} = 0, \forall c, l \in x_l^r \tag{9.38}$$

$$\tau_{c,i} \leq \text{EP}, \forall c, i \tag{9.39}$$

where, variable $\tau_{c,i}$ stands for the arrival time of crew c at point i. Also, parameter $\text{TT}_{i,j}$ is the travel time from point i to j, and parameter SCT_j designates the time required for changing the status of MS at point j. Moreover, binary variable $\beta_{c,i,j}^{\text{route}}$ indicates whether crew c moves from point i to j or not. Equations (9.26) and (9.27) stand for determining the arrival time of a crew to a specific point, which comprises of the travel time between the two so-called points, switch status changing time, and previous actions' time. In this regard, (9.28) ensures that the arrival time of crew c to point j is equal to zero in case the crew is not dispatched to there. As shown in (9.29), crew teams start their ride from the depot. Furthermore, a crew team should not move to a switch location unless he/she inclines to change its status, which is modeled in (9.30). In addition, in case a crew moves to the location of the switch of line l, he/she should leave its location after changing the status as stated in (9.31). Also, (9.32) imposes that at most one crew could be dispatched to the location of each switch. Also, (9.33) and (9.34) ensure that crew teams after reconfiguring the system, i.e., in their last action, are dispatched to the staging locations with the aim of enabling prompt service restoration of the system after occurrence of the disaster. (9.35) and (9.36) are contemplated to determine the time that the status of a switch changes, which is equal to the arrival time of the crew team to the location of the MS plus the time required for changing the status of the MS. Also, (9.37) states that crew c cannot start its travel from point i unless it has been moved to that point previously, except from depot, which is the initial point. In addition, since changing the status of RCSs calls for no crew, crews should not be dispatched to the location of RCSs, which is shown in (9.38). Finally, the arrival time of the crew teams to each point should not exceed the value of EP (9.39).

9.4.2 Mid-Event Monitoring

In this study, it is assumed that during the occurrence of the event, considering the harsh nature of HILP events in order to maintain the safety of the crew teams, the system operators take no actions and only monitor the situation; thus, they aggregate the data associated with the operating condition of different components. Therefore, by taking this strategy and having the real status of the lines in the post-event stage, the system operator will be able to restore the system from the devastated state as soon as possible through reconfiguring the system and dispatching output power of energy sources.

9.4.3 Post-event Restoration Problem

After the occurrence of the event and collecting the real data, now the distribution company is capable of mitigating the consequences of the event by co-optimizing the electric power distribution system and transportation system.

$$\text{Min} \sum_{t,m} \omega_m \times P_{t,m}^{\text{Shed}}.\Delta t \tag{9.40}$$

where, parameter ω_n represents the importance of load at bus m, and variable $P_{t,m}^{Shed}$ denotes the amount of curtailed load at bus m at time t. Also, Δt is the time step. Therefore, as can be observed in (9.40), after the occurrence of the event, the objective of the distribution company is to minimize the curtailed load considering the priorities. The objective function is subject to various constraints as follows:

$$(9.11), (9.15) - (9.25), (9.27) - (9.29), (9.32), (9.33), (9.35) \tag{9.41}$$

$$P_{m,t}^D - P_{m,t}^{\text{Shed}} - P_{m,t}^{\text{ES}} - P_{m,t}^{\text{DG}} - P_{m,t}^{\text{WT}} - P_{m,t}^{\text{PV}} + \sum_t \sum_l P_{t,l}^f = 0, \forall m, t \tag{9.42}$$

$$Q_{m,t}^D - Q_{m,t}^{\text{Shed}} - Q_{m,t}^{\text{ES}} - Q_{m,t}^{\text{WT}} - Q_{m,t}^{\text{PV}} - Q_{m,t}^{\text{DG}} + \sum_t \sum_l Q_{t,l}^f = 0, \forall m, t \tag{9.43}$$

$$\xi_{t,l} = S_{t,l} \times D_{t,l}, \forall t, l \tag{9.44}$$

$$-\xi_{t,l}.\text{Cap}_l \le P_{t,l}^f \le \xi_{t,l}.\text{Cap}_l, \quad \forall t, l \tag{9.45}$$

$$v_{t,n} - v_{t,m} + \frac{r_l . P_{t,l}^f + x_l . Q_{t,l}^f}{V_{t,1}} \geq \left(\xi_{t,l} - 1 \right) M, \forall t, l \tag{9.46}$$

$$v_{t,n} - v_{t,m} + \frac{r_l . P_{t,l}^f + x_l . Q_{t,l}^f}{V_{t,1}} \leq \left(1 - \xi_{t,l} \right) M, \forall t, l \tag{9.47}$$

$$0 \leq P_{m,t}^{\text{Shed}} \leq P_{m,t}^D, \forall m, t \tag{9.48}$$

$$Q_{m,t}^{\text{Shed}} = P_{m,t}^{\text{Shed}} \frac{Q_{m,t}^D}{P_{m,t}^D}, \forall t, m > 1 \tag{9.49}$$

$$\alpha_{m,n,t} + \alpha_{n,m,t} = \xi_{t,l}, \forall t, l \in (m, n), n \in \varphi_m \tag{9.50}$$

$$\sum_{n \in \varphi_m} \alpha_{m,n,t} \leq 1, \forall t, m \notin \varphi_{\text{root}} \tag{9.51}$$

$$\alpha_{1,n,t} = 0, \forall t, m \in \varphi_{\text{root}} \tag{9.52}$$

where, binary parameter $D_{t,l}$ indicates whether line l is damaged at time t or not. Similarly, binary variable $\xi_{t,l}$ signifies the overall status of line l, i.e., including the status of the switch (i.e., open or closed) and status of the line (i.e., whether damaged or not). Also, binary variable $\alpha_{m,n,t}$ indicates whether bus n at time t is the parent of bus m or not. In the above expressions, (9.42) and (9.43) are the active and reactive powers balance at bus m at time t. As shown in (9.44), after the occurrence of the event, the status of a line depends on both the status of the switch of the line (whether closed or not) and status of the conductor (whether damaged ($D_{t,l} = 0$) or not). In this regard, (9.45) indicates the capacity limit of line l. In addition, (9.46) and (9.47) are the modified version of the power flow equations. In (9.48), the curtailed load at bus m at time t is restricted to the total load hosted by the bus at time t. Set of constraints (9.50)–(9.52) are considered for maintaining the radial structure of the distribution system [33]. The following expressions are the modified versions of (9.26), (9.30), (9.31), and (9.34), respectively. They are considered to model the initial and final dispatching of the crew team, since in the post-event stage crews start their travel from the staging locations, and finally they travel to the depot.

$$\sum_{l}\sum_{sl}\beta_{c,st,l}^{\text{route}} \leq 1, \forall c \tag{9.53}$$

$$\sum_{dp}\sum_{i}\beta_{c,i,dp}^{\text{route}} = 1, \forall c \tag{9.54}$$

$$\sum_{j}\beta_{c,dp,j}^{\text{route}} = 0, \forall cr, dp \tag{9.55}$$

$$\sum_{i}\beta_{c,i,j}^{\text{route}} - \sum_{i}\beta_{c,j,i}^{\text{route}} \geq 0, \forall c, j/\{sl\} \tag{9.56}$$

All the above problems are established in MILP fashion which can be solved exploiting the available software packages. The optimal decisions derived from the pre-event model include the output power of various energy sources, switching sequence of the RCSs and MSs via contemplating the crews' travel in the transportation system as well as the prepositioned locations of the crew teams. Also, the decisions derived from the post-event model contain the optimal restoration of the system contemplating the travel of crew teams in the transportation system and output power of the DGs, PVs, WTs, and ESs.

9.5   Simulation Results

Here, the proposed model is applied to a test system, shown in Fig. 9.4, which is taken from [34]. As can be seen in the figure, the system consists of four feeders, 47 buses with a total peak load of 35 MW. Also, the single line diagram of the system is shown in Fig. 9.5. The load profile of the system is illustrated in Fig. 9.6. Furthermore, there are three RCSs and three maneuver points in the system. As can be observed, three WT units are installed on buses 26, 33, and 37, and their output powers are drawn in Fig. 9.7. Additionally, a PV unit is installed on bus 20, and its output power is drawn in Fig. 9.8. Moreover, two DG units are available in the system with a maximum output of 5 MW. Also, three 0.5 MWh ESs are available at buses 19, 24, and 33. In order to consider the traffic issues after the occurrence of the event and likely congestions, traffic congestion factor (TCF) is defined as an emergency condition traveling time over the normal condition traveling time. In this study, it is assumed that the value of TCF is set equal to two. In addition, it is assumed that only two crew teams are available in the system.

Case 1 Here, to simulate the post-event restoration process of the distribution system, it is considered that the system operator takes no actions in the pre-event

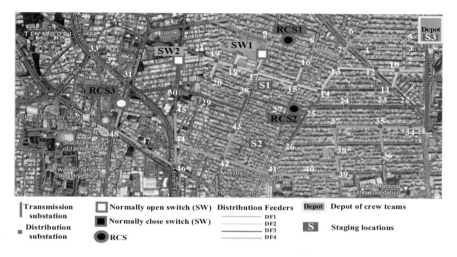

Fig. 9.4 Geographical diagram of the test system

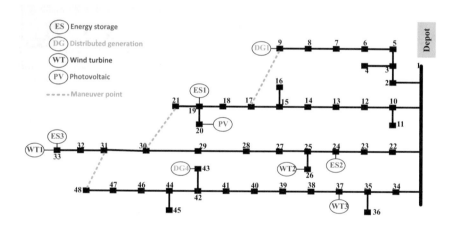

Fig. 9.5 Single line diagram of the test system

stage and recovers the system after the occurrence of the disturbance. In this regard, switching actions associated with recovering the system under this circumstance is provided in Table 9.1.

As can be observed, after the event hits the system, the operator by running the proposed model dispatches the crew teams to the location of MSs. In order to restore the curtailed customers, the system operator closes the switch of line 31–48 remotely within 5 min after the event. Then, the second crew arrives at the location of line 9–17 and closes it within 85 min. Afterward, the first crew closes the switch of line 21–30 after 105 min. It stands to reason that due to traffic issues,

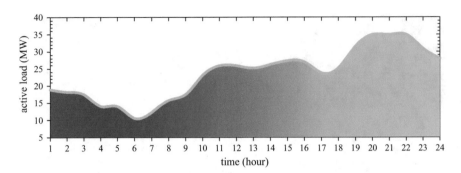

Fig. 9.6 Load profile of the test system

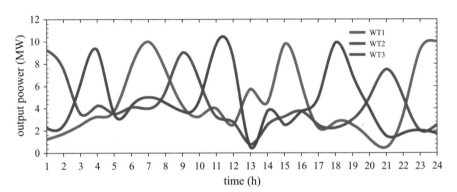

Fig. 9.7 The output power of wind turbines

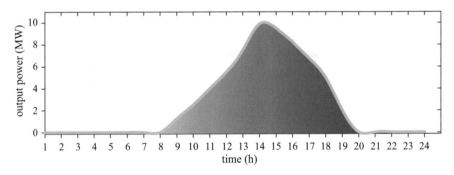

Fig. 9.8 The output power of the photovoltaic unit

travelling of crew teams in the transportation system takes a long time even with considering two for the TCF value, which is a totally normal value considering major natural disasters. The scheduling of the crews and the configuration of the system are illustrated in Fig. 9.9. In addition, the percentage of supplied load during the event and restoration process is depicted in Fig. 9.10. It should be noted that the

Table 9.1 Switching actions after the occurrence of the event: Case 1

Crew/RCS	Path/switch	Action	Required time (min)
RCS	Line 31–48	Close	5
Crew2	Depot → Line 9–17	Close	85
Crew1	Depot → Line 21–30	Close	105
Crew2	Line 9–17 → Depot	–	165
Crew1	Line 21–30 → Depot	–	205

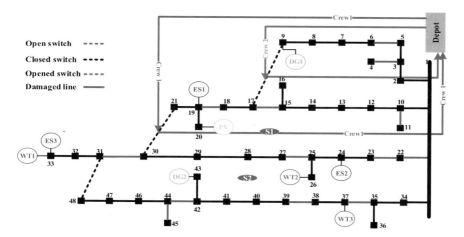

Fig. 9.9 Post-event scheduling of the crew teams: Case 1

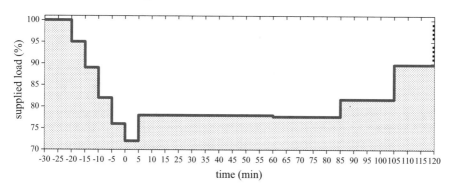

Fig. 9.10 Supplied load during the restoration process: Case 1

point zero in the figure indicates the start point of the post-event stage. As can be seen, after the occurrence of the event, the supplied load level of the system falls to 72% which rises to 78% by closing the switch of line 31–48 within 5 min. Then, it almost stands until the actions of the first crew team which takes place after 85 min by which the supplied load level of the system reaches 81.7%. Finally, closing the

Table 9.2 Switching actions before the occurrence of the event: Case 2

Crew/RCS	Path/switch	Action	Required time (min)
RCS	Line 31–48	Close	5
Crew1	Depot → Line 15–17	Open	35
Crew1	Line 15–17 → Line 9–17	Close	45
Crew2	Depot→ Line 30–31	Open	55
Crew1	Line 9–17 → Line 22–23	Open	80
Crew2	Line 30–31 → Line 21–30	Close	85
Crew1	Line 22–23 → S2	–	100
Crew2	Line 21–30 → S1	–	115

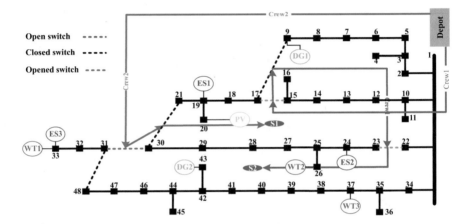

Fig. 9.11 Pre-event scheduling of the crew teams: Case 2

switch of line 21–30, the system operator energized 89.7% of the total load. Also, the amount of curtailed energy within the 2 h after the event is equal to 10.107 MWh.

Case 2 In this case, in order to simulate the pre-event actions of the crew teams, it is assumed that the predictability of the approaching event is 2 h, which means that the system operator has 2 h to reconfigure the system before the disturbance hits the system; thus, acquire a stronger configuration of the system in dealing with the natural calamity. To do so, the value of the EP is set to 2 h in the simulations, and the associated results are provided in Table 9.2. In addition, schedule of the crew teams is depicted in Fig. 9.11. As can be observed, in the first action, switch of line 31–48 is closed remotely within 5 min. Then, the first crew moves to the location of the switch of line 15–17 and opens the switch in 35 min. Afterward, the mentioned crew closes the switch of line 9–17 within 45 min. At the same time, the second crew moves to the location of line 30–31 in order to open the respective switch. Then, 80 min after the occurrence of the event, the first crew opens the switch of line 22–23. Finally, the second crew closes the switch of line 21–30 after 85 min. Furthermore, the percentage of supplied load during the event

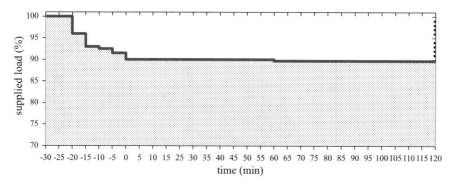

Fig. 9.12 Supplied load during the restoration process: Case 2

Table 9.3 Switching actions before the occurrence of the event: Case 3

Crew/RCS	Path/switch	Action	Required time (min)
RCS	Line 31–48	Close	5
Crew1	Depot → Line 15–17	Open	35
Crew1	Line 15–17 → Line 21–30	Close	55
Crew2	Depot→ Line 30–31	Open	55
Crew1	Line 21–30 → S2	–	90
Crew2	Line 30–31 → S1	–	90

and restoration process is shown in Fig. 9.12. As can be seen, due to proactive actions, the load level of the system drops to 90%, which means that the obtained configuration has been strong enough to confront the disturbance. The amount of curtailed energy within 2 h is equal to 5.21 MWh.

It is worthwhile to mention that no more actions needed to be applied after the event in this case. This is mainly due to the enough time for taking appropriate proactive actions as well as the pretty precise forecasts about the line damages caused by the event. Needless to mention, shorter times for proactive actions and/or erroneous forecasts about potential damages may make some post-event actions necessary.

Case 3 In this case, it is assumed that the system operator can foresee the upcoming disaster 90 min before it hits the system; so, he/she has 90 min to reconfigure the system and preposition the crew teams in order to mitigate the negative consequences of the event and heighten the resilience level of the system. In this regard, the required proactive actions to reconfigure the system are provided in Table 9.3. As can be seen, in the first action, the switch of line 31–48 is closed remotely. Then, after 35 min, the first crew arrives at the location of line 15–17 and opens the switch of the line. Following this action, the crew moves to the location of the switch of line 21–30 and closes the switch within 55 min. Simultaneously, the second crew travels to the location of line 30–31 and opens the switch after 55 min. Afterward,

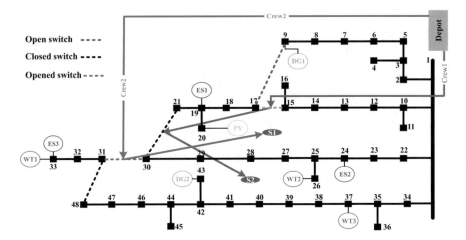

Fig. 9.13 Pre-event scheduling of the crew teams: Case 3

Table 9.4 Switching actions after the occurrence of the event: Case 3

Crew/RCS	Path/switch	Action	Required time (min)
Crew2	S1 → Line 9–17	Close	55
Crew1	S2 → Depot	–	80
Crew1	Line 9–17 → Depot	–	135

both of the crew teams are dispatched to the staging location in order to speed up the post-event actions. In addition, schedule of the crews is depicted in Fig. 9.13.

After the occurrence of the event and collecting the data related to the damage status of the lines, the system operator runs the post-event model in order to reconfigure the system and recover the curtailed customers optimally. Post-event scheduling of the crew teams is provided in Table 9.4. As can be observed, 55 min after the occurrence of the event, the second crew closes the switch of line 9–17. Furthermore, the configuration of the system is illustrated in Fig. 9.14. Also, the amount of supplied energy after the occurrence of the event is shown in Fig. 9.15. It is worthwhile to mention that in this circumstance, the curtailed energy within 2 h after the event is equal to 6.127 MWh.

Finally, the amount of supplied load during the restoration process regarding the provided three cases is compared in Fig. 9.16. As can be seen, applying proactive actions considerably enhances the resilience level of the system. Furthermore, the sooner the system operator could foresee the approaching disaster, the more he/she can mitigate the consequences of the disaster; thus, the system becomes more resilient in dealing with a natural hazard.

Case 4 In this case, it is assumed that the input information regarding the situation of the system changes after the first run. Thus, the system operator by running the model with the updated data changes its strategy regarding the dispatching of the

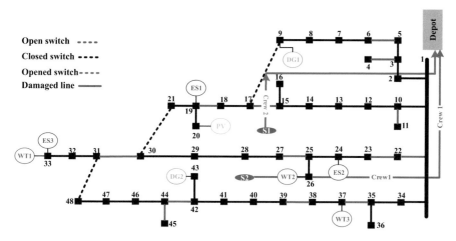

Fig. 9.14 Post-event scheduling of the crew teams: Case 3

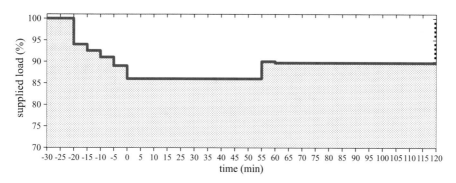

Fig. 9.15 Supplied load during the restoration process: Case 3

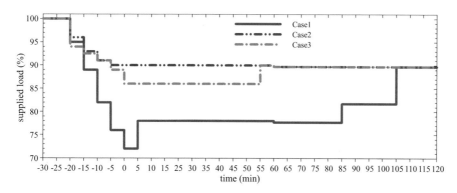

Fig. 9.16 Comparison of the restoration processes in Case 1–3

Table 9.5 Switching actions before the occurrence of the event: Case 4

Crew/RCS	Path/switch	Action	Required time (min)
RCS	Line 31–48	Close	5
Crew1	Depot → Line 15–17	Open	35
Crew1	Line 15–17 → Line 9–17	Close	45
Crew2	Depot→ Line 30–31	Open	55
Crew1	Line 9–17 → S1	–	85
Crew2	Line 30–31 → S2	–	105

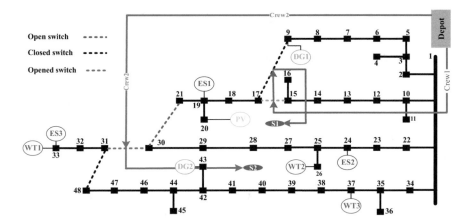

Fig. 9.17 Pre-event scheduling of the crew teams: Case 4

crew teams. To do so, it is assumed that at first the system operator has 2 h before the occurrence of the disaster. So, it dispatches the crew teams as discussed in Case 2. But, after closing the switch of line 9–17 by the first crew, the input data of the navigator regarding the travel time of the crew teams changes. Thus, the system operator in order to preserve the safety of the crew teams and resilience of the system, modifies its strategy. In this regard, the new plan of the system operator in the pre-event stage is given in Table 9.5. Besides, the configuration of the system and crew schedules in this case are shown in Fig. 9.17.

So, after the occurrence of the event and collecting the data associated with the damage status of the lines, the system operator runs the post-event model and recovers the devastated system optimally. In this regard, scheduling of the crew teams is given in Table 9.6. Also, scheduling of the crew teams and the system configuration are illustrated in Fig. 9.18. Moreover, the amount of the supplied load during the restoration process is shown in Fig. 9.19, where the amount of curtailed energy within 2 h after the event is equal to 7.406 MWh. As can be seen, any error in forecasted data can deviate the navigator from the optimal actions. This emphasizes on the importance of developing precise forecasting toolboxes besides working on models for the optimal proactive action scheduling.

Table 9.6 Switching actions after the occurrence of the event: Case 4

Crew/RCS	Path/switch	Action	Required time (min)
Crew1	S1 → Line 21–30	Close	65
Crew2	S2 → Depot	–	80
Crew1	Line 21–30 → Depot	–	165

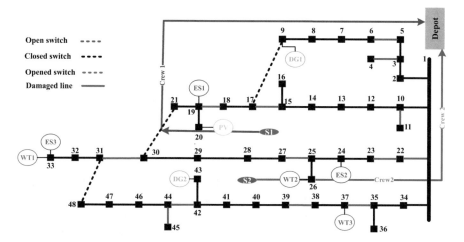

Fig. 9.18 Post-event scheduling of the crew teams: Case 4

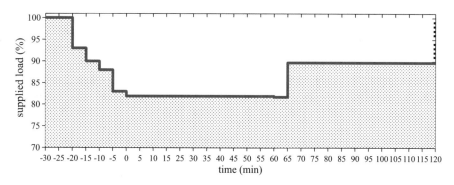

Fig. 9.19 Supplied load during the restoration process: Case 4

Here, to investigate the impact of the traffic issues on the restoration process of the system, a sensitivity analysis is conducted on the value of TCF, and the associated results are illustrated in Fig. 9.20. As can be observed, increasing the value of TCF drastically influences the amount of the curtailed load. It should be noted that the harnessed test system was relatively small and it has only two maneuver points which have MSs. Therefore, it stands to reason that the larger the system becomes, the crucial the impacts of proactive actions, number of crew teams, and traffic issues become. As can be seen in Fig. 9.20, the number of crew

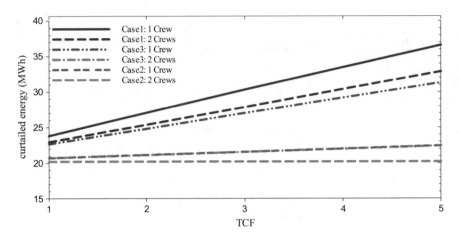

Fig. 9.20 Amount of curtailed energy versus the value of TCF

teams, predictability of the event, and TCF have a drastic impact on the amount of curtailed energy after a major disaster. It is worthwhile to mention that regarding the events which cause massive destructions in the transportation system (e.g., floods), applying the presented model for optimally scheduling proactive actions could improve the resilience level of the system efficiently.

9.6 Conclusion

In this chapter, in order to enhance the resilience of electric power distribution systems in dealing with the natural calamities, a new approach has been presented, wherein, the problem is modeled in an MILP fashion. The presented strategy consists of three stages namely pre-, mid-, and post-event stages. The first stage takes place before the occurrence of an upcoming disaster, wherein the system operator using a navigator, tends to change the configuration of the distribution system in the hope of acquiring a more robust configuration to confront the disaster. To do so, he/she optimizes the network configuration contemplating the manifold parameters including the travel time of the field crews in the transportation system. Also, he/she prepositions the crew teams in the staging locations after reconfiguring the system in the hope of accelerating expected post-event actions. The second stage is for collecting the data related to the operating status of the system components. Then, after the occurrence of the event, the system operator using the navigator restores the curtailed customers as quickly as possible with respect to their criticality. Finally, the presented model is implemented on a 47-bus power distribution system. The simulation results confirm the efficacy of applying proactive actions before the occurrence of the natural calamities, which strengthens the system and enables it to rapidly recover from the devastated state.

References

1. M. Izadi, M. Farajollahi, A. Safdarian, Switch deployment in distribution networks, in *Electric Distribution Network Management and Control*, ed. by A. Arefi, F. Shahnia, G. Ledwich, ch. 8 (Springer, Singapore, 2018), pp. 179–233
2. C. Yuan, M.S. Illindala, A.S. Khalsa, Modified Viterbi algorithm based distribution system restoration strategy for grid resiliency. IEEE Trans. Power Delivery **32**(1), 310–319 (2017)
3. M.R. Kleinberg, K. Miu, H. Chiang, Improving service restoration of power distribution systems through load curtailment of in-service customers. IEEE Trans. Power Syst. **26**(3), 1110–1117 (2011)
4. M. Gholami, J. Moshtagh, L. Rashidi, Service restoration for unbalanced distribution networks using a combination two heuristic methods. Int. J. Electr. Power Energy Syst. **67**, 222–229 (2015)
5. Y. Li, J. Xiao, C. Chen, Y. Tan, Y. Cao, Service restoration model with mixed-integer second order cone programming for distribution network with distributed generations. IEEE Trans. Smart Grid **10**(4), 4138–4150 (2018)
6. R. Romero, J.F. Franco, F.B. Leão, M.J. Rider, E.S. de Souza, A new mathematical model for the restoration problem in balanced radial distribution systems. IEEE Trans. Power Syst. **31**(2), 1259–1268 (2016)
7. Y. Xu, C. Liu, H. Gao, Reliability analysis of distribution systems considering service restoration, in *2015 IEEE Power and Energy Society Innovative Smart Grid Technologies Conference (ISGT)* (2015), pp. 1–5
8. M. Izadi, A. Safdarian, Financial risk evaluation of RCS deployment in distribution systems. IEEE Syst. J. **13**(1), 692–701 (2019)
9. O.K. Siirto, A. Safdarian, M. Lehtonen, M. Fotuhi-Firuzabad, Optimal distribution network automation considering earth fault events. IEEE Trans. Smart Grid **6**(2), 1010–1018 (2015)
10. A. Safdarian, M. Farajollahi, M. Fotuhi-Firuzabad, Impacts of remote control switch malfunction on distribution system reliability. IEEE Trans. Power Syst. **32**(2), 1572–1573 (2016)
11. M. Farajollahi, M. Fotuhi-Firuzabad, A. Safdarian, Optimal placement of sectionalizing switch considering switch malfunction probability. IEEE Trans. Smart Grid **10**(1), 403–413 (2017)
12. M. Izadi, M. Farajollahi, A. Safdarian, M. Fotuhi-Firuzabad, A multistage MILP-based model for integration of remote control switch into distribution networks, in *2016 International Conference on Probabilistic Methods Applied to Power Systems (PMAPS)* (2016), pp. 1–6
13. M. Izadi, A. Safdarian, Financial risk constrained remote controlled switch deployment in distribution networks. IET Gener. Transm. Distrib. **12**(7), 1547–1553 (2017)
14. M. Farajollahi, M. Fotuhi-Firuzabad, A. Safdarian, Simultaneous placement of fault indicator and sectionalizing switch in distribution networks. IEEE Trans. Smart Grid **10**(2), 2278–2287 (2019)
15. M. Izadi, A. Safdarian, A MIP model for risk constrained switch placement in distribution 493 networks. IEEE Trans. Smart Grid **10**(4), 4543–4553 (2018)
16. M. Farajollahi, M. Fotuhi-Firuzabad, A. Safdarian, Sectionalizing switch placement in distribution networks considering switch failure. IEEE Trans. Smart Grid **10**(1), 1080–1082 (2018)
17. M. Izadi, M. Farajollahi, A. Safdarian, Optimal deployment of remote-controlled switches in distribution networks considering laterals. IET Gener. Transm. Distrib. **13**(15), 3264–3271 (2019)
18. E. Office and P. August, White House 2013 - Grid resiliency and economic benefit, no. August (2013)
19. R. Billinton, *Power System Reliability Evaluation* (Gordon and Breach, 1970)
20. North American Electric Reliability Corporation (NERC), Reliability standards for the bulk electric systems of North America, in *NERC Reliability Standard Complete Set* (Atlanta, 2019). https://www.nerc.com/pa/Stand/Reliability%20Standards%20Complete%20Set/RSCompleteSet.pdf

21. Y. Wang, C. Chen, J. Wang, R. Baldick, Research on resilience of power systems under natural disasters—a review. IEEE Trans. Power Syst. **31**(2), 1604–1613 (2015)
22. M.H. Amirioun, F. Aminifar, H. Lesani, Resilience-oriented proactive management of microgrids against windstorms. IEEE Trans. Power Syst. **33**(4), 4275–4284 (2017)
23. C. Wang, Y. Hou, F. Qiu, S. Lei, K. Liu, Resilience enhancement with sequentially proactive operation strategies. IEEE Trans. Power Syst. **32**(4), 2847–2857 (2017)
24. A. Arab, A. Khodaei, Z. Han, S.K. Khator, Proactive recovery of electric power assets for resiliency enhancement. IEEE Access **3**, 99–109 (2015)
25. B. Taheri, A. Safdarian, M. Moeini-Aghtaie, M. Lehtonen, *Distribution Systems Resilience Enhancement Via Pre- and Post-event Actions* (IET Smart Grid, 2019)
26. M. Panteli, D.N. Trakas, P. Mancarella, N.D. Hatziargyriou, Boosting the power grid resilience to extreme weather events using defensive islanding. IEEE Trans. Smart Grid **7**(6), 2913–2922 (2016)
27. M. Panteli, C. Pickering, S. Wilkinson, R. Dawson, P. Mancarella, Power system resilience to extreme weather: fragility modeling, probabilistic impact assessment, and adaptation measures. IEEE Trans. Power Syst. **32**(5), 3747–3757 (2016)
28. M. Panteli, P. Mancarella, D.N. Trakas, E. Kyriakides, N.D. Hatziargyriou, Metrics and quantification of operational and infrastructure resilience in power systems. IEEE Trans. Power Syst. **32**(6), 4732–4742 (2017)
29. X. Zhan, S. Hasan, S.V. Ukkusuri, C. Kamga, Urban link travel time estimation using large-scale taxi data with partial information. Transp. Res. Part C Emerg. Technol. **33**, 37–49 (2013)
30. E. Jenelius, H.N. Koutsopoulos, Travel time estimation for urban road networks using low frequency probe vehicle data. Transp. Res. Part B Methodol. **53**, 64–81 (2013)
31. M.E. Baran, F.F. Wu, Optimal capacitor placement on radial distribution systems. IEEE Trans. Power Delivery **4**(1), 725–734 (1989)
32. M.R. Sarker, M.A. Ortega-Vazquez, D.S. Kirschen, Optimal coordination and scheduling of demand response via monetary incentives. IEEE Trans. Smart Grid **6**(3), 1341–1352 (2014)
33. R.A. Jabr, R. Singh, B.C. Pal, Minimum loss network reconfiguration using mixed-integer convex programming. IEEE Trans. Power Syst. **27**(2), 1106–1115 (2012)
34. M. Rahmani-Andebili, M. Fotuhi-Firuzabad, An adaptive approach for PEVs charging management and reconfiguration of electrical distribution system penetrated by renewables. IEEE Trans. Ind. Inform. **14**(5), 2001–2010 (2017)

Chapter 10
Control of Cooperative Unmanned Aerial Vehicles: Review of Applications, Challenges, and Algorithms

Arman Sargolzaei, Alireza Abbaspour, and Carl D. Crane

Abstract A system of cooperative unmanned aerial vehicles (UAVs) is a group of agents interacting with each other and the surrounding environment to achieve a specific task. In contrast with a single UAV, UAV swarms are expected to benefit efficiency, flexibility, accuracy, robustness, and reliability. However, the provision of external communications potentially exposes them to an additional layer of faults, failures, uncertainties, and cyberattacks and can contribute to the propagation of error from one component to other components in a network. Also, other challenges such as complex nonlinear dynamic of UAVs, collision avoidance, velocity matching, and cohesion should be addressed adequately. Main applications of cooperative UAVs are border patrol; search and rescue; surveillance; mapping; military. Challenges to be addressed in decision and control in cooperative systems may include the complex nonlinear dynamic of UAVs, collision avoidance, velocity matching, and cohesion. In this paper, emerging topics in the field of cooperative UAVs control and their associated practical approaches are reviewed.

Keywords Cooperative unmanned aerial vehicles · Nonlinear dynamics · UAV swarm

Abbreviations

CCUAVs	Cooperative Control Unmanned Aerial Vehicles
CLSF	Constrained Local Submap Filter

A. Sargolzaei (✉) · C. D. Crane
Department of Mechanical and Aerospace Engineering, University of Florida, Gainesville, FL, USA
e-mail: asargolzaei@ufl.edu; ccrane@ufl.edu

A. Abbaspour
Department of Advanced Engineering, Hyundai Mobis, Plymouth, MI, USA
e-mail: aabbaspour@mobis-usa.com

© Springer Nature Switzerland AG 2020
M. H. Amini (ed.), *Optimization, Learning, and Control for Interdependent Complex Networks*, Advances in Intelligent Systems and Computing 1123,
https://doi.org/10.1007/978-3-030-34094-0_10

CML	Concurrent mapping and localization
DDDAS	Dynamic Data-Driven Application System
DDF	Decentralized Data Fusion
DI	Dynamic inversion
DoS	Denial of service
ECM	Electronic Counter-Measure
EJ	Escort Jamming
FDI	Fault Detection and Identification
FTC	Fault Tolerant Controllers
GNN	Grossberg Neural Network
LIDAR	Light detection and ranging
LOS	Line of Sight
LQR	Linear Quadratic Regulator
PC	Probability collective
PDF	Probability Density Function
POMDP	Observable Markov Decision Process
PN	Proportional navigation
PP	Pure pursuit
PRS	Personal Remote Sensing
ROS	Robot Operating System
SAM	Surface-to-Air Missile
SLAM	Simultaneous Localization and Mapping
SWEEP	Swarm Experimentation and Evaluation Platform
TDS	Time delay switch
UAV	Unmanned Aerial Vehicles
UCAVs	Unmanned Combat Air Vehicles
WSN	Wireless Sensor Network

10.1 Introduction

Swarm intelligence deals with physical and artificial systems formed of entities that have internal and external interactions coordinating by incentive or a predefined control algorithm. Flocking of birds, swarming of insects, shoaling of fishes, and herding of quadrupeds were a motive for the cooperated control of UAVs. A group of UAVs can be modeled similar to natural animal cooperation where bodies operate as a system toward reaching mutual benefits. Animals can benefit from swarm performance in defending against predators, food seeking, navigation, and energy saving. Cooperative multi-robots complete a task in a shorter time [1], have synergy [2, 3], and cover a larger area. They are also more cost-effective using smaller, simpler, and more durable robots [4]. Furthermore, they can complete a task more accurately and robustly [5].

Cooperative control is one of the most attractive topics in the field of control systems which has received the attention of many researchers. Cooperative algorithms and utilization are mainly discussed in the recent decade. Many useful surveys have been done to review the recent contributions in this field [6–11]. However, most of them were just focused on the algorithms and on the consensus control theory. A valuable review of the consensus control problem was done by Ren et al. [6]; however, significant contributions have been done thereafter. Anderson et al. [7] also focused on consensus control of the multi-agent systems. Wang et al. [8] and Zhu et al. [10] reviewed most of the consensus control problems; however, other cooperative techniques and the application of these algorithms were not discussed. Senanyake et al. investigated cooperative algorithms for searching and tracking applications [11]; however, the other algorithms and applications of the cooperative system were not considered.

To enhance the current related literature mentioned above and cover most of the applications and algorithms, the recent research studies in the field of cooperative control design will be reviewed. The applications, algorithms, and challenges are considered. The applications are categorized into surveillance, search and rescue, mapping, and military applications; then, the recent developments related to each category are reviewed. Similarly, the algorithms can be categorized into three main classes: consensus control, flocking control, and guidance based cooperative control. The challenges related to the cooperative control and applications of cooperative algorithms are investigated in a separate section. Moreover, the related mathematics of cooperative control algorithms is simplified to make it easier for readers to understand the concepts.

This paper is organized as follows: Sect. 10.2 provides potential applications of cooperative control, and Sect. 10.3 highlights possible challenges when applying cooperative control. Section 10.4 reviews algorithms used in cooperative control design. Finally, Sect. 10.5 provides the summary and conclusion of this work.

10.2 Applications and Literature Review

Cooperative control of multiple unmanned vehicles is one of the topics in control areas that have received increasing interest in the past several years. Single UAVs have been applied for various applications, and recently, investigators have attempted to expand and improve their applications by using a combination of multiple agents. The multiple agents concept has been used for search and rescue [2, 12–16], geographic mapping [17–20], military applications [21–23], etc. In this section, the current and potential applications of the cooperative control of UAVs are surveyed.

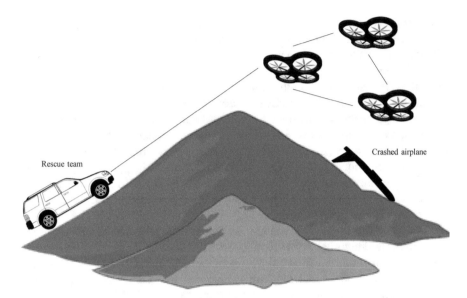

Rescue team

Crashed airplane

Fig. 10.1 A cooperative approach to search for a victim in a hard to access area [25]

10.2.1 Search and Rescue

UAVs have been used for several years for search and rescue purposes since they
are more compact and cost-effective and require less amount of time to deploy than
a plane or helicopter, particularly when multiple numbers of UAVs are required
to accomplish the task. Figure 10.1 displays a scenario for cooperative control
of quadrotors to search for and rescue a patient or missing person in a hard to
access environment. In this kind of operation, search time is the most critical
factor. To satisfy the time constraint, Scherer et al. implemented a distributed
control system in the robot operating system (ROS) of the multiple multi-copters,
to capture the situations and display them as video streams in real-time at base
stations [24]. Since UAVs have their advantages such as agility, swiftness, remote-
controlling, birds eye-vision, and other integrities, they can creditably perform
practical work promptly. However, when those advantageous of UAVs are operated
by a cooperative control algorithm to complete a mission, the requirement of
minimum time delay and other critical constraints can be achieved in searching and
rescuing casualties or victims.

Many types of research and experiments are performed in search and rescue
requiring cooperative control unmanned aerial vehicles (CCUAVs). For example,
Waharte et al. showed that employing multiple autonomous UAVs has excellent
benefits in the search and rescue operations for the corollary of Hurricane Kat-
rina in September 2006. The notable sophistication of their work was that they
divided the real-time approaches into three main categories which were greedy

heuristics, potential-based heuristics, and partially observable Markov decision process (POMDP) based heuristics [25]. In a case of fire, Maza et al. investigated a multi-UAV firefighters monitoring mission in the framework of the AWARE Project using two autonomous helicopters to monitor the firemen's performance and safety in real-time from a simulated situation where firefighters are assisting injured people in front of a burning building. This work has been done based on their previous work's algorithm [26], SIT algorithm, which follows a market-based approach combined with a network of ground cameras and a wireless sensor network (WSN) [27]. Another scenario that CCUAVs can be wholly beneficial is to search and rescue missing persons in a wilderness. It has been many centuries that travelers had been lost in wildernesses such as mountains, oceans, deserts, jungles, rain forests, or any abandoned or uncolonized areas. Some of the missing people could be found and rescued, but many of them were lost from their families forever. Goodrich et al. have shown and identified a set of operational practices for using mini unmanned aerial vehicles (mUAVs) to support wilderness search and rescue (WiSAR) operations. In their work, technical operations such as sequential operations, remote-led operations, and base-led operations have been used to gather and analyze evidence or potential signs of a lost person to simulate a stochastic model of his behavior and a geographic description of a particular region. If the model is well matched to a specific victim, then the location of the missing person would be estimated according to the probability of the area where the lost person could be located [12]. The result of their research shows that the mUAVs could address the limitations of human-crewed aircraft which also upholds the research algorithm of CCUAVs.

10.2.2   Surveillance

Surveillance is one of the applications of UAVs that have been widely used. Figure 10.2 shows an overall scheme of the surveillance application using the cooperative quadrotors system. Bread et al. studied aerial surveillance of fixed-wing multi-UAVs. Fixed-wing aircraft may have a significant advantage in speed. However, the lack of hovering ability would increase their chance of collision when they work in cooperative control mode. To mitigate and overcome this constraint, Bread et al. presented an approach which consists of four significant steps: cooperation objective and constraints, coordination variable and coordination function, centralized cooperation scheme, and consensus building [28].

Ahmadzadeh et al. [16] have studied the cooperative motion-planning problem for a group of heterogeneous UAVs. In their work, the surveillance operations were conducted via the body-fixed cameras equipped on their fixed-wing UAV. They demonstrated multi-UAV cooperative surveillance with spatiotemporal specifications [16]. Besides, they used an integer programming strategy to reduce the computational effort. The main contribution of their study was to generate an appropriate trajectory associated with the complexities of coupling cameras field

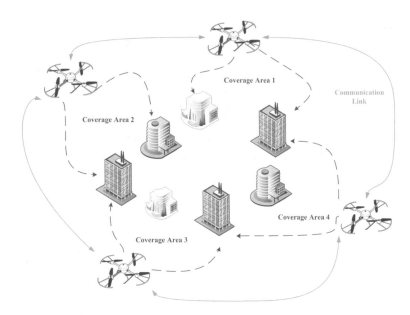

Fig. 10.2 Cooperative surveillance concept for patrolling urban areas

of view with flight paths. Paley et al. designed a glider with a coordinated control system for long-duration ocean sampling using real-time feedback control [23]. In their design, agents were modeled as Newtonian particles to steer a set of coordinated trajectories. However, this model cannot be applied for closed flocking due to the assumption that there is enough space between particles.

In the case of persistent surveillance, Nigam et al. have intensively researched on UAVs for persistent surveillance and their works have been consecutively released in the past few years. Their early efforts focused on investigating techniques for a high-level, scalable, reliable, efficient, and robust control of multiple UAVs [14] and derived an optimum policy with a single UAV [29]. They also suggested that modifications of the existing control policies would improve the system performance under dynamic constraints and proposed multi-agent reactive policy to integrate multiple UAVs and optimized the performance using a real-encode probability collective (PC) optimization framework. In the later works, Nigam et al. have developed algorithms to control multiple UAVs for persistent surveillance and devised a semi-heuristic approach for a surveillance task using multiple UAVs [15]. Their research considered the effect of aircraft dynamics on the performance of the designed cooperative mission and the advantages of their policy's performance was demonstrated by comparing it with other benchmark approaches such as the potential field-like approach, the planning-based approach, and the optimum approach. Paley and Peterson developed their previous research for ocean sampling [23], for environmental monitoring and surveillance [30]. Each UAV was considered as a Newton particle which was incorporated in a gyroscopic steering control system. This design has several drawbacks: first, obstacle avoidance in Newton particle method is not considered; second, all UAVs are moving in the same direction which is not flexible for surveillance and searching tasks; third, each UAV orbit around an inertially fixed point at constant radius which is not an energy efficient method for monitoring and surveillance.

10.2.3   Localization and Mapping

High agility, wide vision, and accessibility are some of the significant factors that made the UAVs a popular tool to map and model lands or terrains [18]. UAVs have been used to map in several types of research [18–20]. Figure 10.3 shows the concept of cooperative 3D mapping by multiple quadrotors. Remondino et al. used UAVs for space-mapping and 3D-modeling in several types of vehicles and techniques [18]. One of the high systems in the mapping technology of UAVs is known as light detection and ranging (LIDAR) was employed by Lin et al. [19]. They have applied the LIDAR-based system on a mini-UAV-borne cooperating with Ibeo Lux and Sick laser scanners and an AVT Pike F-421 CCCD camera to map a local area in Vanttila, Espoo, Finland in a fine-scale.

   As the surveillance and searching algorithms, the cooperative mapping task of UAVs can help to improve the accuracy and reduce the operation time through sharing their responsibilities. Cooperative control of autonomous vehicles can be

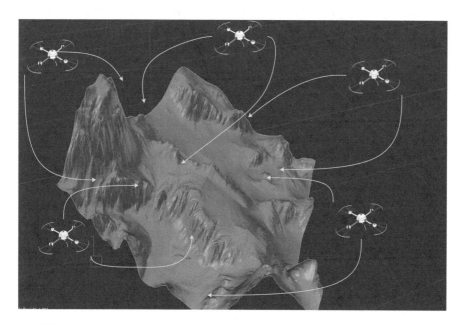

Fig. 10.3 Cooperative three-dimensional mapping using quadrotor UAVs

used to make a map for an unknown environment and 3D-modeling. Fenwick et al. introduced a novel algorithm for concurrent mapping and localization (CML) which combines the information of navigation and sensors of multiple unmanned vehicles [17]. This algorithm is working based on stochastic estimation and to extract landmarks from the mapping area using a feature-based approach. Gktoan et al. developed and demonstrated the multiple sensing nodes of numerous UAV platforms using decentralized data fusion (DDF) algorithm to simultaneously localize and map the flight simulator in real-time [31].

Simultaneous localization and mapping (SLAM) presented by Williams et al. [32] can be used to examine the prospect of the constrained local submap filter (CLSF) algorithm and applied to the multi-UAVs as SLAM algorithm. The advantage of this approach is that it allows the cross-covariance process to be scheduled at convenient intervals and aids in the data association problem.

Localization and mapping in unsafe or obscure places is another critical application of UAVs. Multi-UAV cooperative control has been used for mapping in wild or unknown areas in several types of research [33, 34] such as the continuation of the SLAM algorithm and its applications presented by Bryson and Sukkarieh [33]. Han et al. have introduced personal remote sensing (PRS) multi-UAVs for contour mapping in two scenarios of nuclear radiation [34]. Their work also focused on the costs of the multi-UAVs and the efficiency of atomic radiation detection in a necessary time which were the main advantages over a single UAV mapping. Kovacina et al. also focused on mapping a hazardous substance which was a chemical cloud. To map the chemical cloud, Kovacina et al. used swarm experimentation and evaluation platform (SWEEP) with their developed rule-based, decentralized control algorithm to simulate an air vehicle swarm searching for and mapping a chemical cloud [35].

10.2.4 Military Applications

The cooperative control of UAVs has various practical and potential military applications varying from reconnaissance and radar deception to surface-to-air-missile jamming. It has been demonstrated that a group of low-cost and well-organized UAVs can have better effects than a single high-cost UAV [36]. Generally, the application of cooperative control for the unmanned system in the military can be categorized into two main categories: reconnaissance and penetrating strategies. To achieve these types of applications, UAVs may need to flying near each other with a specific structure. Formation flight control is one of the most straightforward cooperative strategies which consists of a set of aircrafts flying near to each other in a defined distance [37]. One of the advantages of flight formation is a significant reduction in fuel consumption through locating the follower aircraft such that the vortex of the leader aircraft reduces the induced drag of the follower aircraft [38].

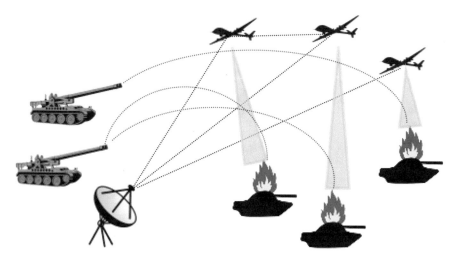

Fig. 10.4 Target detection using multiple cooperative UAVs in a reconnaissance mission [44]

10.2.4.1   Reconnaissance Strategy

A formation or cooperative design of UAVs can be used as reliable radars or reconnaissance tools to detect enemy troops and ballistic missiles [39, 40]. The integration of the UAVs radars will help to identify incursion objects or observe ground activities of an adversary [41, 42]. Ahmadzadeh et al. introduced a cooperative strategy to enable a heterogeneous team of UAVs to gather information for situational awareness [43]. In their work, an overall framework for reconnaissance and an algorithm for cooperative control of UAVs considering collision and obstacle avoidance were presented. Figure 10.4 shows a reconnaissance mission using multiple cooperative UAVs.

10.2.4.2   Penetrating Strategy

The new and robust defense mechanism of rivals makes it difficult to penetrate to their territories. To this aim, various strategies have been designed to deceive the target radar and defense mechanism [45–47].

Being hidden from the enemy radars through electronic counter-measure (ECM) is called radar jamming which is a very important action that is mostly used by unmanned combat air vehicles (UCAVs) to protect or defend themselves from surface-to-air missiles when the vehicles reconnoiter into enemy territories. The radar jamming consists of sending some noise to deceive the enemies radar signal. The radar jamming and deception can be more effective when a group of UCAVs works together. Jongrae et al. focused on the escort jamming (EJ) of the UAVs while a close formation and cooperative control procedure are designed to deceive the

Fig. 10.5 Cooperative radar jamming using multiple UAVs [45]

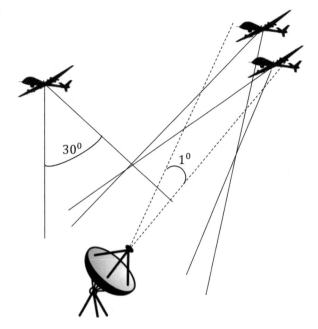

tracking radar of the surface-to-air missile (SAM) [45]. Generally, jamming can be classified into two categories: self or support jamming. Figure 10.5 shows the two mentioned methods of interference, where "D" shows the self-jamming and "A, B, C" UAVs show the support jamming.

The missiles control system is similar to the UAV control system. However, they are not designed to come back to the station. Since penetrating to the high-tech defense mechanism of a target is very complicated, a group of cooperative missiles will have more chance to penetrate a defense mechanism in comparison with being independently operated [46, 47]. Figure 10.6 shows a collective missile attack to a ship target.

10.3 Challenges

Multi-UAV systems have advantages over single UAVs in the impact of failure, scalability, survivability, the speed of the mission, cost, required bandwidth, and range of antennas [48]. However, these systems are complex and hard to coordinate. Gupta and Vaszkun considered three challenges in providing a stable and reliable UAV network: architectural design of networks; routing the packet from an origin to a destination and optimizing the metric; transferring from an out-of-service UAV to an active UAV, and energy conservation [48]. According to a study at MIT, the main challenges associated with the development and testing of cooperative

Fig. 10.6 Cooperative missile attacking concept to a target

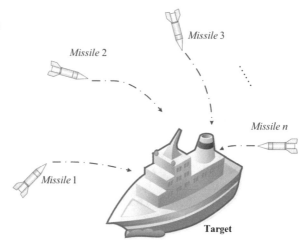

UAVs in dynamic and uncertain situations are real-time planning; designing a robust controller; and using communication networks [23]. Ryan et al. address issues in cooperative UAV control which are aerial surveillance, detection, and tracking which allows vision-based control; collision and obstacle avoidance and formation reconfiguration; high-level control needed for real-time human interfacing; and security of communication links [49]. Oh et al. addressed the problem of modeling the agent's interactions with each other and with the environment which is challenging to predict [50]. The most significant challenges in cooperative control of multi-agent systems can be summarized as below.

1. In cooperative control, instead of developing a control objective for a single system, it is necessary to devise control objectives for several sub-systems. Moreover, the relation between the team goal and agent goal needs to be negotiated and balanced [51].
2. The communication bandwidth and quality of connection among agents in the system are limited and variable. Moreover, the security of communication links in the presence of intruders should be considered in the design [52–55]. The CUAV is vulnerable to a range of cyberattacks such as denial of service (DoS) and time delay switch (TDS) attacks [56–59].
3. The aerodynamic interference of the agents on each other should be considered in the design [50]. Close cooperative flight control or formation has also specific aerodynamic challenges which are called aerodynamic coupling. These aerodynamic interferences are caused by the vortex effect of the leading aircraft and should be modeled and quantified in the controller design to avoid their critical impact on the system stability. Otherwise, unwanted rolling or yawing moment will be generated which can destabilize the overall system [60, 61]. However, incorporating the coupled dynamic in the formation design can help to reduce energy consumption through the mission [62, 63].

4. The controller design of CUAV should include fault tolerable algorithms through software redundancy because hardware redundancy is not an option for mini-UAVs. The fault tolerant control design for one UAV is a challenging task by itself, which has been discussed in the literature [64, 65].

10.4   Algorithms

The cooperative algorithms can be categorized into three main groups based on their methodologies. They are (1) consensus techniques; (2) flocking techniques; and (3) formation based techniques. Figure 10.7 shows the main algorithms that are used for the UAVs system. Algorithms for consensus control, flocking control, and formation control are discussed below, respectively.

10.4.1   Consensus Strategies

In the area of cooperative control, consensus control is an important and complicated problem. In consensus control, a group of agents communicates with each other through a sensing or communication network to reach a common decision. The roots of the consensus control belong to computer science and parallel computing [66, 67]. In the last decade, the research works of Jadbabaie et al. [68] and Olfati-Saber et al. [69] had a considerable impact on other researchers to work on consensus control problems. Generally, Jadbabaie et al. [68] provided a theoretical explanation for the alignment behavior of the dynamic model introduced by Vicsek [70], and Olfati-Saber introduced a general framework to solve consensus control problem of the networks of the integrators [69]. In the following subsection, the basic concepts of the consensus control will be explained; then, recent research works in this area will be reviewed. In the cooperative control, the communications among agents are modeled by undirected graphs. Thus, a basic knowledge of graph

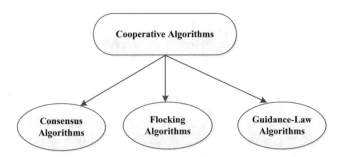

Fig. 10.7 Cooperative algorithms are categorized and explained in three main algorithms

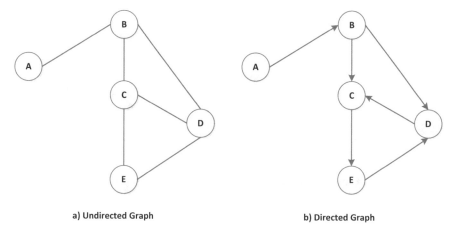

a) Undirected Graph   b) Directed Graph

Fig. 10.8 Directed and undirected graph structure

theory is needed to understand the concept of cooperative algorithms. Therefore, the basic concept of graph theory will be briefly explained, followed by the concept of consensus control theory.

10.4.1.1   Graph Theory Basics in Communication Systems

Communications or sensing among the agents of a team is commonly modeled by undirected graphs. An undirected graph is denoted by $G = (V, \varepsilon, A)$, where $V = \{1, 2, \ldots, N\}$ is the set of N nodes or agents in the network, and $\varepsilon(i, j) \in V \times V$ is set of edges between the ordered pairs of jth and ith agents. $\beta = [a_{ij}] \in R^{N \times N}$ is the adjacency matrix associated with graph G which is symmetric, and $a_{i,j}$ is a positive value if $(i, j) \in \varepsilon$ and $i \neq j$, otherwise $a_{ij} = 0$. Figure 10.8 shows the basic structure of a directed and undirected graph.

For example the adjacency matrix associated with the undirected graph shown in Fig. 10.8 is

$$\beta = \begin{bmatrix} 0 & 1 & 0 & 0 & 0 \\ 1 & 0 & 1 & 1 & 0 \\ 0 & 1 & 0 & 1 & 1 \\ 0 & 0 & 1 & 1 & 0 \end{bmatrix} \tag{10.1}$$

where node A, B, C, D, and E are considered to be nodes 1, 2, 3, 4, and 5, respectively.

10.4.1.2   Consensus Control Theory

The basic concept of the consensus control theory is to stimulate similar dynamics on the state's information of each agent in the group. Based on the communication type, each agent (vehicle) in the system can be modeled based on differential or difference equations. If the bandwidth of the communication network among the agents is large enough to allow continuous communication, then a differential equation can be used to model agent dynamics. Otherwise, the transmitted data among agents should be sent through discrete packets that need difference equations to model the agent dynamics. These are briefly explained here.

- **Continuous-time Consensus:** The most common consensus algorithm used for the dynamics defined by differential equations can be presented as [6, 71, 72]

$$\dot{x}_i(t) = -\sum_{j=1}^{n} a_{ij}(t)\big(x_i(t) - x_j(t)\big), \quad i = 1, \ldots, n \tag{10.2}$$

where $x_i(t)$ is the information state of the ith agent, and $a_{ij}(t)$ is the (i, j) element of the adjacency matrix β which is obtained from the graph G. If $a_{ij} = 0$, it indicates that there is no connection between agents i and j, subsequently, they cannot exchange any information between them. The consensus algorithm shown in Eq. 10.2 can be rewritten in a matrix form as

$$\dot{x}(t) = -L(t)x(t) \tag{10.3}$$

where the Laplacian matrix $L = [l_{ij}] \in R^{N \times N}$ is related to the graph G and can be obtained as follows

$$l_{ij} = \begin{cases} \sum_{j \in N_i}, & i = j \\ -a_{i,j}, & i \neq j \end{cases} \tag{10.4}$$

Since the l_{ij} has zero row sums, an eigenvalue of L is 0, which is associated with an eigenvector of 1. Because L is symmetric, in a connected graph, L has $N - 1$ real eigenvalues on the right side of the imaginary plane. Thus, N eigenvalues of L can be defined as follows

$$0 = \lambda_1 < \lambda_2 \leq \lambda_2 \ldots \leq \lambda_N \tag{10.5}$$

Based on this condition and the fact that L is symmetric, the diagonalized L can be obtained by orthogonal transformation matrix as

$$L = P J P^T \tag{10.6}$$

where P consists of the eigenvectors of the L and J is a diagonal form of L which are defined as follows

$$P = [r_1 \ r_2 \ \ldots \ r_n]$$

$$J = \begin{bmatrix} 0 & 0_{1 \times (N-1)} \\ 0_{(N-1) \times 1} & \gamma \end{bmatrix}$$

where γ is a matrix with diagonal form which contains $N - 1$ eigenvalues of L which have positive values, and r_i, $i \in \{1, 2, \ldots, N\}$ describes the eigenvectors of L where $r_i^T r_i = 1$ [6].

It can be claimed that *consensus* is achieved for a team of agents for all $x_i(0)$ and all $i, j = 1, \ldots, n$, if $lim_{t \to \infty} |x_i(t) - x_j(t)| = 0$ [6].

- **Discrete-time Consensus:** The discrete-time consensus is used when the communication bandwidth among the agents in the team is weak or occurs at discrete instants. In this case, the information states are updated through difference equations. The following form commonly presents the discrete-time consensus [73–76]

$$x_i[k+1] = \sum_{j=1}^{n} d_{ij}[k] x_j[k], \quad i = 1, \ldots, n \tag{10.7}$$

where k is the solving step associated to the communication event; $d_{ij}[k]$ is the (i, j) element of the stochastic matrix $D = [d_{ij}] \in R^{n \times n}$. The discrete-time consensus algorithm in Eq. 10.7 can be rewritten in a matrix form as

$$x[k+1] = D[k] x[k] \tag{10.8}$$

where $D = [d_{ij}] > 0$, if $i \neq j$ and the information flows from the agent j to i, otherwise $d_{ij}[k] = 0$ [75]. Similarly, a discrete-time *consensus* is achieved for a team of agents for all $x_i[0]$ and all $i, j = 1, \ldots, n$, if we have $lim_{k \to \infty} |x_i[k] - x_j[k]| = 0$ [75].

10.4.1.3 Consensus Recent Researches

The consensus control algorithm, which is based on graph theory, has received a growing interest among researchers [77, 78]. Jamshidi et al. developed a testbed and a consensus technique for cooperative control of UAVs [77]. Rezaee and Abdollahi proposed a consensus protocol for a class of high-order multi-agent systems [78]. They showed how agents achieve consensus on the average of any shared quantities using their relative positions. Li presented a geometric decomposition approach for cooperative agents [79]. Under topology adjustments, decomposing a system into sufficiently simple sub-systems facilitates subsequent analyses and provides the flexibility of choice. Liang et al. introduced an observer-based discrete consensus control system. The nonlinear observer was used to obtain the states of the agents, and a feedback control law was designed based on the data received from the

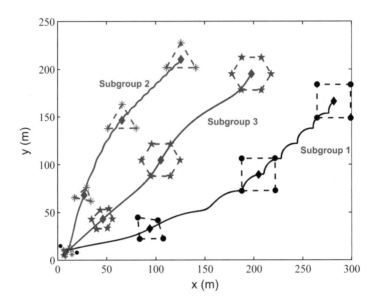

Fig. 10.9 Multi-consensus control of three subgroups by Han et al. [82]

observer [80]. Xia et al. introduced an optimal design for consensus control of agents with double-integrator dynamics with collision avoidance considerations [81]. Han et al. introduced a nonlinear multi-consensus control strategy for multi-agent systems [82]. In their research, both of the switching and fixed topology were considered, and their consensus controller could control three subgroups, as shown in Fig. 10.9. They were also compared their research work with their previous work [83] in which they could reduce the convergence time in consensus control. Shoja et al. introduced an estimator based consensus control scheme for agents with nonlinear and nonidentical dynamic systems [84]. In their design, they used an undirected graph model for their communication system among the agents, and multiple leaders were considered in their design. A sliding mode consensus control design for double-integrator multi-agent systems and 3-DoF helicopters was introduced by Hou et al. [85]. The advantage of their proposed method was achieving synchronization in the presence of disturbances and the ability to be implemented on 3-DoF model of helicopters.

Taheri et al. introduced an adaptive fuzzy wavelet network approach for consensus control of a class of a nonlinear second-order multi-agent system [86]. The adaptive laws were obtained using the Lyapunov theory to maintain the nonlinear dynamic stability. Then, an adaptive fuzzy wavelet network was used to compensate for the effect of unknown dynamics and time delay in the system. However, the authors did not address the design of a consensus control design for a second-order multi-agent system with a directed graph. Neural networks and robust control techniques have been used in [87] and [88] to design a consensus controller for higher-order multi-agent systems and their semi-global boundedness of consensus error was ensured by choosing sufficiently large control gains. Consensus fault tolerant controllers (FTC) with the ability to tolerate faults in the actuators of agents in a multi-agent system were also investigated [89–91]. Gallehdari et al. introduced an online redistributed control reconfiguration approach that employed the nearest neighbor information and the internal fault detection and identification (FDI) of the agent to keep the consensus control in the presence of faults in the actuators. They used the first-order dynamic model for their agents, and their proposed controller was designed based on minimizing the cost of faulty agent performance index which led to optimizing the performance index of the team. Later, they developed their work to optimize all the agents in the consensus FTC system [90]. Hua et al. introduced a consensus FTC design for time-varying high-order linear systems which could tolerate faults in the actuators [91].

Wang et al. introduced a new smooth function-based adaptive consensus control approach for multi-agent systems with nonlinear dynamic, unknown parameters, and uncertain disturbances without the need for the assumption of linearly parameterized reference trajectory [92]. Their approach was based on the premise of transmitting data among the agents based on an undirected graph model. Later, they extended their work for directed graph model as well [93].

10.4.2 Flocking Based Strategies

Flocking can be defined as a form of collective behavior of a group of interacting agents with mutual objectives. Flocking algorithms are inspired by a flock of birds and developed based on Reynolds rules. Reynolds modeled the steering behavior of each agent based on the positions and velocities of nearby flock-mates, using three terms of separation (collision avoidance), alignment (velocity matching), and cohesion (flock centering) [94].

10.4.2.1 Flocking Control Theory

Similar to consensus algorithms, flocking algorithms are based on graph theory. Unlike the formation strategies that require the group of agents to be in a particular shape, the group of agents in the flocking is not necessarily in a rigid shape or form. In other words, in flocking control, as long as the flock goals are satisfied, transition in the shape of the flock is allowed, e.g., it can be transformed from a rectangular shape to a triangular shape.

Several flocking algorithms have been devised for multi-agent systems with a second-order dynamic model [95–97]. The following equation of motion can present a group of agents with a second-order dynamic model.

$$\begin{cases} \dot{q}_i = p_i \\ \dot{p}_i = u_i \end{cases} \tag{10.9}$$

where q_i is the position of agent (node) i and p_i is the velocity. $p_i, q_i, u_i \in R^m$ and $i \in V = 1, 2, \ldots, N$ (set of N nodes or agents in the network). Flocking algorithms consists of three terms: (1) a gradient-based term, (2) a consensus term, and (3) a navigational feedback term, and can be presented as follows [96]

$$u_i = \underbrace{\sum_{j \in N_i} \phi_\alpha(\|q_j - q_i\|)n_{ij}}_{\text{gradient-based term}} + \underbrace{\sum_{j \in N_i} a_{ij}(q)(p_j - p_i)}_{\text{consensus term}} + \underbrace{f_i^\gamma(q_i, p_i, q_r, p_r)}_{\text{Navigational-based term}}$$

$$\tag{10.10}$$

where $\phi(\bullet)$ is a potential function, and $n_{ij} = \sigma_\epsilon(q_j - q_i) = (q_j - q_i / \sqrt{1 + \epsilon \|q_j q_i\|^2})$ is a vector along the line connecting q_i to q_j in which $\epsilon \in (0, 1)$ is a constant parameter of the norm in σ-norm. The pair $(p_r, q_r) \in R^m \times R^m$ is the state of a γ agent. The navigational feedback term f_i^γ is given as follows

$$f_i^\gamma(q_i, p_i, q_r, p_r) = -c_1(q_i - q_r) - c_2(p_i - p_r), \quad c_1, c_2 > 0 \tag{10.11}$$

The flocking algorithm in Eq. 10.10 can be developed by using some updating terms to tackle the problem of uncertainties in the flock control. One major problem with flocking control is its incapability of covering a large area. Thus, a semi-flocking algorithm was introduced to tackle this problem [98]. In the semi-flocking algorithm, the navigation feedback term is modified to make each agent able to decide whether to track a target or to search for a new one.

10.4.2.2 Flocking Recent Researches

Moshtagh and Jadbabaie introduced a novel flocking and velocity alignment algorithm to control the kinematic agents using graph theory [95]. In their design which was capable of flocking control in two and three dimensions, they used a geodesic control to minimize the misalignment potential which leads to flocking and velocity alignment. They also demonstrated that their method could keep the flocking even when the topology of proximity graph changes, and as long as the joint connectivity is well-maintained, the algorithm will be successful in consensus control. However, to guarantee the flocking success, still, one problem has to be solved, and that is how to keep the connectivity condition in the proximity graph. Olfati-Saber introduced a systematic approach for the generation of cost functions for flocking [96]. In these cost functions, the deviation from flock objects will be penalized. They demonstrated that a peer-to-peer network of agents could be used for the migration of flocks and the need for a single leader for the flock can be eliminated. The simulation results for flocking hundreds of agents in 2-D and 3-D, squeezing, and split/reuniting maneuvers were provided that showed the success of the proposed algorithm in the presence of obstacles. Saif et al. introduced a linear quadratic regulator (LQR) controller for a flock of UAVs which is independent of the number of agents in the flock [97]. This control strategy can satisfy the Reynolds rules, and independent of the number of UAVs in the flock it allows designing an LQR controller for each of the UAVs. Chapman and Mesbahi designed an optimal controller for UAV flocking in the presence of wind gusts, using a consensus-based leader-follower system to improve velocity tracking [99].

Tanner et al. introduced a control law for flocking of multi-agent systems with double-integrator dynamics and arbitrary switching in the topology of agent interaction network [100]. The non-smooth analysis was used to accommodate arbitrary switching the agent's network, and they demonstrated that their control law is robust against arbitrary changes in the agent communication network as long as they are connected in their maneuvers. Hung and Givigi developed a model-free reinforcement learning approach to flocking of small fixed-wing UAVs in a leader-follower topology [4]. In their study, agents experience disturbances in a stochastic environment. The advantage of their online learning design is that their model is not dependent on the environment; hence, it can be implemented in a different environment without any information about the plant and disturbances in the system. This characteristic increases the adaptability of the system to unforeseen situations. However, the learning rate and convergence speed of flocking are two factors that still need to be solved. Quintero et al. introduced a leader-follower design for flocking control of multiple UAVs to conduct a sensing task [101]. The UAVs were considered as fixed-wing airplanes flying at a constant speed with fixed altitude which limits its movement in a 2-D planar surface. In their strategy, each of the followers is controlled using a stochastic optimal control problem where the cost function is the heading and distance toward the leader. This algorithm was successfully applied and implemented in three UAVs equipped with cameras; however, the offline solving the optimization problem cannot guarantee the flocking behavior of the system in the presence of nonlinear behavior of flock and its agents.

McCune et al. introduced a framework based on a dynamic data-driven application system (DDDAS) to predict, control, and improve decision making artificial swarms using repeated simulations and synergistic feedback loops [5]. Using this strategy helps to improve the decision making in the process of swarm control; however, the time frame for the real-time application of this strategy has not been considered which can affect the effectiveness of this approach. Martin et al. [102] considered a system of agents with second-order dynamics. They determined conditions to ensure that agents agree on a common velocity to achieve system flocking. The significance of their design was the allowance for disconnected communication links that were unnecessary for flocking. Practical bounds for two different communication rules were investigated; first, the agents communicate within the radius of communication bound; second, agents communicate with each other with different and randomly communication radiuses. Overall, they concluded that by choosing a proper initial velocity disagreement or by setting a small enough time step, flocking can be achieved with random communication radiuses. One of the drawbacks of their approach was an asymmetric requirement in the interaction among the agents. Generally, other types of interaction (i.e., assuming that agents interact with the nearest neighbors with a fixed parameter which is called topological interaction rule) can happen in the flock. Riehl et al. introduced a receding-horizon search algorithm for cooperative UAVs [2]. In order to find a target in the minimum time, each of the UAVs was equipped with a gimbal sensor which could be rotated to observe the nearby target; then by gathering information on a potential location for the target, they could find it. The algorithm helps to minimize the expected time for finding the target by controlling the position of UAVs and their sensors. The optimization process is a receding-horizon algorithm based on a graph with variable target probability density function (PDF). This algorithm was successfully tested using two small UAVs equipped with gimbaled video cameras.

10.4.3 Guidance Law Based Cooperative Control

This subsection is separated from the other cooperative control techniques because they do not deal with the guidance system in their design. In order to achieve a formation, the acceleration and angular velocity of each agent in the formation group should be calculated separately [103]. To this aim, guidance law techniques are used to obtain the desired acceleration and angular velocities. Pure pursuit (PP) guidance algorithm is one of the most practical leader-follower guidance techniques in the formation control. This algorithm was initially implemented on ground-attack missile systems that aim to hit the target [104]. Later by introducing the concept of the virtual leader (or target) it has been developed for the formation of flight control which the followers keep their line of sight (LoS) in-line with the leader movement. In other words, the velocity direction of the agents should be aligned with the velocity of the leader [103].

In the PP algorithm, between the follower speed vector $\vec{V}$ and the virtual leader $\vec{R}$ the following equation is maintained:

$$\vec{V}_f \times \vec{R} = 0 \tag{10.12}$$

Figure 10.10 shows the geometry between the virtual leader and the follower in the PP algorithm. In this figure, $d_{x_{ref}}$, $d_{y_{ref}}$, and $d_{z_{ref}}$ represent the distance between the leader and the virtual leader in the longitudinal axis, lateral axis, and the vertical axis, respectively. The required acceleration in the follower aircraft to reach the virtual leader can be calculated as follows [105, 106]

$$\vec{A}_f = \frac{N(\vec{V}_f \times \vec{R}) \times \vec{V}_f}{\|\vec{V}_f\| \, \|\vec{R}\|} \tag{10.13}$$

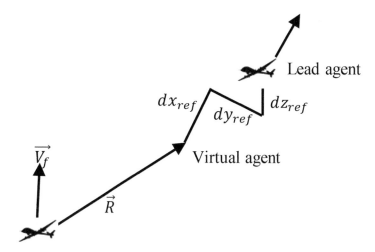

Fig. 10.10 Geometry of the PP guidance algorithm [105]

where N is the navigational constant which is usually chosen between 0.3 and 0.5. Proportional navigation (PN) guidance is another candidate that can be applied in the formation control design; however, because when the closing velocity is negative (the leader velocity is higher than the follower aircraft), the PN guidance is likely to guide the follower away from the leader [106]. In contrast, the PP guidance does not depend on the leader velocity and always guides the follower in the direction of the leader. Thus, we discussed the PP guidance laws application in the control design of the flight formation systems.

10.4.3.1   Guidance Law Based Recent Researches

Gu et al. [107] introduced a nonlinear leader-follower based formation control law. A two-loop controller was designed where nonlinear dynamic inversion (DI) was used to design the velocity and position tracker in the outer-loop, and a linear controller was used to track the leader attitude in the inner-loop. This two-loop design is based on the difference in the changing rate of the inner-loop and outer-loop dynamic parameters. The introduced controller was experimentally tested on two WVU YF-22 aircrafts as leader and follower. The experimental results demonstrated the effectiveness of their proposed formation control law. Yamasaki et al. introduced a PP guidance based formation control system for a group of UAVs [106]. Their proposed control system uses a PP guidance algorithm and a velocity controller based on the DI control technique to avoid a collision. The attitude controller of the follower aircraft was designed based on a two-loop DI controller. Sadeghi et al. improve the Yamasaki work [106] and introduced a new approach to integrating the guidance and control system through a PID control design [37].

Their proposed approach could improve the PP guidance algorithm accuracy and the maneuverability of the formation group.

Zhu et al. introduced a least-squares method for the estimation of the leader location, then, a guidance law based on sliding mode control was designed to control the heading rate of the follower aircrafts toward the leader estimated location [108]. Ali et al. presented a guidance law for lateral formation control of UAVs based on sliding mode theory [109]. Two sliding surfaces were integrated into series to improve the control response in the formation design. A new approach for UAVs formation control considering obstacle/collision avoidance using modified Grossberg neural network (GNN) was developed by Wang et al. [110]. In order to track the desired trajectory, a model predictive controller was used. They simulated their collision/obstacle avoidance design in a 3-D environment. A LOS guidance law approach for formation control of a group of under-actuated vessels is studied in [111]. In their approach, a nonlinear synchronization controller was combined with the LOS-based path following controller to make the overall system more robust and controllable under the under-actuation situation.

10.5 Summary and Conclusion

In this chapter, the algorithms and applications of cooperative control techniques for UAVs are reviewed. By categorizing the recent researches to applications and methods, each was discussed separately. The latest studies in the field of cooperative control of UAVs have been investigated and the advantages and disadvantages of methods were discussed. Applications of cooperative UAVs mission in various fields have been explored. Although some studies in the cooperative field may have been missed in this survey, it is hoped that this survey would be helpful for researchers to overview the major achievements in cooperative control of UAVs.

Acknowledgements The authors would like to thank Dr. Kang Yen for his guidance, contribution, and effort toward this paper.

Bibliography

1. L. Lin, M. Roscheck, M.A. Goodrich, B.S. Morse, Supporting wilderness search and rescue with integrated intelligence: autonomy and information at the right time and the right place. in *Association for the Advancement of Artificial Intelligence* (2010)
2. J.R. Riehl, G.E. Collins, J.P. Hespanha, Cooperative search by UAV teams: a model predictive approach using dynamic graphs. IEEE Trans. Aerosp. Electron. Syst. **47**(4), 2637–2656 (2011)
3. J. How, Y. Kuwata, E. King, Flight demonstrations of cooperative control for UAV teams, in *AIAA 3rd "Unmanned Unlimited" Technical Conference, Workshop and Exhibit* (2004), p. 6490
4. S.-M. Hung, S.N. Givigi, A q-learning approach to flocking with UAVS in a stochastic environment. IEEE Trans. Cybern. **47**(1), 186–197 (2017)

5. R.R. McCune, G.R. Madey, Control of artificial swarms with DDDAS. Proc. Comput. Sci. **29**, 1171–1181 (2014)
6. W. Ren, R.W. Beard, E.M. Atkins, Information consensus in multivehicle cooperative control. IEEE Control Syst. **27**(2), 71–82 (2007)
7. B.D. Anderson, C. Yu, J.M. Hendrickx, et al., Rigid graph control architectures for autonomous formations. IEEE Control Syst. **28**(6), 48–63 (2008)
8. Q. Wang, H. Gao, F. Alsaadi, T. Hayat, An overview of consensus problems in constrained multi-agent coordination. Syst. Sci. Control Eng. Open Access J. **2**(1), 275–284 (2014)
9. K.-K. Oh, M.-C. Park, H.-S. Ahn, A survey of multi-agent formation control. Automatica **53**, 424–440 (2015)
10. B. Zhu, L. Xie, D. Han, X. Meng, R. Teo, A survey on recent progress in control of swarm systems. Sci. China Inf. Sci. **60**(7), 070201 (2017)
11. M. Senanayake, I. Senthooran, J.C. Barca, H. Chung, J. Kamruzzaman, M. Murshed, Search and tracking algorithms for swarms of robots: a survey. Robot. Auton. Syst. **75**, 422–434 (2016)
12. M.A. Goodrich, J.L. Cooper, J.A. Adams, C. Humphrey, R. Zeeman, B.G. Buss, Using a mini-UAV to support wilderness search and rescue: practices for human-robot teaming, in *IEEE International Workshop on Safety, Security and Rescue Robotics, 2007. SSRR 2007* (IEEE, Piscataway, 2007), pp. 1–6
13. J.Q. Cui, S.K. Phang, K.Z. Ang, F. Wang, X. Dong, Y. Ke, S. Lai, K. Li, X. Li, F. Lin, et al., Drones for cooperative search and rescue in post-disaster situation, in *IEEE 7th International Conference on Cybernetics and Intelligent Systems (CIS) and IEEE Conference on Robotics, Automation and Mechatronics (RAM), 2015* (IEEE, Piscataway, 2015), pp. 167–174
14. N. Nigam, I. Kroo, Control and design of multiple unmanned air vehicles for a persistent surveillance task, in *12th AIAA/ISSMO Multidisciplinary Analysis and Optimization Conference* (2008), p. 5913
15. N. Nigam, S. Bieniawski, I. Kroo, J. Vian, Control of multiple UAVS for persistent surveillance: algorithm and flight test results. IEEE Trans. Control Syst. Technol. **20**(5), 1236–1251 (2012)
16. A. Ahmadzadeh, A. Jadbabaie, V. Kumar, G.J. Pappas, Multi-UAV cooperative surveillance with spatio-temporal specifications, in *45th IEEE Conference on Decision and Control, 2006* (IEEE, Piscataway, 2006), pp. 5293–5298
17. J.W. Fenwick, P.M. Newman, J.J. Leonard, Cooperative concurrent mapping and localization, in *IEEE International Conference on Robotics and Automation, 2002. Proceedings. ICRA'02*, vol. 2 (IEEE, Piscataway, 2002), pp. 1810–1817
18. F. Remondino, L. Barazzetti, F. Nex, M. Scaioni, D. Sarazzi, UAV photogrammetry for mapping and 3d modeling–current status and future perspectives. Int. Arch. Photogramm. Remote. Sens. Spat. Inf. Sci. **38**(1), C22 (2011)
19. Y. Lin, J. Hyyppa, A. Jaakkola, Mini-UAV-borne LIDAR for fine-scale mapping. IEEE Geosci. Remote Sens. Lett. **8**(3), 426–430 (2011)
20. L. Zongjian, UAV for mapping low altitude photogrammetric survey, in *International Archives of Photogrammetry and Remote Sensing, Beijing, China*, vol. 37 (2008), pp. 1183–1186
21. M.J. Mears, Cooperative electronic attack using unmanned air vehicles, in *Proceedings of the 2005 American Control Conference, 2005* (IEEE, Piscataway, 2005), pp. 3339–3347
22. Z. Junwei, Z. Jianjun, Target distributing of multi-UAVs cooperate attack and defend based on DPSO algorithm, in *Sixth International Conference on Intelligent Human-Machine Systems and Cybernetics (IHMSC), 2014*, vol. 2 (IEEE, Piscataway, 2014), pp. 396–400
23. D.A. Paley, F. Zhang, N.E. Leonard, Cooperative control for ocean sampling: the glider coordinated control system. IEEE Trans. Control Syst. Technol. **16**(4), 735–744 (2008)
24. J. Scherer, S. Yahyanejad, S. Hayat, E. Yanmaz, T. Andre, A. Khan, V. Vukadinovic, C. Bettstetter, H. Hellwagner, B. Rinner, An autonomous multi-UAV system for search and rescue, in *Proceedings of the First Workshop on Micro Aerial Vehicle Networks, Systems, and Applications for Civilian Use* (ACM, New York, 2015), pp. 33–38

25. S. Waharte, N. Trigoni, Supporting search and rescue operations with UAVS, in *International Conference on Emerging Security Technologies (EST), 2010* (IEEE, Piscataway, 2010), pp. 142–147
26. A. Viguria, I. Maza, A. Ollero, Distributed service-based cooperation in aerial/ground robot teams applied to fire detection and extinguishing missions. Adv. Robot. **24**(1–2), 1–23 (2010)
27. I. Maza, F. Caballero, J. Capitan, J. Martinez-de Dios, A. Ollero, Firemen monitoring with multiple UAVs for search and rescue missions, in *IEEE International Workshop on Safety Security and Rescue Robotics (SSRR), 2010* (IEEE, Piscataway, 2010), pp. 1–6
28. R.W. Beard, T.W. McLain, D.B. Nelson, D. Kingston, D. Johanson, Decentralized cooperative aerial surveillance using fixed-wing miniature UAVS. Proc. IEEE **94**(7), 1306–1324 (2006)
29. N. Nigam, I. Kroo, Persistent surveillance using multiple unmanned air vehicles, in *IEEE Aerospace Conference, 2008* (IEEE, Piscataway, 2008), pp. 1–14
30. D.A. Paley, C. Peterson, Stabilization of collective motion in a time-invariant flowfield. J. Guid. Control. Dyn. **32**(3), 771–779 (2009)
31. A.H. Goktogan, E. Nettleton, M. Ridley, S. Sukkarieh, Real time multi-UAV simulator, in *IEEE International Conference on Robotics and Automation, 2003 Proceedings. ICRA'03*, vol. 2 (IEEE, Piscataway, 2003), pp. 2720–2726
32. S.B. Williams, G. Dissanayake, H. Durrant-Whyte, Towards multi-vehicle simultaneous localisation and mapping, in *IEEE International Conference on Robotics and Automation, 2002. Proceedings. ICRA'02*, vol. 3 (IEEE, Piscataway, 2002), pp. 2743–2748
33. M. Bryson, S. Sukkarieh, Co-operative localisation and mapping for multiple UAVs in unknown environments, in *IEEE Aerospace Conference, 2007* (IEEE, Piscataway, 2007), pp. 1–12
34. J. Han, Y. Xu, L. Di, Y. Chen, Low-cost multi-UAV technologies for contour mapping of nuclear radiation field. J. Intell. Robot. Syst. **70**, 401–410 (2013)
35. M.A. Kovacina, D. Palmer, G. Yang, R. Vaidyanathan, Multi-agent control algorithms for chemical cloud detection and mapping using unmanned air vehicles, in *IEEE/RSJ International Conference on Intelligent Robots and Systems, 2002*, vol. 3, pp. 2782–2788 (IEEE, Piscataway, 2002)
36. I.-S. Jeon, J.-I. Lee, M.-J. Tahk, Homing guidance law for cooperative attack of multiple missiles. J. Guid. Control. Dyn. **33**(1), 275–280 (2010)
37. M. Sadeghi, A. Abaspour, S.H. Sadati, A novel integrated guidance and control system design in formation flight. J. Aerosp. Technol. Manag. **7**(4), 432–442 (2015)
38. E. Lavretsky, F/a-18 autonomous formation flight control system design, in *AIAA Guidance, Navigation, and Control Conference and Exhibit* (2002), p. 4757
39. L. Liu, D. McLernon, M. Ghogho, W. Hu, J. Huang, Ballistic missile detection via micro-Doppler frequency estimation from radar return. Digital Signal Process. **22**(1), 87–95 (2012)
40. M.L. Stone, G.P. Banner, Radars for the detection and tracking of ballistic missiles, satellites, and planets. Lincoln Lab. J. **12**(2), 217–244 (2000)
41. C. Schwartz, T. Bryant, J. Cosgrove, G. Morse, J. Noonan, A radar for unmanned air vehicles. Lincoln Lab. J. **3**(1), 119–143 (1990)
42. Y. Wang, S. Dong, L. Ou, L. Liu, Cooperative control of multi-missile systems. IET Control Theory Appl. **9**(3), 441–446 (2014)
43. A. Ahmadzadeh, G. Buchman, P. Cheng, A. Jadbabaie, J. Keller, V. Kumar, G. Pappas, Cooperative control of UAVS for search and coverage, in *Proceedings of the AUVSI Conference on Unmanned Systems*, vol. 2 (2006)
44. R.M. Murray, Recent research in cooperative control of multivehicle systems. J. Dyn. Syst. Meas. Control. **129**(5), 571–583 (2007)
45. J. Kim, J.P. Hespanha, Cooperative radar jamming for groups of unmanned air vehicles, in *43rd IEEE Conference on Decision and Control, 2004. CDC*, vol. 1 (IEEE, Piscataway, 2004), pp. 632–637
46. J.-I. Lee, I.-S. Jeon, M.-J. Tahk, Guidance law using augmented trajectory-reshaping command for salvo attack of multiple missiles, in *UKACC International Control Conference* (2006), pp. 766–771

47. I.-S. Jeon, J.-I. Lee, M.-J. Tahk, Impact-time-control guidance law for anti-ship missiles. IEEE Trans. Control Syst. Technol. **14**(2), 260–266 (2006)
48. L. Gupta, R. Jain, G. Vaszkun, Survey of important issues in UAV communication networks. IEEE Commun. Surv. Tutorials **18**(2), 1123–1152 (2016)
49. A. Ryan, M. Zennaro, A. Howell, R. Sengupta, J.K. Hedrick, An overview of emerging results in cooperative UAV control, in *43rd IEEE Conference on Decision and Control, 2004. CDC*, vol. 1 (IEEE, Piscataway, 2004), pp. 602–607
50. H. Oh, A.R. Shirazi, C. Sun, Y. Jin, Bio-inspired self-organising multi-robot pattern formation: a review. Robot. Auton. Syst. **91**, 83–100 (2017)
51. F.L. Lewis, H. Zhang, K. Hengster-Movric, A. Das, *Cooperative Control of Multi-Agent Systems: Optimal and Adaptive Design Approaches* (Springer, Berlin, 2013)
52. S. Martini, D. Di Baccio, F.A. Romero, A.V. Jiménez, L. Pallottino, G. Dini, A. Ollero, Distributed motion misbehavior detection in teams of heterogeneous aerial robots. Robot. Auton. Syst. **74**, 30–39 (2015)
53. B. Khaldi, F. Harrou, F. Cherif, Y. Sun, Monitoring a robot swarm using a data-driven fault detection approach. Robot. Auton. Syst. **97**, 193–203 (2017)
54. A. Abbaspour, M. Sanchez, A. Sargolzaei, K. Yen, N. Sornkhampan, Adaptive neural network based fault detection design for unmanned quadrotor under faults and cyber attacks, in *25th International Conference on Systems Engineering (ICSEng)* (2017)
55. A. Abbaspour, K.K. Yen, S. Noei, A. Sargolzaei, Detection of fault data injection attack on UAV using adaptive neural network. Proc. Comput. Sci. **95**, 193–200 (2016)
56. A. Sargolzaei, K.K. Yen, M.N. Abdelghani, Preventing time-delay switch attack on load frequency control in distributed power systems. IEEE Trans. Smart Grid **7**(2), 1176–1185 (2015)
57. A. Sargolzaei, K. Yen, M.N. Abdelghani, Delayed inputs attack on load frequency control in smart grid, in *Innovative Smart Grid Technologies 2014* (IEEE, Piscataway, 2014), pp. 1–5
58. A. Sargolzaei, K.K. Yen, M.N. Abdelghani, S. Sargolzaei, B. Carbunar, Resilient design of networked control systems under time delay switch attacks, application in smart grid. IEEE Access **5**, 15,901–15,912 (2017)
59. A. Sargolzaei, A. Abbaspour, M.A. Al Faruque, A.S. Eddin, K. Yen, Security challenges of networked control systems, in *Sustainable Interdependent Networks* (Springer, Berlin, 2018), pp. 77–95
60. F. Giulietti, M. Innocenti, M. Napolitano, L. Pollini, Dynamic and control issues of formation flight. Aerosp. Sci. Technol. **9**(1), 65–71 (2005)
61. R.L. Pereira, K.H. Kienitz, Tight formation flight control based on h approach, in *24th Mediterranean Conference on Control and Automation (MED), 2016* (IEEE, Piscataway, 2016), pp. 268–274
62. R.J. Ray, B.R. Cobleigh, M.J. Vachon, C. St. John, Flight test techniques used to evaluate performance benefits during formation flight, in *NASA Conference Publication, NASA* (1998, 2002)
63. J. Pahle, D. Berger, M.W. Venti, J.J. Faber, C. Duggan, K. Cardinal, A preliminary flight investigation of formation flight for drag reduction on the c-17 aircraft (2012)
64. A. Abbaspour, K.K. Yen, P. Forouzannezhad, A. Sargolzaei, A neural adaptive approach for active fault-tolerant control design in UAV. IEEE Trans. Syst. Man Cybern. Syst. Hum. **99**, 1–11 (2018)
65. A. Abbaspour, P. Aboutalebi, K.K. Yen, A. Sargolzaei, Neural adaptive observer-based sensor and actuator fault detection in nonlinear systems: application in UAV. ISA Trans. **67**, 317–329 (2017)
66. V. Borkar, P. Varaiya, Asymptotic agreement in distributed estimation. IEEE Trans. Autom. Control **27**(3), 650–655 (1982)
67. J. Tsitsiklis, D. Bertsekas, M. Athans, Distributed asynchronous deterministic and stochastic gradient optimization algorithms. IEEE Trans. Autom. Control **31**(9), 803–812 (1986)
68. A. Jadbabaie, J. Lin, A.S. Morse, Coordination of groups of mobile autonomous agents using nearest neighbor rules. IEEE Trans. Autom. Control **48**(6), 988–1001 (2003)

69. R. Olfati-Saber, R.M. Murray, Consensus problems in networks of agents with switching topology and time-delays. IEEE Trans. Autom. Control **49**(9), 1520–1533 (2004)
70. T. Vicsek, A. Czirók, E. Ben-Jacob, I. Cohen, O. Shochet, Novel type of phase transition in a system of self-driven particles. Phys. Rev. Lett. **75**(6), 1226 (1995)
71. G. Xie, L. Wang, Consensus control for a class of networks of dynamic agents. Int. J. Robust Nonlinear Control **17**(10–11), 941–959 (2007)
72. F. Xiao, L. Wang, Asynchronous consensus in continuous-time multi-agent systems with switching topology and time-varying delays. IEEE Trans. Autom. Control **53**(8), 1804–1816 (2008)
73. X. Chen, F. Hao, Event-triggered average consensus control for discrete-time multi-agent systems. IET Control Theory Appl. **6**(16), 2493–2498 (2012)
74. H. Zhang, H. Jiang, Y. Luo, G. Xiao, Data-driven optimal consensus control for discrete-time multi-agent systems with unknown dynamics using reinforcement learning method. IEEE Trans. Ind. Electr. **64**(5), 4091–4100 (2017)
75. D. Ding, Z. Wang, B. Shen, G. Wei, Event-triggered consensus control for discrete-time stochastic multi-agent systems: the input-to-state stability in probability. Automatica **62**, 284–291 (2015)
76. J. Jin, N. Gans, Collision-free formation and heading consensus of nonholonomic robots as a pose regulation problem. Robot. Auton. Syst. **95**, 25–36 (2017)
77. M. Jamshidi, A.J. Betancourt, J. Gomez, Cyber-physical control of unmanned aerial vehicles. Sci. Iran. **18**(3), 663–668 (2011)
78. H. Rezaee, F. Abdollahi, Average consensus over high-order multiagent systems. IEEE Trans. Autom. Control **60**(11), 3047–3052 (2015)
79. W. Li, Unified generic geometric-decompositions for consensus or flocking systems of cooperative agents and fast recalculations of decomposed subsystems under topology-adjustments. IEEE Trans. Cybern. **46**(6), 1463–1470 (2016)
80. H. Liang, H. Zhang, Z. Wang, Distributed-observer-based cooperative control for synchronization of linear discrete-time multi-agent systems. ISA Trans. **59**, 72–78 (2015)
81. Y. Xia, X. Na, Z. Sun, J. Chen, Formation control and collision avoidance for multi-agent systems based on position estimation. ISA Trans. **61**, 287–296 (2016)
82. T. Han, Z.-H. Guan, M. Chi, B. Hu, T. Li, X.-H. Zhang, Multi-formation control of nonlinear leader-following multi-agent systems. ISA Trans. **69**, 140–147 (2017)
83. G.-S. Han, Z.-H. Guan, X.-M. Cheng, Y. Wu, F. Liu, Multiconsensus of second order multiagent systems with directed topologies. Int. J. Control, Autom. Syst. **11**(6), 1122–1127 (2013)
84. S. Shoja, M. Baradarannia, F. Hashemzadeh, M. Badamchizadeh, P. Bagheri, Surrounding control of nonlinear multi-agent systems with non-identical agents. ISA Trans. **70**, 219 (2017)
85. H. Hou, Q. Zhang, Finite-time synchronization for second-order nonlinear multi-agent system via pinning exponent sliding mode control. ISA Trans. **65**, 96–108 (2016)
86. M. Taheri, F. Sheikholeslam, M. Najafi, M. Zekri, Adaptive fuzzy wavelet network control of second order multi-agent systems with unknown nonlinear dynamics. ISA Trans. **69**, 89–101 (2017)
87. H. Zhang, F.L. Lewis, Adaptive cooperative tracking control of higher-order nonlinear systems with unknown dynamics. Automatica **48**(7), 1432–1439 (2012)
88. S. El-Ferik, A. Qureshi, F.L. Lewis, Neuro-adaptive cooperative tracking control of unknown higher-order affine nonlinear systems. Automatica **50**(3), 798–808 (2014)
89. Z. Gallehdari, N. Meskin, K. Khorasani, A distributed control reconfiguration and accommodation for consensus achievement of multiagent systems subject to actuator faults. IEEE Trans. Control Syst. Technol. **24**(6), 2031–2047 (2016)
90. Z. Gallehdari, N. Meskin, K. Khorasani, Distributed reconfigurable control strategies for switching topology networked multi-agent systems. ISA Trans. **71**, 51–67 (2017)
91. Y. Hua, X. Dong, Q. Li, Z. Ren, Distributed fault-tolerant time-varying formation control for high-order linear multi-agent systems with actuator failures. ISA Trans. **71**, 40–50 (2017)

92. R. Wang, X. Dong, Q. Li, Z. Ren, Distributed adaptive formation control for linear swarm systems with time-varying formation and switching topologies. IEEE Access **4**, 8995–9004 (2016)

93. W. Wang, C. Wen, J. Huang, Distributed adaptive asymptotically consensus tracking control of nonlinear multi-agent systems with unknown parameters and uncertain disturbances. Automatica **77**, 133–142 (2017)

94. C.W. Reynolds, Flocks, herds and schools: a distributed behavioral model. ACM SIGGRAPH Comput. Graph. **21**(4), 25–34 (1987)

95. N. Moshtagh, A. Jadbabaie, Distributed geodesic control laws for flocking of nonholonomic agents. IEEE Trans. Autom. Control **52**(4), 681–686 (2007)

96. R. Olfati-Saber, Flocking for multi-agent dynamic systems: algorithms and theory. IEEE Trans. Autom. Control **51**(3), 401–420 (2006)

97. O. Saif, I. Fantoni, A. Zavala-Rio, Flocking of multiple unmanned aerial vehicles by LQR control, in *International Conference on Unmanned Aircraft Systems (ICUAS), 2014* (IEEE, Piscataway, 2014), pp. 222–228

98. S.H. Semnani, O.A. Basir, Semi-flocking algorithm for motion control of mobile sensors in large-scale surveillance systems. IEEE Trans. Cybern. **45**(1), 129–137 (2015)

99. A. Chapman, M. Mesbahi, UAV flocking with wind gusts: adaptive topology and model reduction, in *American Control Conference (ACC), 2011* (IEEE, Piscataway, 2011), pp. 1045–1050

100. H.G. Tanner, A. Jadbabaie, G.J. Pappas, Flocking in fixed and switching networks. IEEE Trans. Autom. Control **52**(5), 863–868 (2007)

101. S.A. Quintero, G.E. Collins, J.P. Hespanha, Flocking with fixed-wing UAVS for distributed sensing: a stochastic optimal control approach, in *American Control Conference (ACC), 2013* (IEEE, Piscataway, 2013), pp. 2025–2031

102. S. Martin, A. Girard, A. Fazeli, A. Jadbabaie, Multiagent flocking under general communication rule. IEEE Trans. Control Netw. Syst. **1**(2), 155–166 (2014)

103. W. Naeem, R. Sutton, S. Ahmad, R. Burns, A review of guidance laws applicable to unmanned underwater vehicles. J. Navig. **56**(1), 15–29 (2003)

104. C.-F. Lin, *Modern Navigation, Guidance, and Control Processing*, vol. 2 (Prentice Hall, Englewood Cliffs, 1991)

105. N.A. Shneydor, *Missile Guidance and Pursuit: Kinematics, Dynamics and Control* (Elsevier, Amsterdam, 1998)

106. T. Yamasaki, K. Enomoto, H. Takano, Y. Baba, S. Balakrishnan, Advanced pure pursuit guidance via sliding mode approach for chase UAV, in *Proceedings of AIAA Guidance, Navigation and Control Conference, AIAA*, vol. 6298 (2009), p. 2009

107. Y. Gu, B. Seanor, G. Campa, M.R. Napolitano, L. Rowe, S. Gururajan, S. Wan, Design and flight testing evaluation of formation control laws. IEEE Trans. Control Syst. Technol. **14**(6), 1105–1112 (2006)

108. S. Zhu, D. Wang, C.B. Low, Cooperative control of multiple UAVS for moving source seeking. J. Intell. Robot. Syst. **74**(1–2), 333–346 (2014)

109. S.U. Ali, R. Samar, M.Z. Shah, A.I. Bhatti, K. Munawar, U.M. Al-Sggaf, Lateral guidance and control of UAVs using second-order sliding modes. Aerosp. Sci. Technol. **49**, 88–100 (2016)

110. X. Wang, V. Yadav, S. Balakrishnan, Cooperative UAV formation flying with obstacle/collision avoidance. IEEE Trans. Control Syst. Technol. **15**(4), 672–679 (2007)

111. E. Børhaug, A. Pavlov, E. Panteley, K.Y. Pettersen, Straight line path following for formations of underactuated marine surface vessels. IEEE Trans. Control Syst. Technol. **19**(3), 493–506 (2011)

Chapter 11
An Optimal Approach for Load-Frequency Control of Islanded Microgrids Based on Nonlinear Model

Fatemeh Jamshidi, Mohammad Reza Salehizadeh, Fatemeh Gholami, and Miadreza Shafie-khah

Abstract Due to the increased environmental and economic challenges, in recent years, renewable based distribution generation has been developed. More penetrations from the side of consumers caused a new concept called microgrids which are able to stand with or without connection to the bulk power system. Control of microgrids in islanded mode is very crucial for decreasing the amplitude of frequency deviations as well as damping speed. This chapter aims to propose an optimal combination of FOPD and fuzzy pre-compensated FOPI approach for load-frequency control of microgrids in islanded mode. The optimization parameter of the control scheme is designed by the differential evolution (DE) algorithm which has been improved by a fuzzy approach. In the optimization, control effort is considered as a constraint. Due to the robustness and flexibility of the proposed method, the simulation results have been improved substantially. Robust performance of the proposed control method is examined through sensitivity analysis.

Keywords Microgrid · Intelligent control · Improved differential evolution (DE) · Fuzzy fractional order PID controller

Nomenclature

K_{BESS}	BESS gain
$G_{BESS}(s)$	BESS Linear Transfer Function

F. Jamshidi
Department of Electrical Engineering, Faculty of Engineering, Fasa University, Fasa, Iran

M. R. Salehizadeh (✉) · F. Gholami
Department of Electrical Engineering, Marvdasht Branch, Islamic Azad University, Marvdasht, Iran
e-mail: salehizadeh@miau.ac.ir

M. Shafie-khah
School of Technology and Innovations, University of Vaasa, Vaasa, Finland

© Springer Nature Switzerland AG 2020
M. H. Amini (ed.), *Optimization, Learning, and Control for Interdependent Complex Networks*, Advances in Intelligent Systems and Computing 1123,
https://doi.org/10.1007/978-3-030-34094-0_11

T_{BESS}	BESS time constant
u	Control effort
$G_{DEG}(s)$	DEG Linear Transfer Function
T_T	DEG time constant
K_D	Derivative gain of FOPD
P_{DEG}	Electrical power of DEG
$G_{FC}(s)$	FC Linear Transfer Function
T_{IN}	FC time constant
T_{IC}	FC time constant
T_G	FC time constant
T_{FC}	FC time constant
K_{FESS}	FESS gain
$G_{FESS}(s)$	FESS Linear Transfer Function
T_{FESS}	FESS time constant
K_{p2}	FOPI proportional gain
β	Fractional order of Integral
α	Fractional order of the derivative
K_I	Integral gain of FOPI
P_L	Load power
Δf	Microgrid frequency deviations
P_{BESS}	Output electrical power of BESS
P_{FC}	Output electrical power of FC
P_{FESS}	Output electrical power of FESS
P_{PV}	Output electrical power of PV
K_{p1}	Proportional gain of FOPD
$G_{PV}(s)$	PV Linear Transfer Function
P_{sol}	The solar heat power (light intensity)
$G_{load}(s)$	Transfer function of the model of load disturbance
$G_{sol}(s)$	Transfer function of solar power generation model
$G_{wind}(s)$	Transfer function of wind power generation model
P_W	Wind mechanical power
P_{WTG}	WTG electrical power
K_W	WTG Gain
$G_{WTG}(s)$	WTG Linear Transfer Function
T_W	WTG time constant

11.1 Introduction

Overview Due to the increased environmental and economic challenges, in recent years, renewable based distribution generation has been developed. More penetrations from the side of consumers caused a new concept called microgrids which are able to stand with or without connection to the bulk power system. Control of microgrids in islanded mode is very crucial for decreasing the amplitude of frequency deviations as well as damping speed. This chapter aims to propose an optimal combination of FOPD and fuzzy pre-compensated FOPI approach for load-frequency control of microgrids in islanded mode. The optimization parameter of the control scheme is designed by the differential evolution (DE) algorithm which has been improved by a fuzzy approach. In the optimization, control effort is considered as a constraint. Due to the robustness and flexibility of the proposed method, the simulation results have been improved substantially. Robust performance of the proposed control method is examined through sensitivity analysis.

After the era of post-restructuring, bulk power systems have been faced with different challenges such as, but not limited to the lack of available capacity due to transmission congestion, environmental concerns, as well as new problems related to energy market. Environmental concerns, high cost of installing new power plants, and barriers in front of transmission expansion planning motivate proliferation of distributed generation units in power industry. Thanks to the development of new small-scale generation technologies, newborn stand-alone grid, called microgrid, has been developed. Microgrids are small power systems that work in low voltage and consist of conventional and renewable power generations, controllable and uncontrollable loads. Microgrids could be operated in two modes: on-grid and off-grid (with or without connection to the upper-level networks). One of the most advantages of a microgrid is the islanding capability and independent operation. The advantage of islanding of a microgrid is increasing the reliability of the consumers connected to the microgrid. In both operation modes reacting to the rapid changes in power consumption is very necessary. In the on-grid mode, frequency and power of microgrid depend on the main grid. However, in the islanding mode, frequency and voltage of microgrid oscillate and independent control is required. By disconnecting from the main grid, operation and duties of the microgrid's resources change. The new duties include voltage and frequency control, power-sharing between the resources, and appropriate response to the variation of load and the other disturbances. The power system consists of different components which transmit electrical energy in large-scale. In the case of variation in loads or generated power of the resources, frequency deviation occurs. A lack of sufficient attention to this problem might cause frequency instability. If generated power is less than the consumed power, frequency decreases. The larger is the system, load variation affects less on the frequency of the systems. In this regard, effective control approaches play a very important role to maintain frequency within its acceptable range with less oscillation. The goal of designing load-frequency controller is to decrease frequency oscillation and damping of the disturbance in frequency from the viewpoint of domain or time in the normal operation and the case of disturbance [1–5].

Various literature addressed different load-frequency control in the microgrid. The most commonly used control approach for load-frequency control is the proportional–integral–derivative (PID) controller [6]. In [7, 8], PI or PID controllers have been used. Optimal tuning of PID parameters is very important for getting the best dynamic performance. To this purpose, in the literature, different methods such as Harmonic Search (HS) [9], Social-Spider Optimizer (SSO) [10], and Particle

Swarm Optimization (PSO) [11] have been used in load-frequency control of micro-grid. On the other hand, the performance of the classic PID controller is deteriorated while the operating conditions change. To tackle this challenge, the fuzzy approach is an effective solution to determine the PID parameters. The performance of the fuzzy approach depends on its membership. In [12], fuzzy control has been used and its coefficients have been optimized using PSO, simultaneously with load change. In [13], a fuzzy PSO-based controller has been adopted for the sake of frequency control. In the optimization approach frequency deviations, without paying attention to the control effort, are considered. In [14], an adaptive fuzzy P-PID controller is applied whose parameters have been optimized using an objective function including frequency deviations and eigenvalues. In addition to the PID controller, more advanced control schemes are used for the sake of load-frequency control. In [15], model-predictive control (MPC) has been used. In [16], multiple prediction control has been deployed for load-frequency control in the microgrid. In [17], a model-predictive coordinated control of wind turbine blades and the hybrid electric vehicle has been used for reducing power and frequency oscillation. In [18], in the proposed load-frequency control approach, wind turbine blades and PHEVs are controlled using MPC. In the above-mentioned control approaches, a linear model of the microgrid has been used for load-frequency control. Although the small-signal model can be used to study the dynamic behavior of microgrids, there are several phenomena in microgrids whose nonlinear nature must be taken into account in load-frequency control. Nonlinear factors include non-linearity of the output power of some distributed generation sources (such as wind turbine and solar cell) in terms of inputs, limitations of power conversion rates in energy sources and storage, limitation of energy capacity in storage systems, and saturation phenomena [19–21]. In power system literature, numerous references can be found in which nonlinear phenomena (such as time delay of communication systems, limitations of power rate variations, and saturation phenomena) in load-frequency control have been considered [22–24]. A nonlinear model-based load-frequency control has been proposed in some researches. In [25], the robust (H∞) method has been used and the limitation of power reserve rate variations has been considered. In [26–29], PI or PID controllers—whose parameters are determined in different ways—are used to control the load-frequency in a nonlinear model-based microgrid. In [26] fractional order PID (FOPID) controller, in [27], fractional order fuzzy control based PID (FOFCPID), and in [28, 29], type II fuzzy system have been used. The nonlinear limitations of the generators in [26, 27] are considered. Different control strategies have been proposed to improve system performance despite uncertainties. These include optimal control in [30], sliding mode control in [31], intelligent control in [32], and robust control in [25]. PID control is the most commonly used commercial controller with three design parameters: proportional, integral, and derivative coefficients. On the other hand, fractional calculus has received more attention in recent years. In recent years, fractional calculus has been used for system modeling and controller design. Fractional order PID (FOPID) is the most famous fractional order controller. Some efforts have been made to apply FOPID for load-frequency control purposes. As an instant, in [33], a combination of FOPID and ICA optimization approach is used. Frequency deviations is considered in the objective function. The approach is very prospective for load-frequency control. However, control effort is not considered in the proposed unconstraint objective function. Hence, the application of an improved FOPID control approach seems very necessary.

For the above-mentioned purpose, in this chapter, the optimal combination of FOPD and fuzzy pre-compensated FOPI approach is proposed and applied for load-frequency control. Here, nonlinear phenomena have been added to the linear model of microprocessor components in dynamic network modeling, including saturation and rate limiting. The flexible and robust proposed control scheme is able to improve frequency deviations. Note that optimizing the parameters of this control approach is more complicated in comparison with the conventional fuzzy PID controller. For this purpose, a meta-heuristic approach, differential evolution (DE) algorithm which has been improved by the fuzzy approach, is applied. The purpose of control is to minimize frequency deviations with limited control effort. Microgrid simulation is done in MATLAB environment. The dynamic model of various disturbances, including load changes, changes in wind speed, and changes in sunlight, are considered. Results with PID controller, fractional order PID controller and FOPD+FFOPI controller are compared, while their coefficients are optimized by fuzzy DE. The proposed controller is observed to perform better in damping of oscillations. Also, with the same number of iterations and with the same initial population, better results have been obtained by fuzzy differential evolution algorithm. The simulations have been carried out considering and without considering the uncertainty of the parameters of the dynamic grid model. Considering the considerable variations of the model parameters, the robustness of the proposed control scheme is confirmed.

The chapter is organized as follows: after providing an introduction in Sect. 11.1, the dynamic model of the microgrid is provided in Sect. 11.2. The proposed control method is presented in Sect. 11.3. Simulation results are presented in Sect. 11.4. Finally, Sect. 11.5 concludes the chapter.

11.2   Dynamic Model of Microgrid

A microgrid consists of DGs which electrifies local loads in two operational modes, connected to the network and islanded. Because of the intermittent nature of small-scale renewable resources that are connected to the microgrids, energy storage is required to help the system to stand in a stable way. In this chapter, information related to the examined microgrid is obtained from [12]. This microgrid includes diesel engine generators (DEG), photovoltaic (PV), wind turbine generator (WTG), fuel cells (FC), battery energy storage system (BESS), and flywheel energy storage system (FESS). Table 11.1 shows the transfer functions of the mentioned components.

Table 11.1 Model of the microgrid's component

System	Model
WTG	$G_{WTG}(s) = \frac{K_W}{1+sT_W} = \frac{P_{WTG}}{P_W}$
PV	$G_{PV}(s) = \frac{1}{(1+sT_{IN})(1+sT_{IC})} = \frac{P_{PV}}{P_{sol}}$
DEG	$G_{DEG}(s) = \frac{1}{(1+sT_G)(1+sT_T)} = \frac{P_{DEG}}{u-R^{-1}\Delta f}$
FC	$G_{FC}(s) = \frac{1}{(1+sT_{FC})(1+sT_{IN})(1+sT_{IC})} = \frac{P_{FC}}{u}$
BESS	$G_{BESS} = \frac{K_{BESS}}{1+sT_{BESS}} = \frac{P_{BESS}}{\Delta f}$
FESS	$G_{FESS} = \frac{K_{FESS}}{1+sT_{FESS}} = \frac{P_{FESS}}{\Delta f}$

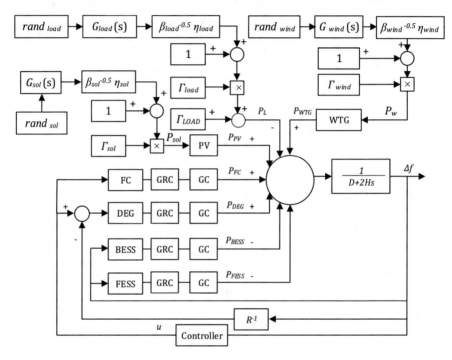

Fig. 11.1 Nonlinear model of the test microgrid

It is mentioned that, in Table 11.1, K and T represent gain and time constant, respectively. Figure 11.1 shows the nonlinear model of the test microgrid which is to be controlled. In the practical applications, the generated power of FESS, BESS, DEG, and FC and their derivatives are limited. In this regard, power generation constraint (GC) and power generation rate constraint (GRC) blocks are considered in the nonlinear model of the test microgrid. For input P, their output, q, is as follows:

$$q = \begin{cases} P_{min} & P < P_{min} \\ P & P_{min} \leq P \leq P_{max} \\ P_{max} & P > P_{max} \end{cases} \qquad (11.1)$$

The input of the GRC is a derivative of generated power, and the input of the GC is generated power.

11.3   The Proposed Intelligent Control Method

Figure 11.2 shows the proposed control approach used in this chapter. It is noticed that this control structure has been used in [34]. The input and output membership functions of the fuzzy system and its fuzzy rules are obtained from [34] and are shown in Fig. 11.3 and Table 11.2, respectively.

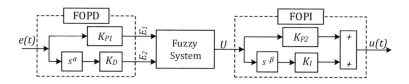

Fig. 11.2   Combination of FOPD and fuzzy pre-compensated FOPI

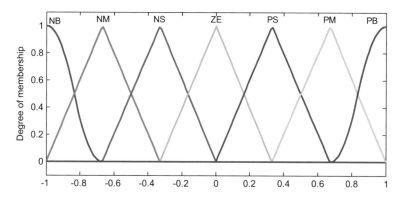

Fig. 11.3   Membership functions of controller's fuzzy system [34]

Table 11.2   Rules of controller's fuzzy system [34]

E2→ E1↓	NL	NM	NS	ZR	PS	PM	PL
PL	ZR	PS	PM	PL	PL	PL	PL
PM	NS	ZR	PS	PM	PL	PL	PL
PS	NM	NS	ZR	PS	PM	PL	PL
ZR	NL	NM	NS	ZR	PS	PM	PL
NS	NL	NL	NM	NS	ZR	PS	PM
NM	NL	NL	NL	NM	NS	ZR	PS
NL	NL	NL	NL	NL	NM	NS	ZR

Fig. 11.4 Fuzzy DE
flowchart

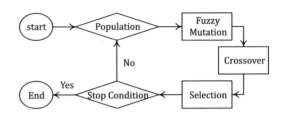

In the industrial implementation and simulation, fractional order of s is approximated with a transfer function of integer order. The most popular approximation for s^α, in the frequency bound $[\omega_L, \omega_H]$, is a filter of the order of $2N + 1$ [35]:

$$s^\alpha = \omega_H^\alpha \prod_{k=-N}^{N} \frac{s - z_k}{s - p_k}, 0 < \alpha < 1 \tag{11.2}$$

where $z_k = -\omega_L (\omega_H / \omega_L)^{\frac{2k+2N+1+\alpha}{2(2N+1)}}$, and $p_k = -\omega_L (\omega_H / \omega_L)^{\frac{2k+2N+1-\alpha}{2(2N+1)}}$.

Design parameters of this controller are proportional gain (K_{P1}), integral gain (K_I), derivative gain (K_D), integral order (β), and derivative order (α). The value of these parameters affects substantially the quality of the system response. In order to achieve the desired performance, fuzzy DE is used. Figure 11.4 shows the flowchart of fuzzy DE.

Fuzzy DE algorithm has been introduced in [36]. In fuzzy DE algorithm, the initial population consists of n random vectors which include d decision variables. The ith member of the population set in the Gth generation, P_i^G, is considered as

$$P_i^G = \begin{bmatrix} p_{i,1}^G & p_{i,2}^G & \cdots & p_{i,d}^G \end{bmatrix}, i = 1, 2, \ldots, n \tag{11.3}$$

In each generation, the population of the next generation is produced using three operators of fuzzy mutation, crossover, and selection. In mutation for each P_i^G, three random vectors $P_{r_1}^G$, $P_{r_2}^G$, and $P_{r_3}^G$ are chosen and mutation vector, M_i^G, is produced as follows:

$$M_i^G = P_{r_1}^G + F \left(P_{r_2}^G - P_{r_3}^G \right), r_1 \neq r_2 \neq r_3 \neq i \tag{11.4}$$

where F is the mutation factor.

Crossover operation combines mutation vector, M_i^G, and parent vector, P_i^G, in order to constitute the polar vector, Z_i^G, as follows:

$$z_{j,i}^G = \begin{cases} m_{j,i}^G \ if \ r_i \leq C_r \ \ or \ \ j = J_r \\ p_{j,i}^G \qquad \text{otherwise} \end{cases}, j = 1, 2, \ldots, d \tag{11.5}$$

where r_i is a random variable between 0 and 1 and J_r ensures $M_i^G \neq P_i^G$. The section operator chooses the best vector between the parent vector and test vector as

$$P_i^{G+1} = \begin{cases} Z_i^G \ if \ f\left(U_i{}^G\right) \leq f\left(X_i{}^G\right) \\ P_i^G \qquad \text{otherwise} \end{cases} \qquad (11.6)$$

Two points should be considered for choosing F. First, at the primary iterations of the DE algorithm, a big value should be assigned to F to speed up the exploration. By increasing the iterations, the smaller values should be assigned to F for speeding up the exploitation. Second, the less is the relative distance of the population members, the less is the effective F and vice versa. Population diversity is defined as

$$\text{diversity}(G) = \frac{\sum_{a=1}^{n-1} \sum_{b=a+1}^{n} \left| \frac{P_a^G - P_b^G}{U - L} \right|}{2d\,(n-1)\,n} \qquad (11.7)$$

where, L and U are the vectors containing the lower and upper bound of the population members. The above equation evaluates the average normal distance between the population members. In this regard, this criterion is effective for representing population diversity. For improving DE performance, fuzzy logic is used to determine F based on the number of generation and population diversity. The Block diagram of the controlled microgrid with FOPD+FFOPI controller based on fuzzy DE algorithm is shown in Fig. 11.5. The related membership function and fuzzy rules of fuzzy DE are shown in Fig. 11.6 and Table 11.3, respectively.

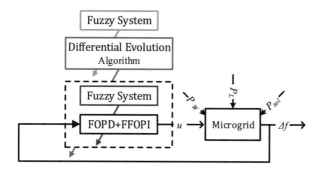

Fig. 11.5 Block diagram of the microgrid by the proposed control approach

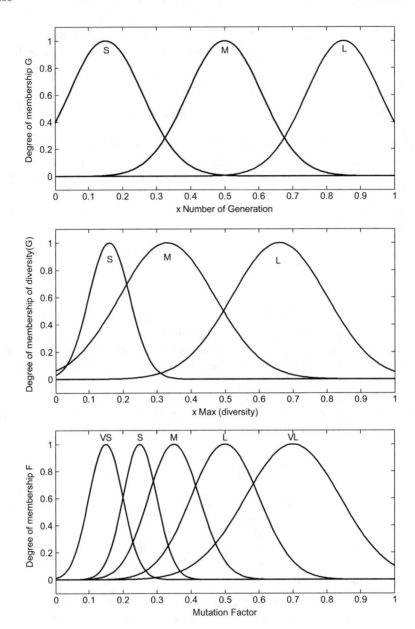

Fig. 11.6 The membership functions of fuzzy DE [36]

Table 11.3 The fuzzy rules of fuzzy DE [36]

		diversity(G)		
		S	M	L
G	S	M	L	VL
	M	S	M	L
	L	VS	S	M

11.4 Simulation and Results

In the load-frequency control problem, minimization of the frequency deviations is desired. On the other hand, control efforts are limited. The objective function of the optimization is the mean absolute of frequency deviations defined as follows:

$$J = \frac{\int_{T_1}^{T} |\Delta f| \, dt}{T - T_1} \tag{11.8}$$

The closed form of the optimization problem is as follows:

$$\begin{aligned} &\text{Min} \quad J(X) \\ &\text{s.t.} \quad \begin{cases} L \leq X \leq U \\ |u| \leq u_{max} \end{cases} \end{aligned} \tag{11.9}$$

Vector X consists of the following decision variables:

$$X = \begin{bmatrix} K_{p1} & K_{p2} & K_I & K_D & \beta & \alpha \end{bmatrix} \tag{11.10}$$

The uncontrolled inputs, wind power, solar thermal power, and variable load, are generated using the models of [37]. The parameters of the system are similar to that of [37, 38] and are shown in Table 11.4.

In this chapter, the parameters of Eq. (11.2) are chosen as $\omega_H = 100$, $N = 3$, and $\omega_L = 0.01$.

For the microgrid of Fig. 11.1, PID controller ($\frac{U(s)}{\Delta F(s)} = K_P + \frac{K_I}{s} + K_D s$), FOPID controller ($\frac{U(s)}{\Delta F(s)} = K_P + \frac{K_I}{s^\beta} + K_D s^\alpha$), and a combination of FOPD and fuzzy pre-compensated FOPI controller of Fig. 11.2 are designed and the related parameters of them are optimized using fuzzy DE. In this regards, these controllers are named FDE-PID, FDE-FOPID, and FDE-FOPD+FFOPI, respectively. Also, FOPD + FFOPI controller is designed, and its parameters are optimized using DE. This controller is named DE-FOPD+FFOPI. In Fig. 11.7, the frequency deviations of the microgrid after implementation of the designed controllers are shown and compared.

In all optimization algorithms, the same number of iterations is considered as the terminal condition. The controllers are sorted based on the value of the oscillation frequency of the frequency deviations as follows: FDE-PID, DE-FOPD-FFOPI, FDE-FOPD-FFOPI, and FDE-FOPID controller. In this list, FDE-PID has

Table 11.4 The parameter values of the test microgrid

Parameter	Value	Parameter	Value
K_w	1	R	3
$K_{FESS} = K_{BESS}$	1	T_w	1.5s
$T_{FESS} = T_{BESS}$	0.1s	T_{IN}	0.04s
Max (P_{FESS}) = Max (P_{BESS})	0.11	T_{IC}	0.004s
Max (P_{FC})	0.48	K_{FC}	1
Max (P_{DEG})	0.45	T_{FC}	0.26
Max (P'_{FESS}) = Max (P'_{BESS})	0.05	T_G	0.08
Max (P'_{FC})	1	T_T	0.4
Max (P'_{DEG})	0.5	D	0.015
Γ_{LOAD}	0.9u (t) + 0.03u $(t - 110)$ + 0.03u $(t - 130)$ + 0.03u $(t - 150)$ + 0.15u $(t - 170)$ + 0.1u $(t - 190)$	H	1/12
Γ_{load}	0.02u (t)	η_{load}	0.9
G_{load} (s)	$1 - 300/(300s + 1) - 1/(1800s + 1)$	B_{load}	10
Γ_{wind}	0.24u (t) − 0.04u $(t - 140)$	η_{wind}	0.8
G_{wind} (s)	$10^4 s/(10^4 s + 1)$	β_{wind}	10
Γ_{sol}	0.05u (t) − 0.02u $(t - 180)$	η_{sol}	0.1
G_{sol} (s)	$10^4 s/(10^4 s + 1)$	B_{sol}	10

the highest and FDE-FOPID has the least oscillation frequency. Negative frequency variations indicate a decrease in frequency and positive indicate an increase in frequency. The maximum values of increase and decrease of frequency variations are observed with FDE-FOPID, DE-FOPD-FFOPI, FDE-FOPD-FFOPI, and FDE-PID controllers, respectively. The FDE-FOPID controller has the lowest oscillation frequency and the highest frequency reduction, while the FDE-PID controller has the highest oscillation frequency and the lowest frequency reduction. In other words, the improvement of the oscillation frequency of frequency deviations is inconsistent with its improvement of the maximum increase and decrease. It can be concluded that the proposed controller has created a favorable trade-off between the oscillation frequency and the maximum value of increasing and decreasing of the frequency variations.

In Fig. 11.8, log 10 of the values of the objective function in the sequential iterations is compared for DE-FOPD+FFOPI and FDE-FOPD+FFOPI. In order to compare DE and fuzzy DE in determining the parameters of FOPD-FFOPI controller rationality and fairly, the random initial population of them are selected identical and their terminal condition is chosen the same number of iterations. As it is expected by applying the fuzzy system, at the primary iterations of the DE algorithm, the exploration speeds up and at the last iterations, the exploitation speeds up. In other words, the fuzzy system improves the performance of the DE algorithm from the viewpoint of convergence rate and more precise searching, considerably.

In Table 11.5, the performance of the controllers is more precisely compared with quantitative indices J, max($|u|$), and max($|\Delta f|$). It is observed that the mean

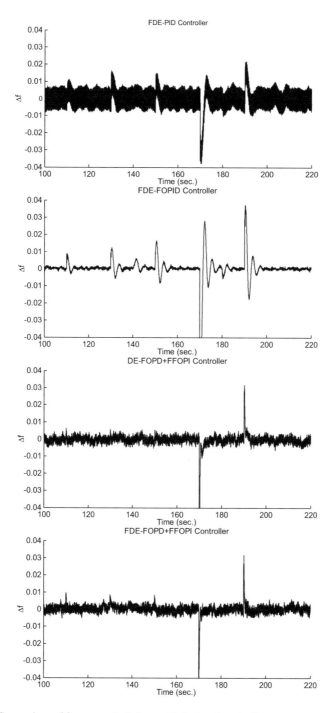

Fig. 11.7 Comparison of frequency deviations with designed controllers

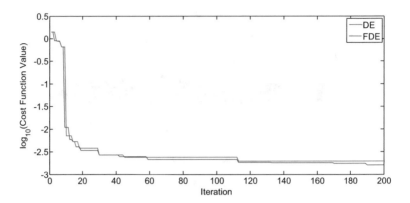

Fig. 11.8 Comparison of the objective function of fuzzy DE and DE

Table 11.5 Quantitative comparison of the performance with designed controllers

Index → Controller ↓	$J \times 10^{-3}$	max($|u|$)	max ($|\Delta f|$) $\times 10^{-2}$
FDE-PID	4.8414	30.218	3.7857
FDE-FOPID	2.489	7.5941	7.395
DE-FOPD+FFOPI	1.9617	70.797	7.324
FDE-FOPD+FFOPI	1.6097	38.665	4.7752

absolute of frequency deviations of the FDE-FOPID controller is less than that of FDE-PID and the performance of DE-FOPD + FFOPI in decreasing the mean absolute of frequency deviations is better than that of FDE-PID controller. The results show that by using the FDE-FOPID controller there is a good trade-off between these three quantitative and conflict indices. This demonstrates the capability of the fractional calculus. There is no logical relationship between max($|u|$) and the other two indices, but the constraint of limited amplitude of the control effort is satisfied. The least mean absolute of frequency deviations is pertaining to the FDE-FOPD + FFOPI controller. These results confirm the superiority of the proposed control method with the fuzzy DE optimization method.

It is noteworthy that considering the variations of the system parameters is essential when evaluating the performance of a load-frequency control system. Ignoring the parameter uncertainties can, in practice, lead to undesirable system behavior and the failure in achieving the desired control objectives. In this regard, the uncertainty of the parameters D, H, R, TFC, TG, and TT are considered and the frequency deviations of the microgrid are assessed under the variation of them for evaluating the robustness of the proposed approach. The related results are shown in Figs. 11.9 and 11.10 and Table 11.6. The results confirm the robustness of the proposed approach.

As shown in Figs. 11.9 and 11.10, in spite of the substantial change in the parameters of the system, frequency deviations are retained in an acceptable range and the control performance is still acceptable. The quantitative results in Table 11.6

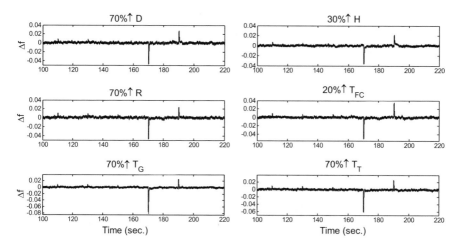

Fig. 11.9 Robust performance under increasing the parameters with FDE-FOPD+FFOPI

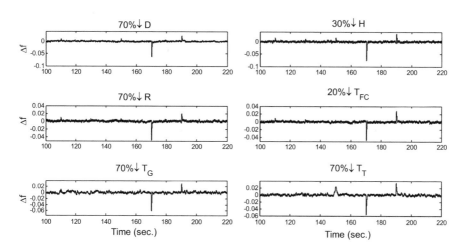

Fig. 11.10 Robustness performance under decreasing the parameters with FDE-FOPD+FFOPI controller

Table 11.6 Robust performance under variation the parameters with FDE-FOPD+FFOPI

	$J \times 10^{-3}$	$\max (|\Delta f|) \times 10^{-2}$	$J \times 10^{-3}$	$\max (|\Delta f|) \times 10^{-2}$
	Increase		Decrease	
70%D	1.5808	4.7672	1.7043	6.2765
30%H	1.2551	3.6343	2.3250	7.6134
70%R	1.6785	4.7939	1.6312	4.8079
20%T_{FC}	1.6333	5.0586	1.6194	4.7472
70%T_G	1.6867	8.2170	2.5443	6.1443
70%T_T	1.6060	6.1088	2.3178	5.7546

demonstrate that variation of the parameters affects the maximum of the frequency variation (max($|\Delta f|$)) more than J. By increasing the parameters, J does not change dramatically and the maximum of the frequency variation changes more. However, by decreasing the parameters both indices, J and max($|\Delta f|$), change. It is also observed that by decreasing H and increasing TG, max($|\Delta f|$) is maximized.

11.5 Conclusion

In this chapter, an optimal combination of FOPD and fuzzy pre-compensated FOPI approach was designed for reducing the frequency deviations of an islanded microgrid. Nonlinear phenomena such as saturation and rate-limiting power have been added to the linear model of the microgrid components. The dynamic model of various perturbations, including load changes, changes in wind speed, and changes in sunlight, were considered. The fractional order structure was proposed because of increasing the controller degrees of freedom and its robust performance. The parameters of the controller were determined via differential evolution (DE) algorithm which was improved by the fuzzy system. In the proposed optimization approach, the objective function was to optimize the average magnitude of the frequency variations while the control effort amplitude does not exceed the pre-determined value. Control effort has been limited by including a relevant constraint in the optimization model. Simulations were performed using MATLAB, and the performance of the proposed control approach was compared with the performance of fractional order PID and PID controllers whose parameters were determined by fuzzy differential evolution algorithm. Frequency deviations had the smallest magnitude with PID controller and it had the least frequency of variations with FOPID controller. Using FDE-FPD-FFOPI controller, a good trade-off was made between the two conflict criteria: variation's frequency and magnitude of frequency deviations. The same number of iterations was chosen as the terminal condition of the fuzzy differential evolution algorithm to determine the parameters of all three controllers. The simulation results showed improvement of frequency deviations via implementing the proposed control schema. To illustrate the superior performance of the fuzzy differential evolution algorithm compared to the differential evolution algorithm, the FPD-FFOPI controller parameters were determined using both algorithms. For fair and rational comparison, the initial population was selected random and identical and the same number of iterations was chosen as the terminal condition in both optimization. Applying the fuzzy system, at the primary iterations of the DE algorithm, the exploration and at the last ones, the exploitation speeded up. In another words, the fuzzy system improves the performance of the DE algorithm from the viewpoint of convergence rate and more precise searching, considerably.

For quantitative analysis of simulation results, the maximum amplitude of the control effort and frequency variations were calculated. Since it is important to evaluate the performance of the load-frequency control system under the parameter uncertainties. The performance of the proposed controller is investigated by considering significant changes in the parameters T, D, H, R, TFC, TG, and TT. The quantitative and qualitative results of this investigation showed that despite considerable variations in system parameters, the frequency variations remained within the acceptable range. A performed sensitivity analysis confirmed the robustness of the proposed approach.

References

1. M.R. Salehizadeh, A. Rahimi-Kian, K. Hausken, A leader–follower game on congestion management in power systems, in *Game Theoretic Analysis of Congestion, Safety and Security* (Springer, Cham, 2015), pp. 81–112
2. M.R. Salehizadeh, A. Rahimi-Kian, M. Oloomi-Buygi, A multi-attribute congestion-driven approach for evaluation of power generation plans. Int. Trans. Electr. Energy Syst. **25**(3), 482–497 (2015)
3. J.-J. Ma, G. Du, B.-C. Xie, CO2 emission changes of China's power generation system: input-output subsystem analysis. Energy Policy **124**, 1–12 (2019)
4. A. Goldthau, N. Sitter, Regulatory or market power Europe? EU leadership models for international energy governance, in *New Political Economy of Energy in Europe* (Palgrave Macmillan, Cham, 2019), pp. 27–47
5. M Oloomi-Buygi, M.R. Salehizadeh. *Toward fairness in transmission loss allocation*, in 2007 Australasian Universities Power Engineering Conference. IEEE, 2007
6. P.K. Ray, S.R. Mohanty, N. Kishor, Proportional–integral controller based small-signal analysis of hybrid distributed generation systems. Energy Convers. Manag. **52**(4), 1943–1954 (2011)
7. R. Dhanalakshmi, S. Palaniswami, Load frequency control of wind diesel hydro hybrid power system using conventional PI controller. Eur. J. Sci. Res. **60**, 630–641 (2011)
8. P.K. Ray, S.R. Mohanty, N. Kishor, Proportional–integral controller based small-signal analysis of hybrid distributed generation systems. Energy Convers. Manag. **52**(4), 1943–1954 (2011)
9. G. Shankar, V. Mukherjee, Load frequency control of an autonomous hybrid power system by quasi-oppositional harmony search algorithm. Int. J. Electr. Power Energy Syst. **78**, 715–734 (2016)
10. M.A. El-Hameed, A.A. El-Fergany, Efficient frequency controllers for autonomous two-area hybrid microgrid system using social-spider optimiser. IET Gener. Transm. Distrib. **11**(3), 637–648 (2017)
11. S.K. Pandey, S.R. Mohanty, N. Kishor, J.P.S. Cataleo, Frequency regulation in hybrid power systems using particle swarm optimization and linear matrix inequalities based robust controller design. Int. J. Electr. Power Energy Systems **63**, 887–900 (2014)
12. H. Bevrani, F. Habibi, P. Babahajyani, M. Watanabe, Y. Mitani, Intelligent frequency control in an AC microgrid: online PSO-based fuzzy tuning approach. IEEE Trans. Smart Grid **3**(4), 1935–1944 (2012)
13. F. Jamshidi, S.L. Emamzadehei, M.M. Ghanbarian, Using fuzzy PI controller optimized by PSO for frequency control of island microgrids. J. Soft Comput. Inf. Technol. **6**(1), 36–43 (2017)
14. H. Shayeghi, A. Ghasemi, Improvement of frequency fluctuations in microgrids using an optimized fuzzy P-PID controller by modified multi objective gravitational search algorithm. Iranian J. Electr. Electron. Eng. **12**(4), 241–256 (2016)
15. A. Parisioa, E. Rikos, G. Tzamalis, L. Glielmo, Use of model predictive control for experimental microgrid optimization. Appl. Energy **115**(15), 37–46 (2014)

16. S.R. Cominesi, M. Farina, L. Giulioni, B. Picasso, R. Scattolini, A Two-Layer Stochastic Model Predictive Control Scheme for Microgrids. IEEE Trans. Control Syst. Technol. **26**(1), 1–13 (2018)

17. J. Pahasa, I. Ngamroo, PHEVs bidirectional charging/discharging and SoC control for microgrid frequency stabilization using multiple MPC. IEEE Trans. Smart Grid **6**(2), 526–533 (2015)

18. J. Pahasa, I. Ngamroo, Coordinated control of wind turbine blade pitch angle and PHEVs using MPCs for load frequency control of microgrid. IEEE Syst. J. **10**(1), 97–105 (2016)

19. C. Wang, X. Li, L. Guo, Y.W. Li, A nonlinear-disturbance-observer-based DC-bus voltage control for a hybrid AC/DC microgrid. IEEE Trans. Power Electron. **29**(11), 6162–6177 (Nov. 2014)

20. H.R. Baghaee, M. Mirsalim, G.B. Gharehpetian, Performance improvement of multi-DER microgrid for small- and large-signal disturbances and nonlinear loads: novel complementary control loop and fuzzy controller in a hierarchical droop-based control scheme. IEEE Sys. J. **12**(1), 444–451 (2018)

21. M. Chaoxu, W. Liu, X. Wei, M.R. Islam, Observer-based load frequency control for island microgrid with photovoltaic power. Int. J. Photoenergy **2017**, 2851436 (2017). https://doi.org/10.1155/2017/2851436, 11 pages

22. Y. Tang, Y. Bai, C. Huang, B. Du, Linear active disturbance rejection-based load frequency control concerning high penetration of wind energy. Energy Convers. Manag. **95**, 259–271 (2015)

23. A. Ahmadi, M. Aldeen, Robust overlapping load frequency output feedback control of multi-area interconnected power systems. Int. J. Electr. Power Energy Syst. **89**, 156–172 (2017)

24. H. Cai, G. Hu, Distributed nonlinear hierarchical control of AC microgrid via unreliable communication. IEEE Trans. Smart Grid **9**(4), 2429–2441 (2018)

25. V.P. Singh, S.R. Mohanty, N. Kishor, P.K. Ray, Robust H-infinity load frequency control in hybrid distributed generation system. Int. J. Electr. Power Energy Syst. **46**, 294–305 (2015)

26. I. Pan, S. Das, Fractional order AGC for distributed energy resources using robust optimization. IEEE Trans. Smart Grid **7**(5), 2175–2186 (2016)

27. I. Pan, S. Das, Fractional order fuzzy control of hybrid power system with renewable generation using chaotic PSO. ISA Trans. **62**, 19–29 (2016)

28. M.-H. Khooban, T. Niknam, F. Blaabjerg, P. Davari, T. Dragicevic, A robust adaptive load frequency control for micro-grids. ISA Trans. **65**, 220–229 (2016)

29. M.H. Khooban, T. Niknam, F. Blaabjerg, T. Dragičević, A new load frequency control strategy for micro-grids with considering electrical vehicles. Electr. Power Syst. Res. **143**, 585–598 (2017)

30. Y. Hain, R. Kulessky, G. Nudelman, Identification-based power unit model for load-frequency control purposes. IEEE Trans. Power Syst. **15**(4), 1313–1321 (2000)

31. Y. Mi, F. Yang, D. Li, C. Wang, P.C. Loh, P. Wang, The sliding mode load frequency control for hybrid power system based on disturbance observer. Int. J. Electr. Power Energy Syst. **74**, 446–452 (2016)

32. İ. Kocaarslan, E. Çam, Fuzzy logic controller in interconnected electrical power systems for load-frequency control. Int. J. Electr. Power Energy Syst. **27**(8), 542–549 (2005)

33. F. Jamshidi, M.M. Ghanbarian, Robust frequency control of islanded microgrids: ICA-based FFOPID control approach. Comput. Intell. Electr. Eng. **8**(1), 51–62 (2017)

34. S. Das, I. Pa, S. Das, A. Gupta, A novel fractional order fuzzy PID controller and its optimal time domain tuning based on integral performance indices. Eng. Appl. Artif. Intell. **25**(2), 430–442 (2012)

35. A. Oustaloup, F. Levron, B. Mathieu, F.M. Nanot, Frequency-band complex noninteger differentiator: characterization and synthesis. IEEE Trans. Circuits Syst. I Fund. Theory Appl. **47**(1), 25–39 (2000)

36. M. Salehpour, A. Jamali, A. Bagheri, N. Nariman-zadeh, A new adaptive differential evolution optimization algorithm based on fuzzy inference system. Eng. Sci. Technol. Int. J. **20**(2), 587–597 (2017)

37. D.C. Das, A. Roy, N. Sinha, GA based frequency controller for solar thermal–diesel–wind hybrid energy generation/energy storage system. Electr. Power Energy Syst. **43**, 262–279 (2012)

38. H. Wang, G. Zeng, Y. Dai, D. Bi, J. Sun, X. Xie, Design of a fractional order frequency PID controller for an islanded microgrid: a multi-objective extremal optimization method. Energies **10**(10), 1502 (2017)

Chapter 12
Photovoltaic Design for Smart Cities and Demand Forecasting Using a Truncated Conjugate Gradient Algorithm

Isa S. Qamber and Mohamed Y. Al-Hamad

Abstract Worldwide, global warming is a very important concern. This refers to climate change caused by human activities, which affect the environment. Climate change presents a serious threat to the natural world. This is likely to affect our future unless action is taken to avoid such phenomena. In addition, without ambitious mitigation efforts, global temperature rises will occur in this century. In recent years, countries all over the world have had their own vision directed toward renewable energy, which is a clean option to will help to avoid the results of global warming. One of these energy sources is solar energy. The idea of solar energy has been raised to improve sustainability in individual countries and in the energy sector. Various countries have made decisions to develop renewable energy projects. Solar energy plans have become important in recent years. Integration of variable energy resources into an electricity grid can use solar photovoltaics as a main resource. These variable energy resources, as new resources, are currently envisioned to be either wind or solar photovoltaics. However, the output of these types of resources can be highly variable and depend on weather fluctuations such as wind speed and cloud cover. Since photovoltaic power generation is highly dependent on weather conditions, photovoltaic power generation operates differently in different regions. In particular, solar irradiance affects photovoltaic power generation. This means that solar power forecasting becomes an important tool for optimal economic management of the electric power network. In this chapter, an artificial intelligence technique is recommended to calculate the number of solar power panels required to satisfy a given estimated daily electricity load for five countries: the Kingdom of

I. S. Qamber (✉)
MBSE, Bahrain Society of Engineers, Electrical Engineering and Energy, Former University of Bahrain, Isa Town, Kingdom of Bahrain

M. Y. Al-Hamad
Power Trade Senior Executive Market Operations, GCC Interconnection Authority, Dammam, Kingdom of Saudi Arabia
e-mail: mhamad@gccia.com.sa

© Springer Nature Switzerland AG 2020
M. H. Amini (ed.), *Optimization, Learning, and Control for Interdependent Complex Networks*, Advances in Intelligent Systems and Computing 1123, https://doi.org/10.1007/978-3-030-34094-0_12

277

Bahrain, Egypt, India, Thailand, and the UK. Such artificial intelligence techniques play an important role in modeling and prediction in renewable energy engineering. The main focus of this chapter is the design of photovoltaic solar power plants, which help to reduce carbon dioxide emissions where they are connected to the national electricity grid in order to feed the grid with the extra electricity they generate. In this case, the power plant becomes more efficient than a combined cycle plant. At the same time, modeling and prediction in renewable energy engineering helps engineers to make predictions regarding future required loads.

Keywords Demand forecast · Optimization · Smart cities · Photovoltaics · Rule-based neural networks · Solar power · Renewable energy

Abbreviations

ANFIS	Adaptive neuro-fuzzy inference system
ANN	Artificial neural network
APopPV	Actual power output of a photovoltaic panel
Bapco	Bahrain Petroleum Company
BHD	Bahraini dinars
CombE	Combined efficiency
CPV	Concentrated photovoltaic
CSP	Concentrated solar power
EnPby1PD	Energy produced by a 1-peak-watt panel in a day
FS	Feature selection
GCC	Gulf Cooperation Council
GWh	Gigawatt-hour
ICT	Information and communications technology
IPP	Independent power producer
IRENA	International Renewable Energy Agency
KISR	Kuwait Institute for Scientific Research
kWh	Kilowatt-hour
kWp	Peak kilowatt
MENA	Middle East and North Africa
MLR	Multiple linear regression
NhrsPD	Number of hours per day
NoUnits	Number of units
NREAP	National Renewable Energy Action Plan
NSP	Number of solar power panels required to satisfy a given estimated daily electricity load
OpF	Operating factor
Ophrs	Operating hours
PEndU	Power used at the end use [it is less because of lower combined efficiency of the system]

PGF	Panel generation factor
PPR	Peak power rating
PV	Photovoltaic
QSE	Qatar Solar Energy
REPDO	Renewable Energy Projects Development Office
REQP	Rating of the equipment
RES	Renewable energy source
SEIA	Solar Energy Industries Association
STEEB	Solar Technology Energy and Environment in Bahrain
TRL	Total required load (total connected load)
TWh	Terawatt-hour
TWhrR	Total watt-hour rating of the system
UAE	United Arab Emirates
VER	Variable energy resource
W_p	Peak watt

12.1 Introduction

A solar energy project involves several factors that affect the selection of the area for that project. Thus, site selection is a critical issue for building a solar energy plant. Solar energy is a natural and renewable energy source (RES). The main resource for solar energy is the sun, which rises every day. Nowadays, many countries are moving toward using renewable energy because it is clean. In recent years, the Gulf Cooperation Council (GCC) countries (Kuwait, the Kingdom of Saudi Arabia, the Kingdom of Bahrain, Qatar, the United Arab Emirates, and the Sultanate of Oman) have been considering moving toward renewable energy, especially solar energy. It is well known that the Arabian Gulf region is the site of the highest summer temperatures ever recorded. This means that it is among the most productive solar regions, and it has been presented in the literature as an example of this. These countries are pursuing their own solar energy projects, aimed at the most economical targets, as they have very high electricity consumption per capita. The reasons for this target are rapid population growth and industrial expansion in the GCC area. Many relevant studies have been carried out, as discussed in the present chapter. It is just a matter of how much is enough to justify the cost of solar energy over other energy options. Even areas with large amounts of rain can easily provide solar energy, albeit less than other areas with more sunlight hours. All of this has been prompted by global climate change, which presents a serious threat to the nature of the world we are living in [1, 2].

Solar irradiation has been investigated as a source of energy, as it is easily obtained at the earth's surface and has various useful applications. Solar energy can be used as renewable energy in many fields such as power generation, in the form of solar photovoltaic (PV) systems. Solar energy estimation includes accurate measurement of solar irradiation, which depends on different atmospheric variables that can influence the energy yield to a large extent. The largest amount of energy is generated in areas with high solar irradiation [3].

12.2   Objectives and Targets

The objective of the present mission is to achieve a clean environment, which entails several factors that need to be satisfied [4], such as:

1. Reducing emissions of greenhouse gases into the atmosphere—for example, by planting trees, which absorb carbon dioxide and give off oxygen
2. Isolating greenhouse gas emissions from the atmosphere

At the same time, climate change due to human behavior is occurring because of the building of factories without due consideration regarding the risks they pose. These risks are raised by ignoring the outcome of the operation of factories and even power stations. There would be substantial benefits from reducing the waste produced by both factories and power stations. Burning of fuels in gas power stations results in discharge of waste gas (flue gas) into the air. This discharge contains various different gases, such as carbon dioxide, nitrogen, methane, and water vapor, in addition to other pollutants such as sulfur oxides. By reducing the waste, the following outcomes can be achieved:

3. Sustainable electric power use
4. Sustainable electric power management
5. Reduction, prevention, and control of air pollution

12.3   Literature Review

The development of a smart grid anywhere is a step toward improved generation of electricity. Improvements are achieved by diversification and conservation in electric power generation. Smart grids manage strong and healthy energy demand and reduce the influences of climate change and global warming. In the growth of any electricity network in any country or region, it is helpful and recommended to include renewable energy, which helps to provide clean energy. This aim drives the selection of smart grid technology. Selection of RES has increased in recent years. The ongoing challenges facing governments, utilities, and even commercial companies are (1) to obtain clean energy that avoids discharge of waste gases into the air and (2) to replace use of thermal power stations to generate electricity with a clean power generation source. This means that the clean energy that is generated is required to provide reliable power to meet the demand.

The process of finding and selecting the most useful features for the generated power is called feature selection (FS), which is suitable for economic consumption and is used as a procedure to pick a suitable prediction model for the future. Eseye et al. [5] proposed use of machine learning to find the most relevant model not characterized by repetition, for accurate short-term load demand forecasting in distributed power systems. The suggested approach improved the quality and efficiency of the estimated selection with minimal selection for accuracy. The suggested model was trained using a 2-year set of hourly data. Then, the results were tested with another 1-year set of hourly data. Finally, the obtained results verified that the feedforward artificial neural network model forecast training feature had reached an annual 1.96% mean absolute percentage error, which is a very acceptable value for electrical load demand forecasting in small-scale, decentralized power systems.

Mito et al. [6] reviewed numerous studies that focused on using mature RES, and presented the state of the art in both wind and solar PV power. These RES can be used to drive reverse osmosis plants on a small scale. Direct coupling of a reverse osmosis plant to a RES requires variable-speed operation and/or modular operation to match the needed load to the power available in the network. This review of the use of wind and solar PV to drive reverse osmosis took into consideration the plant configuration, operational strategy, control system, and methods followed to tailor the plant selection for the relevant RES. The performance of membrane desalination is helping the operation of wind and solar power to be economically viable in RES-powered reverse osmosis plants.

The Bahrain government has provided coordination to help protect Bahrain's environment by supporting sustainable investment in clean technologies. This will help to minimize pollution and conserve national resources. Albuflasa [7] published a paper highlighting the average annual solar irradiation in Bahrain, which is approximately 2600 kWh/m^2/year, and the achievable generation of electricity using the heat energy from the sun is approximately 33 TWh/year. Bahrain's target for renewable energy is 5% by 2025 with generation of around 480 GWh/year of clean energy. By the year 2035, the target will be raised to 10%, which is double that for 2025. This means that generation of 1460 GWh/year of clean energy is required. This is the National Renewable Energy Action Plan (NREAP). The growth projection is around 50% from the years 2025 to 2035. Albuflasa [7] mentioned that the first renewable energy grid in Bahrain is a 5 MW PV project owned by the Bahrain Petroleum Company (Bapco). The project was commissioned in the year 2012 and installed at three locations: Awali, the Refinery, and the University of Bahrain.

For evaluation of the performance of a solar energy conversion system, data on terrestrial solar irradiation are essential. In addition, the evaluation of solar energy must take into account the location and the efficiency of the solar system. In a study by Pareek and Gidwani [3], a curve-fitting method was used to find a modeling equation and estimate the horizontal global solar irradiation. The authors [3] presented a comparison between software and instruments for measuring solar irradiation using different techniques and methodology. Artificial neural network (ANN) models approximate daily solar irradiation at locations where solar irradiation data are not collected.

A new method is needed to develop renewable-energy-based grids that balance the load and generation using efficient energy storage models. This was discussed by Sidorov et al. [8], who proposed a new mathematical model for load forecasting using deep learning and support vector regression models. Their research included various features such as the average daily temperature. Germany's electrical grid was selected, which has many RES, with approximately 6.7 GW of pumped storage power plants. The proposed models are efficient.

Ferwatia et al. [9] concentrated on green building rating systems in the development of neighborhood sustainability assessment tools. Qatar selected some of these tools. In their study, the authors developed the Qatar Sustainability Assessment System. Their study was divided into three phases. The first phase identified eight sustainability criteria. The second phase was used to design the model and apply an analytical network process method to find the weights of the criteria. The developed model was applied to a real case.

The objective of a study by Al-Hamad and Qamber [10] was a 20% reduction of the load flow through distribution and transmission equipment to help relieve loaded equipment in all networks. In addition, many projects are starting to be developed in the GCC countries. These developments are helping to maximize the benefits from involvement of RES in networks. It is well known that the GCC countries have a good location for solar energy, with a high intensity of solar irradiation. In addition, the sky is clear of cloud cover in the GCC countries throughout the year. The opportunities for this region are being explored to create sustainable energy resources. Moreover, the target of this research is to engage PV technology in such a way that reduces the overload on current equipment and meets the electricity demand on the consumer side. PV systems could be used to add electricity generation units in cities and other areas. The roofs of houses could be used to install PV units to handle a portion of the local demand. Annually, the GCC countries are spending millions of dollars to support their networks to satisfy the needs for reinforcement.

Albuflasa [11] has studied Bahrain residents' needs for sustainable economic development. Household energy demand accounts for 51% of the total electricity consumption in Bahrain. The optimal scenario is that future energy systems will be sustainable, integrated, smart energy systems. The challenge in designing smart energy technologies is achieving full understanding of the energy consumption. The research carried out by Albuflasa [11] assesses the actual deployment of smart energy solutions and energy supply systems in Bahrain, and highlights the opportunities for using information and communications technology (ICT), such as smart meters to enable customers to make smart decisions regarding their power consumption. At the same time, it helps utility providers to reshape the overall energy profile.

A research article by Qamber and Al-Hamad [12] presented a way of connecting PV panels to the grid to reduce the demand on the power authority and help achieve a reduction of load for the authority. Since the intensity of the solar irradiation in this region is high, this means the production of electricity from the sun will increase in these locations. In their study, the location of the six GCC countries was considered

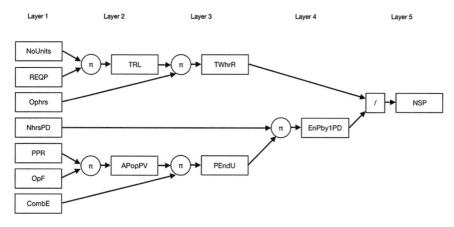

Fig. 12.1 Proposed model for solar power panels

a strategic location; thus, the GCC countries have a new competitive energy resource and can be engaged in both distribution and transmission systems. Qamber and Al-Hamad derived their model's solar cell design from five countries with the same house specifications. In addition, a panel generation factor (PGF) was included in the study to achieve the specification of the cells. The PGF depends on the climate. Therefore, renewable energy provides many benefits for the climate, the economy, and health. The total (peak kilowatts (kWp)) of the PV panel capacity for the five countries, the number of PV panels needed to design a 110-peak-watt (110 Wp) PV module, and the solar charge controller rating were calculated on the basis of the PGF of each country, following the proposed model shown in Fig. 12.1.

Al-Hamad and Qamber [13] developed a suitable model to calculate electricity demand forecasting as requested by the relevant decision-makers. They dealt with the electrical long-term peak load demand forecasting using a developed adaptive neuro-fuzzy inference system (ANFIS) and multiple linear regression (MLR) methods. The neuro-fuzzy training used previous sets of data. The obtained results had an acceptable level of mean errors, encouraging the GCC countries to further explore solar energy for the future. The novelty of this research was that it avoided an increase in generation capacity in both medium-term and long-term plans to help the GCC countries avoid load shedding. In addition, the developed models helped to find optimum times for electrical energy trading. It was concluded that the neuro-fuzzy technique was a more accurate technique than MLR and could be used in power system planning and development.

The NREAP of Bahrain [14–18] was presented on the occasion of STEEB [Solar Technology Energy and Environment in Bahrain] 2017, organized by the Bahrain Solar Industry Association in the capital of Bahrain. The planned PV capacity by the year 2025 is 255 MW. This amount of power is sufficient to cover approximately 5% of Bahrain's power load, and by the year 2030 it is aimed to reach a capacity of 700 MW of renewable energy power generation. In addition, the target of the Askar project [14–18] is a 100 MW solar power plant located at the Askar landfill site, which is in the Southern Governorate of Bahrain.

The 50 MW Shagaya concentrated solar power (CSP) plant is part of the first phase of the 2 GW Kuwait Shagaya Renewable Energy Park [19]. (CSP is a type of solar technology that uses mirrors to direct sunlight onto a receiver. It is discussed in more detail later in the text.) This project has been developed by the Kuwait Institute for Scientific Research (KISR). The concept of the project is based on a micro gas turbine. The capacity of the project is 100 kW, with a compact heliostat (tracking mirror) field. The heliostats are close to the tower. The 2 GW Shagaya project has a three-phase master plan. The first phase of the Shagaya Power Plant has a 70 MW renewable energy capacity (50 MW CSP, 10 MW PV, and 10 MW wind power) [19]. The second phase is a 930 MW extension of the plant to reach a total of 1000 MW installed renewable energy capacity (including CSP, concentrated PV (CPV), PV, and/or wind power). However, although the target year for completion of the expansion is 2030, the latest analysis by KISR illustrates that the CSP capacity will not exceed 400 MW by then [20].

Solar energy is one of the only renewable energy resources available in Oman. The solar irradiation in Oman varies between 4.51 and 6.09 kWh/m^2/day [21]. This means that it varies between 1640 and 2200 kWh/year. The Fahud location in Oman is taken as an example. Its global insolation varies between 2 and 5 kWh/day during January and between 5 and 7 kWh/day between July and September. The average variation annually varies between 4 kWh/m^2/day during January and about 6.5 kWh/m^2/day in May [21]. The other locations can be obtained by referring to SEM/NBP/KF/SAJ [21] and Al-Mahrouqi and Amin [22]. Oman, in most of its regions, possesses a statistical high predictability of insolation during summer. The solar energy procurement includes the 500 MW Ibri II solar independent power project (IPP)) scheduled for the year 2021 [23]. Al Hatmi and Tan [24] performed a comprehensive review of energy sources in Oman, where the population is growing. Furthermore, the spacing between housing units in Oman has become a major challenge to the government.

Several research projects at centers in Qatar are focused on renewable energy. KAHRAMAA has planned to generate 200 MW of solar energy at approximately 60 sites across the country by the year 2020. Its first PV project is the 220 MW Duhail Power Plant, and the second project is an increase of 10 MW at the same plant. This is based on a MEED Middle East Business Intelligence (MENA Business Intelligence) report [25]. The solar energy potential has an ambitious target of a 2% renewable energy contribution to the national energy mix by the year 2022, reaching 220 MW [26]. Qatar Solar Energy (QSE) was selected as the provider for the project because of its ability to produce higher-efficiency solar solutions at a lower cost through a contract with both the Japanese and Thai markets [27].

In the Gulf region, the Saudi Arabia power sector is considered to have the biggest installed generating capacity. The capacity was over 61 GW in the year 2014. It has the greatest potential for renewable energy in the Middle East and North Africa (MENA) region. Seven PV solar projects are planned as part of Saudi Arabia's NREAP [28]. These projects will have a combined total capacity of 1.51 GW, and each project will be developed under construction: 50 MW from the Medina IPP, 45 MW from the Rafha IPP, 200 MW from the Qurayyat IPP, 600 MW from the Al-Faisaliah IPP, 300 MW from the Rabigh IPP, 300 MW from the Jeddah IPP, and 20 MW from the Mahd al-Dahab IPP. The kingdom is aiming at an ambitious 27.3 GW of clean energy by the year 2024 and 58.7 GW by the year 2030. The year 2024 target is 20 GW of PV solar capacity, while the year 2030 target includes 40 GW of PV solar capacity [28]. The Renewable Energy Projects Development Office (REPDO) was set to issue tenders for 2225 MW of solar power projects in the year 2019 [29].

In terms of renewable energy, the International Renewable Energy Agency (IRENA) has published a roadmap report envisaging that by the year 2050, 86% of the world's power demands could be met by renewable energy. The United Arab Emirates (UAE) aims to obtain 30% of its energy from renewable sources by the year 2030, using CSP technology with extremely huge mirrors. The sunlight directed onto the receiver by the mirrors is converted to heat. Several types of mirrors are used: parabolic troughs, rounded dishes, and power towers. It has been found that CSP is a lot more effective than solar PV technology, which needs sunlight to operate, whereas CSP does not. The UAE has had CSP since 2013. The capacity installed at that time was 100 MW, generating energy of 261 GWh. Masdar states that its 10 MW and 1 MW solar power plant and rooftop panels can power 500 homes for a year [30, 31].

The Sweihan PV power station [31–34] started commercial operations at the 1177 MW Noor Abu Dhabi PV power project in the UAE on June 30, 2019. The project was financed by eight commercial banks and sponsored by an AED 3.2 billion (US$871.2 million) loan agreement signed in May 2017. This amount is expected to reduce Abu Dhabi's carbon dioxide emissions by one million tons. This is equivalent to taking 200,000 cars off the roads. The station is providing enough power capacity to meet the demands of 90,000 people. The project will allow Abu Dhabi to increase its production of renewable energy and reduce its dependence on natural gas for electricity generation.

Boyle [35], in his book, provides a comprehensive overview of the principal types of renewable energy, explaining the underlying physical and technological principles of renewable energy and discussing the environmental impact and prospects of different energy sources. The book includes more than 350 illustrations and more than 50 tables of data, with some case studies also being illustrated [35].

Turner [36] highlights the USA as an example of renewable resource use, taking into consideration the number of renewable systems and discussing energy payback issues, carbon dioxide abatement, energy storage, and the implementation of hydrogen technologies in the energy infrastructure.

Johansson et al. [37] highlight the Energy Technology Options Conference, opening with a presentation by Johansson on the prospects for renewable energy in a global context, based on a study commissioned by the United Nations Solar Energy Group for Environment and Development. The study includes sources for fuels and electricity, including reports by specialists on a long list of renewable technologies.

12.4   Rule-Based Neural Network Structure

The rule-based neural network detailed in the present chapter uses a multilayer network structure with transfer functions and links representing a fuzzy logic system [13]. In minimizing the neuro-fuzzy weight, an updating vector needs to be found, such as:

$$\alpha^x = \frac{\alpha_r}{\sum_{i=0}^{m} a_i} \tag{12.1}$$

A popular truncated method is the Newton algorithm to obtain a step size and direction in weight space that drives a cost function toward its minimum. In addition, using Taylor's expansion, the cost function JN (.) can be approximated by the quadratic function:

$$J_N\,(w + \Delta w) = J_N(w) + \Delta w\,\frac{d\,J_N(w)}{dw} + \frac{1}{2}\,\Delta w^T\,\frac{d^2 J_N(w)}{dw^2}\,\Delta w \tag{12.2}$$

where Δw is the weight vector update.

Differentiating Eq. (12.2) with respect to Δw, and with the result set to zero, setting the equation to zero will minimize the equation:

$$\frac{d\,J_N(w)}{dw} = \frac{d^2 J_N(w)}{dw^2}\,\Delta w \tag{12.3}$$

$$g = -H\,\Delta w \tag{12.4}$$

where g and H represent the gradient and Hessian of $J_N(w_k)$, respectively.

A truncated conjugate gradient algorithm will help us to find the solution for Newton's equation. This will ensure that the weight vector lies in a trust region. This algorithm can be summarized as follows:

(i) $k = 0$.
(ii) Set initial weight vector, w_k.
(iii) Calculate gradient $g(w_k)$ and Hessian $H(w_k)$.
(iv) If $||g(w_k)|| < \epsilon$ then terminate.
(v)   Solve $H(w_k)\,\Delta w_k + g(w_k)$ using a conjugate gradient, ensuring $||\,w_{k+1}|| \le D_k$.
(vi) Starting with $\lambda_k = 1$, find λ_k where $w_{k+1} = w_k + \lambda_k \Delta w_k$.
(vii) Adjust the trust region radius (D_{k+1}) with the following heuristic:

$$D_{k+1} = \begin{cases} 2\,D_k & if\ \lambda_k \ge 1 \\ \frac{D_k}{3} & if\ \lambda_k < 1 \end{cases} \tag{12.5}$$

(viii) Go to step (iv).

In a sequence to implement the previous algorithm, the gradient and Hessian of the cost function $J_N(w_k)$ are required. The weight vector **w** needs to be taken into consideration. These derivatives can be constructed as follows:

$$\frac{\partial J_N(w)}{\partial w_p} = \frac{1}{N} \sum_{k=1}^{N} 2 \frac{\partial \hat{y}(w(k), w)}{\partial w_p} \left[y(k) - \hat{y}(w(k), w) \right] \tag{12.6}$$

$$\frac{\partial^2 J_N(w)}{\partial w_p \partial w_q} = \frac{1}{N} \sum_{k=1}^{N} \left[2 \frac{\partial^2 \hat{y}(y(k), w)}{\partial w_p \partial w_q} y(k) - \hat{y}(w(k), w) \right.$$
$$\left. + 2 \frac{\partial \hat{y}(w(k), w)}{\partial w_p} \frac{\partial \hat{y}(w(k), w)}{\partial w_q} \right] \tag{12.7}$$

The gradient of the model output with respect to the weights is represented as follows:

$$\frac{\partial \hat{y}(x, w)}{\partial w_p} = \begin{cases} \left[\prod_{u=1, u \neq k}^{U} \sum_{i=1}^{P_u} \mu_{A^i}^{u} w_i^{u} \right] \mu_{A^j}^{k} & \mu_{A^i}^{u} \end{cases} \tag{12.8}$$

w_p refers to the ith weight of the uth tensor model, and w_p refers to the jth weight of the kth submodel.

$$\frac{\partial^2 \hat{y}(x, w)}{\partial w_p \partial w_q} = \begin{cases} \left[\prod_{u=1, u \neq m \neq k}^{U} \sum_{i=1}^{P_u} \mu_{A^i}^{u} w_i^{u} \right] \mu_{A^j}^{k} \mu_{A^l}^{m} \\ 0 \end{cases} \tag{12.9}$$

w_p and w_p refer to the jth weight of the kth submodel and the lth weight of the mth submodel.

The dependency of $\mu_{A^i}^{u}(x_i)$ on the input vector x_i has been dropped to reduce the notation.

12.5   The Proposed Model

The developed model illustrated in Fig. 12.1 represents the model proposed in the present chapter. The final step of the proposed model is reaching the number of solar power panels required to satisfy a given estimated daily electricity load (NSP). Furthermore, the presented model illustrated in Fig. 12.1 is a neuro-fuzzy system to calculate the NSP. The model is the corresponding equivalent ANFIS architecture. The fuzzy system uses a learning algorithm derived from neural network theory, where the objective is to calculate its parameters by processing the input data. The represented neuro-fuzzy system is viewed as a five-layer feedforward neural network. The first layer represents the input variables, where the second, third, and fourth layers (multiplications) represent fuzzy rules. Every node from the second to the fourth layer is fixed, whose output is the product of the inputs. At the same time, every node is adaptive. In addition, the fifth layer represents the output variable. The fuzzy sets are encoded as fuzzy connection weights.

Therefore, following the steps needed to calculate the number of solar panels required for the solar PV system design for each considered country, the study considers the five countries presented in Qamber et al. [12]. The following steps were followed to meet the target of the study as presented by Qamber et al. [12]:

$$TRL = NoUnits.REQP \tag{12.10}$$

$$APopPV = PPR.OpF \tag{12.11}$$

$$TWhrR = TRL.Ophrs \tag{12.12}$$

$$PEndU = APopPV.CombE \tag{12.13}$$

$$EnPby1PD = PEndU.NhrsPD \tag{12.14}$$

$$NSP = \frac{TWhrR}{EnPby1PD} \tag{12.15}$$

where $APopPV$ is the actual power output of a photovoltaic panel, $CombE$ is the combined efficiency, $EnPby1PD$ is the energy produced by a 1-peak-watt panel in a day, $NhrsPD$ is the number of hours per day, $NoUnits$ is the number of units, OpF is the operating factor, $Ophrs$ is the operating hours, $PEndU$ is the power used at the end use [it is less because of lower combined efficiency of the system], PPR is the peak power rating, $REQP$ is the rating of the equipment, TRL is the total required load (total connected load), and $TWhrR$ is the total watt-hour rating of the system.

12.6   Results and Discussion

On the basis of the model proposed in the present chapter, it is obvious from Eqs. (12.10–12.15) that the neuro-fuzzy model shown in Fig. 12.1 could be followed to obtain the required results. The results shown in Table 12.1 were found for five countries [12]: Bahrain, Egypt, India, Thailand, and the UK. If the GCC countries are considered, the results obtained for Bahrain will be the same as those for the other GCC countries or very close to them because their climates are almost the same. Furthermore, carbon dioxide emissions in the GCC countries are causing climate extremes and temperature rises. In addition, urbanization and economic

Table 12.1 Total photovoltaic (PV) panel capacity needed in five countries [12]

Country	Total PV panel capacity needed (kWp)	Total PV panel capacity needed (%)
Bahrain	53.5313	14.73429101
Egypt	39.5725	10.89218328
India	72.3664	19.91858215
Thailand	91.1436	25.08693654
UK	106.6972	29.36800702
Total	363.3110	100

kWp peak kilowatts

growth are leading to air pollution. At the same time, the heat of the islands is raised with the increase in urbanization, which pushes the country toward greater energy consumption. Power consumption for cooling increases the temperatures in the region, which are related to emissions. GCC investments have started in several sectors, including modernization, renewable energy, technological investments, and high technology.

Figure 12.1 illustrates the proposed neuro-fuzzy system model to calculate the NSP. The model consists of interconnected layers of processing factors. The required factors are passed through the layer's interconnections. Data are passed through the network from layer to layer via the junction between two neurons. In addition, an activation function is associated to limit the amplitude of the output of a neuron. To satisfy the relationship between the input and output of the neuro-fuzzy schematic diagram, the connection weights and the activation functions must be derived. The method has been derived previously and is known as supervised training. The proposed model, when implemented, is trained with respect to data sets until it learns the patterns used as inputs. Once it is trained, new patterns may be obtained for the number of solar panels, then for prediction of the electricity load. The ANN can automatically learn the recognition patterns in data from real systems or from physical models. The model can obtain many inputs and produce results. Figure 12.1 has five layers, starting with an input layer and reaching the output layer. The incoming connection has seven factors. The output is a function of the target required, which is the NSP.

It can be concluded that the processing is an interconnection formed by links (synapses) with weights. Furthermore, the neuro-fuzzy approach has been developed for prediction of the index or what might be called a factor and has a symbol (NSP). An adaptive ANN and hybrid models for prediction of solar panels are proposed and discussed in the present chapter. The model consists of a combination of ANN and fuzzy logic, forming an ANFIS. The proposed new neuro-fuzzy-based model is used for predicting the design of a PV system. Finally, the application for sizing a PV model is presented on the basis of the data generated by this model. The study uses fuzzy logic to obtain and assess the solar panel model in the five countries, where the evaluation using fuzzy logic is based on different factors as mentioned earlier.

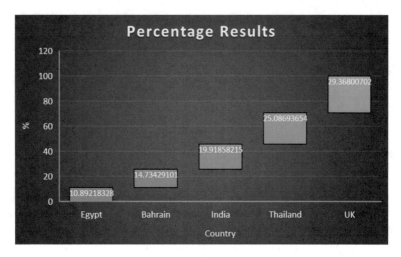

Fig. 12.2 Waterfall chart of the total photovoltaic (PV) panel capacity (expressed in peak kilowatts (kWp)) needed in five countries

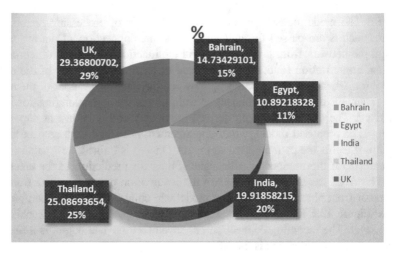

Fig. 12.3 Pie chart of the total photovoltaic (PV) panel capacity (expressed in peak kilowatts (kWp)) needed in five countries

Figures 12.2, 12.3, 12.4, 12.5, and 12.6 show different charts for the five considered countries. The results for Bahrain are almost identical to those for the other GCC countries. The waterfall chart in Fig. 12.2 shows the positive amounts for the studied countries that have influenced the total amount, based on consideration of Bahrain for the starting value.

The results are plotted as a pie chart in Fig. 12.3, which displays the data in a circular graph form. The pieces of the graph (with country as a category) are

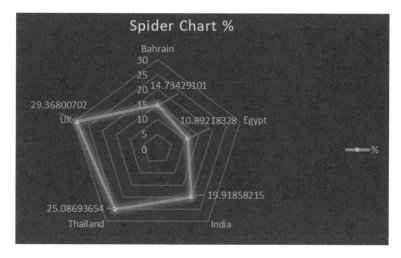

Fig. 12.4 Radar chart of the total photovoltaic (PV) panel capacity (expressed in peak kilowatts (kWp)) needed in five countries

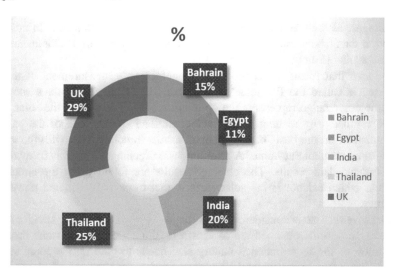

Fig. 12.5 Doughnut chart of the total photovoltaic (PV) panel capacity (expressed in peak kilowatts (kWp)) needed in five countries

proportional to the fraction of the whole. In addition, each slice is relative to the size of that category. Each pie slice represents a portion of the whole.

The radar chart shown in Fig. 12.4 simplifies the reading of the results and makes the total PV panel capacity easy to see. In percentage terms, the UK has the highest total PV panel capacity and Egypt has the lowest percentage capacity. All axes in Fig. 12.4 are arranged radially with equal distances between them, and the scale is the same between all axes, where the grid lines are used as a guide.

Fig. 12.6 Photovoltaic (PV) solar surface area for five countries

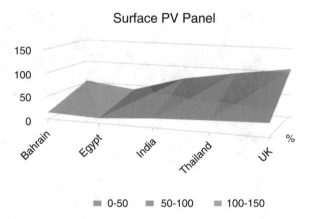

The doughnut chart in Fig. 12.5 shows the relationship of the parts to the whole shape. The results show the proportion of each country's percentage versus the summation of the countries (which is 100%), and the center of the doughnut chart can be selected to extract additional information such as the total of all data percentages, as well as the data value of the slice being examined. In addition, execution data labels and data values are not separately required. This means that both are available on the chart.

The area that forms the basis of the manufacturer's measurement of module efficiency is called the PV solar surface area (Fig. 12.6). The values shown in the same color range represents the optimum results. The size of the solar panel can be estimated by reading the surface chart. The surface area of the panel is recommended on the basis of the obtained results in comparison with the garden or roof area available at home. A two-dimension rectangular grid is formed after obtaining the data results. The obtained results are found using the simulation program, which calculates the solar surface area from the calculated power and efficiency.

If an inverter size is considered in the range of a 25–30% bigger size for safety purposes, the results obtained are as shown in Table 12.1. At the same time, the calculation of the recommended battery size needs to take into account several factors. These factors are the total watt-hours per day, the expected efficiency of the battery (which is assumed to be approximately equal to 0.85), the depth of discharge (which is assumed to be 0.6), and the nominal battery voltage (which is equal to 12 Volts DC). When there is no power produced by the PV panels, autonomy days are assumed, where these days are the number of days when the system needs to operate when no power is produced by the PV panels. The autonomy days are assumed to be equal to 3 days.

12.7 Conclusion

Worldwide, CSP systems are being used to generate solar power. This type of power is generated using mirrors or lenses. The mirrors and lenses concentrate a large area of sunlight, or solar thermal energy, onto a small area. The resulting concentrated heat is used to eventually spin a turbine and generate electricity. CSP systems depend on moving parts, whereas photovoltaic systems do not.

According to the Solar Energy Industries Association (SEIA), CSP plants use different formations of mirrors to concentrate the heat from the sun to drive traditional steam turbines or engines that can generate electricity. Therefore, the thermal energy concentrated in a CSP plant can be stored and used to produce electricity day or night whenever it is needed.

Obtaining a suitable model for the required PV panels (NSP) using artificial intelligence is the main goal of the present study. After finding the required NSP model, some factors may be adjusted for more suitable design of the solar panels. The calculated total PV panel capacity varies between 39.5725 and 106.6972 kWp. The main difference between the present study and the previous ones is the NSP. In the present study, the calculated NSP is helpful in the design of PV panels, taking into account the considered factors. The NSP is the main advantage of the present study.

The study carried out by Al-Hamad and Qamber [13] developed a model to find the forecast electricity demand for the six GCC countries, using ANFIS and MLR methods. ANFIS was the most accurate technique according to the comparison in their study [13]. On the basis of the plan for each GCC country using renewable energy, the model for each country could be studied and investigated. In addition, the investment saving could be targeted.

The main purpose of a solar energy power plant is summarized in the following points:

1. Carbon dioxide emissions are reduced, which is considered one of the environmental benefits.
2. The power plant becomes more efficient than a conventional combined cycle plant.
3. Solar power plants directly feed the national electricity grid.

The development of CSP helps to lower costs and increase efficiency. In addition, CSP provides more reliable performance than current technologies. The proposed model shows new concepts in the collector, where the technologies will lower operation and management costs. The GCC countries are conducting their own solar energy projects, including Bahrain, which is one of the five countries considered in our analysis. Furthermore, in the present analysis, artificial intelligence techniques have been used to help calculate the number of PV solar panels required by each considered country. The calculation of the number of solar panels is based on the required power produced by the solar cells designed for each purpose. Furthermore, these artificial intelligence techniques play an important role in modeling and prediction of renewable energy engineering.

Load forecasting, as a future study, is recommended, using the proposed artificial intelligence model, and can be determined for the proposed five countries or for any other country on the basis of the power output produced by the solar cells designed for such purposes. The combination of the models from these studies [12, 13] will allow us to devise a novel model to assist the design and estimation load for any country. In addition, the cost of each solar panel, with its power output under the required specification, can be found.

References

1. H. Al Guarni, A. Awasthi, Solar PV power plant site selection using a GIS-AHP based approach with application in Saudi Arabia. Appl. Energy **206**, 1225–1240 (2017)
2. H. Al Garni, A. Kassem, A. Awasthi, D. Komljenovic, K. Al-Haddad, A multicriteria decision making approach for evaluating renewable power generation sources in Saudi Arabia. Sustain. Energy Technol. Assess. **16**, 137–150 (2016)
3. A. Pareek, L. Gidwani, Solar irradiation data measurement analysing techniques, in Proceedings of International Conference on Renewable Energy and Sustainable Environment—RESE 15, Dr. Mahalingam College of Engineering and Technology, Pollachi, August 10–13, 2015
4. European Bank for Reconstruction and Development (EBRD), Implementing the EBRD Green Economy Transition (GET), Technical Guide for Consultants, Climate Policy Initiative, Version 2, May 2018, London
5. A.T. Eseye, M. Lehtonen, T. Tukia, S. Uimonen, R.J. Millar, *Machine Learning Based Integrated Feature Selection Approach for Improved Electricity Demand Forecasting in Decentralized Energy Systems* (IEEE Access, 2019). https://doi.org/10.1109/ACCESS.2019.2924685
6. M.T. Mito, X. Maa, H. Albuflasab, P.A. Davies, Reverse osmosis (RO) membrane desalination driven by wind and solar photovoltaic (PV) energy: state of the art and challenges for large-scale implementation. Renew. Sust. Energ. Rev. **112**, 669–685 (2019). https://doi.org/10.1016/j.rser.2019.06.008
7. H. Albuflasa, Renewable energy in Bahrain: background paper (2019). https://doi.org/10.13140/RG.2.2.23998.64320
8. D. Sidorov, Q. Tao, I. Muftahov, A. Zhukov, D. Karamov, A. Dreglea, F. Liu, Energy Balancing Using Charge/Discharge Storages Control and Load Forecasts in a Renewable-Energy-Based Grids (2019). https://arxiv.org/pdf/1906.02959v1.pdf
9. M.S. Ferwatia, M. AlSaeeda, A. Shafaghat, A. Keyvanfar, Qatar sustainability assessment system (QSAS)—neighborhood development (ND) assessment model: coupling green urban planning and green building design. J. Build. Eng. **22**, 171–180 (2019)
10. M.Y. Al-Hamad, I.S. Qamber, Smart PV grid to reinforce the electrical network, in World Renewable Energy Congress-17, E3S Web of Conferences 23, 01002 (2017). https://doi.org/10.1051/e3sconf/20172301002
11. H. Albuflasa, Smart Energy Management Systems for Households in Bahrain, in *Sustainability and Resilience Conference: Mitigating Risks and Emergency Planning*, KnE Engineering, pp. 135–149. https://doi.org/10.18502/keg.v3i7.3078
12. I.S. Qamber, M.Y. Al-Hamad, Novel PV panels design modeling to support smart cities. Int. J. Comput. Digit. Syst. **8**(2), 125–130 (2019)
13. M.Y. Al-Hamad, I.S. Qamber, GCC electrical long-term peak load forecasting modeling using ANFIS and MLR methods. Int. Arab J. Basic Appl. Sci. **26**(1), 269–282 (2019). https://doi.org/10.1080/25765299.2019.1565464
14. E. Bellini, *Bahrain's Run to Sun Begins with 255 MW Plan* (2017). https://www.pv-magazine.com/2017/09/19/bahrains-run-to-sun-begins-with-255-mw-plan/

15. E. Bellini, *Bahrain Launches 100 MW PV Tender* (2019). https://www.pv-magazine.com/2019/01/14/bahrain-launches-100-mw-pv-tender/
16. Amin Al Yaquob, *Askar Landfill Site Solar Independent Power Plant. Gulf Cooperation Council's Association for Renewable Energy and Sustainability* (2019). https://www.protenders.com/projects/askar-landfill-site-solar-independent-power-plant
17. Electricity and Water Authority (EWA) of Bahrain, *Bahrain EWA's Askar Solar PV Project* (2019). http://www.theenergyinfo.com/detail.php?project=ktXHj8g7iWs8Ttw1kJs7 ito1iZQx&sample=1
18. M. Willuhn, *Bahrain Launches 3 MW Rooftop Tender* (2019). https://www.pv-magazine.com/2019/01/31/bahrain-launches-3-mw-rooftop-tender/
19. O. Baudson, *Kuwait 1st Concentrated Solar Power Plant 50 MW Shagaya Solar Completed this year, Micro-CSP (Concentrated Solar Power)*, Solar Thermal Energy News, Helioscsp (2015)
20. K. Chamberlain, *Kuwait Eyes up to 400 MW of Concentrated Solar Power, Helioscsp* (2019). http://helioscsp.com/kuwait-eyes-up-to-400-mw-of-concentrated-solar-power/
21. SEM/NBP/KF/SAJ, *Study on Renewable Energy Resources: Oman, in Authority for Electricity Regulation of Oman*, final report, number 66847-*1-1*, (2008)
22. M.A. Al-Mahrouqi, A.Z. Amin, *Sultanate of Oman: Renewable Readiness Assessment* (International Renewable Energy Agency (IRENA), Abu Dhabi, 2014)
23. O. Power, Water Procurement Co. (SAOC), OPWP's 7-year statement (2018–2024), Issue 12, Sultanate of Oman (2018)
24. Y. Al Hatmi, C.S. Tan, Issues and challenges with renewable energy in Oman. Int. J. Res. Eng. Technol. **2**(7), 212–219 (2013)
25. M.T. Davies, T. Armsby, J. Palmer, S. Ahmad, *Developing Renewable Energy Projects: A Guide to Achieving Success in the Middle East* (PWC, Eversheds, 2016)
26. S. Zafar, *Solar Energy in Qatar, Dubai's Journey Towards a Sustainable City, Middle East, Renewable Energy, Solar Energy, Technology*, 2018. https://www.ecomena.org/solar-energy-in-qatar/
27. Intelligent Word Recognition (IWR), *Qatar Solar Energy Puts Online 300 MW Solar PV Module Factory* (2014). http://www.renewable-energy-industry.com/news/world/article-4099
28. MEED, *Saudi Arabia Launches Requalification for Seven Solar Projects*. Power Technology (2019). https://www.power-technology.com/comment/saudi-arabia-solar-projects/
29. J.S. Hill, *Saudi Arabia Planning 2.6 Gigawatt Solar Project Near Mecca* (2019). https://cleantechnica.com/2019/03/26/saudi-arabia-planning-2-6-gigawatt-solar-project-near-mecca/
30. S.A. Rahman, *Is UAE Leading the Way for Concentrated Solar Power in GCC?* (2019). http://www.ipsnews.net/2019/05/uae-leading-way-concentrated-solar-power-gcc/
31. Power Technology, *Noor Abu Dhabi Solar Power Plant in UAE Begins Operations* (2019). https://www.power-technology.com/news/noor-abu-dhabi-solar-power-plant-uae/
32. Apicorp Energy Research, Solar Energy in the UAE: Impressive Progress, vol. 3(2) (2017). http://www.apicorp-arabia.com/Research/EnergyReseach/2017/APICORP_Energy_Research_V03_N02_2017.pdf
33. C. Weatherby, B. Eyler, R. Burchill, *UAE Energy Diplomacy: Exporting Renewable Energy to the Global South* (Trends Research and Advisory, Stimson Center, Washington, 2018)
34. ENGIE (French Multinational Electric Utility Company), *Solar Outlook Report 2019* (Middle East Solar Industry Association, Dubai, 2019)
35. G. Boyle, *Renewable Energy* (Oxford University Press, Oxford, 2004), p. 456
36. J.A. Turner, A realizable renewable energy future. Science **285**(5428), 687–689 (1999)
37. T.B. Johansson, H. Kelly, A.K.N. Reddy, R.H. Williams, *Renewable Energy: Sources for Fuels and Electricity* (1993)

Index

© Springer Nature Switzerland AG 2020
M. H. Amini (ed.), *Optimization, Learning, and Control for Interdependent
Complex Networks*, Advances in Intelligent Systems and Computing 1123,
https://doi.org/10.1007/978-3-030-34094-0

Printed in the United States
By Bookmasters